# TOPICS IN
# STATISTICAL
# MECHANICS

**Second Edition**

# Advanced Textbooks in Physics

ISSN: 2059-7711

The *Advanced Textbooks in Physics* series explores key topics in physics for MSc or PhD students.

Written by senior academics and lecturers recognised for their teaching skills, they offer concise, theoretical overviews of modern concepts in the physical sciences. Each textbook comprises of 200–300 pages, meaning content is specialised, focussed and relevant.

Their lively style, focused scope and pedagogical material make them ideal learning tools at a very affordable price.

*Published*

*Topics in Statistical Mechanics (Second Edition)*
    by Brian Cowan

*Physics of Electrons in Solids*
    by Jean-Claude Tolédano

*Astronomical Spectroscopy: An Introduction to the Atomic and Molecular Physics of Astronomical Spectroscopy (Third Edition)*
    by Jonathan Tennyson

*A Guide to Mathematical Methods for Physicists: Advanced Topics and Applications*
    by Michela Petrini, Gianfranco Pradisi & Alberto Zaffaroni

*Quantum States and Scattering in Semiconductor Nanostructures*
    by Camille Ndebeka-Bandou, Francesca Carosella & Gérald Bastard

*An Introduction to Particle Dark Matter*
    by Stefano Profumo

*Studying Distant Galaxies: A Handbook of Methods and Analyses*
    by François Hammer, Mathieu Puech, Hector Flores & Myriam Rodrigues

*Trapped Charged Particles: A Graduate Textbook with Problems and Solutions*
    edited by Martina Knoop, Niels Madsen & Richard C Thompson

Advanced Textbooks in Physics

# TOPICS IN STATISTICAL MECHANICS

**Second Edition**

### Brian Cowan

Royal Holloway, University of London, UK

**World Scientific**

NEW JERSEY · LONDON · SINGAPORE · BEIJING · SHANGHAI · HONG KONG · TAIPEI · CHENNAI · TOKYO

*Published by*

World Scientific Publishing Europe Ltd.

57 Shelton Street, Covent Garden, London WC2H 9HE

*Head office:* 5 Toh Tuck Link, Singapore 596224

*USA office:* 27 Warren Street, Suite 401-402, Hackensack, NJ 07601

**Library of Congress Cataloging-in-Publication Data**
Names: Cowan, B. P., 1951–   author.
Title: Topics in statistical mechanics / Brian Cowan, Royal Holloway, University of London, UK.
Description: Second edition. | New Jersey : World Scientific, [2022] |
    Series: Advanced textbooks in physics, 2059-7711 |
    Includes bibliographical references and index.
Identifiers: LCCN 2020051830 | ISBN 9781786349781 (hardcover) |
    ISBN 9781786349903 (paperback) | ISBN 9781786349798 (ebook for institutions) |
    ISBN 9781786349804 (ebook for individuals)
Subjects: LCSH: Statistical mechanics.
Classification: LCC QC174.8 .C69 2022 | DDC 530.13--dc23
LC record available at https://lccn.loc.gov/2020051830

**British Library Cataloguing-in-Publication Data**
A catalogue record for this book is available from the British Library.

For any available supplementary material, please visit
https://www.worldscientific.com/worldscibooks/10.1142/Q0288#t=suppl

Desk Editors: Christina Ramalingam/Michael Beale/Shi Ying Koe

Typeset by Stallion Press
Email: enquiries@stallionpress.com

# PREFACE

Ludwig Boltzmann, who spent much of his life studying
statistical mechanics, died in 1906, by his own hand. Paul
Ehrenfest, carrying on the work, died similarly in 1933.
Now it is our turn to study statistical mechanics.

Perhaps it will be wise to approach the subject
cautiously.

— David Goodstein, *States of Matter*, 1975, Dover NY

## The First Edition

Statistical Mechanics, more than any other branch of physics, is beset with
problems of methodology and presentation. Philosophers have argued
over the meaning of probability, particularly when applied to a single
"event". Mathematicians have neatly side-stepped the issue by stripping
away physical interpretation and treating probability simply as a "measure" accompanied by a set of rules. But disengaging from reality in this
way is of scant use to physics. For physicists, probability and the statistical
method continue to cause their share of anguish. The statistical method
was a contributory factor to Boltzmann's suicide and possibly to that of
Paul Ehrenfest. Even today the conundrums of Quantum Mechanics have
aspects of probability at their heart.

In Statistical Mechanics, the operational approach of the mathematicians is paralleled by the Information Theory approach of E. T. Jaynes.
True, this method is championed by some outstanding pedagogues, but

I confess to finding it distinctly unappealing. Certainly it might be an expedient way to obtain results, but (to my mind) it obscures *understanding*. And understanding is central to the physicist's endeavours. By contrast, the ensemble formalism of scholars such as T. L. Hill provides maximum clarity and physical meaning. Some might find this overly formal; I find it distinctly appealing.

The second structural issue is the relationship between Statistical Mechanics and the older discipline of Thermodynamics. Should these be developed independently or are they better treated in a unified manner? Landau was a strong supporter of the latter view and I have mostly been persuaded by his arguments. At the undergraduate level this philosophy benefited from the magnificent exposition of Reif (also his volume in the Berkley Physics series). It is probable that this had a major impact on the intellectual formation of many of today's professional physicists. It was certainly an ideal preparation for the study of the appropriate Landau and Lifshitz volume(s). Nevertheless, there is no doubt that classical thermodynamics has its strengths. Most important is the model independence of its results. Aspects of its abstract formalism may deter some and indeed parts of its deeper logic may be obscure: witness the gradual conversion to the Carathéodory viewpoint by Zemansky through the many editions of his Thermodynamics textbook. These days the logical aspects of Classical Thermodynamics have been elegantly clarified and reformulated by Callen, but this is probably not appropriate at an undergraduate level.

Regardless of the deeper philosophical issues, this second matter has been mostly resolved through necessity. The increasing pressure on the undergraduate curriculum means that in most UK universities there is no longer the space available to present self-contained courses on Classical Thermodynamics and Statistical Mechanics. Indeed, this is the case at Royal Holloway University of London. However, the introduction of the four-year undergraduate integrated Master's degree, the M.Sci. or M.Phys., has helped and allowed the incorporation of some more advanced material into the degree programmes.

Within the University of London, King's, Queen Mary, Royal Holloway, and University Colleges collaborate in the joint teaching of the fourth year of the M.Sci. degree, with lectures held in central London. This has allowed a wide range of courses to be provided, many of which are at the cutting edge of the subject. And at the same time this has given the opportunity for a more detailed coverage of some material latterly squeezed out of the traditional three-year B.Sc. degrees.

This book has arisen out of such an intercollegiate course in Statistical Mechanics that I have taught to M.Sci. fourth year students over the last 10 or so years. Such intercollegiate teaching presents its own challenges. At the completion of their third year, students are required to have reached a common standard and level for embarking on the intercollegiate fourth year, although presentation and flavour will have been different in each college. This is particularly the case in the area of thermal physics for the reasons outlined in the first few paragraphs above. And, because of this, the learning material for this course provided to students included some more elementary "foundation" material, albeit presented from a slightly more mature standpoint. Approximately half the material of Chapters 1 and 2 falls into this category.

This was a "paperless" course, whereby students were provided with lecture notes and other learning material on the web. That included the lecture calendar, problem exercises, etc., all of which students could access remotely from their home college or from elsewhere. I was alerted to the wider appeal of the course material when I started to receive requests from students, outside London and around the world, for answers to problem exercises (often with deadlines!) and queries about unclear portions of the notes. This became strikingly poignant on occasions when the College web service was interrupted. Then I would often find frantic email enquiries asking where the material had gone. So, when I was approached by the Imperial College Press to consider producing a book version of the course, I eventually agreed. This book comprises a reworking of the course material, incorporating changes and suggestions from the various cohorts of students, colleagues and reviewers commissioned by ICP.

Electromagnetic units have traditionally been the cause of many problems to students learning their Physics from a range of sources. It was expected that such difficulties would have been eliminated through the adoption of the SI system. But there has been resistance. Kittel's *Solid State Physics* book has settled on a compromise whereby many equations are quoted in both their Gaussian and SI forms. I have adopted, almost exclusively, SI units although I am unhappy about aspects of the $B - H$ controversy that this can lead to. However, readers should note that I use the symbol $M$ to represent total magnetic moment rather than magnetic moment per unit volume. I also confess to the eccentricity of representing the complex dynamical susceptibility as $\chi(\omega) = \chi'(\omega) + i\chi''(\omega)$ rather than the more common complex conjugate form.

# ABOUT THE AUTHOR

**Brian Cowan** was educated at the then new University of Sussex, receiving a B.Sc. in Physics and Mathematics in 1973 and a D.Phil. in 1976. For his doctoral research, he applied the technique of Nuclear Magnetic Resonance to the study of helium films at low temperatures under the supervision of Michael Richards. On leaving Sussex, he moved to Paris, working at the French Atomic Energy Commission in the Laboratory of Anatole Abragam. Following a short spell as a post-doctoral fellow with the NMR group at Nottingham University, he was appointed a Lecturer in Physics at London University's Bedford College in 1978, joining the Low Temperature Physics group of Roland Dobbs.

In 1985, Bedford College merged with Royal Holloway College and Brian moved to new purpose-built laboratories in Egham, Surrey. Apart from sabbatical stays at the Technion, Israel Institute of Science and Technology in Haifa, and at the University of Florida in Gainesville, he has remained at Royal Holloway ever since. During the Bedford–Holloway merger, he was involved in the reorganisation of the teaching of Thermal Physics. In 1993, the London Intercollegiate Physics M.Sci. degree was proposed. He was a member of the Intercollegiate Board, which designed the new degree programmes; he then served as its Chair from 1996 to 2014. Between 1996 and 2011, he chaired the Physics Department's Curriculum Development Group, redesigning the Royal Holloway physics undergraduate programmes.

Over the years, Brian has taught most undergraduate physics subjects. He pioneered the introduction of Computer Algebra into the physics curriculum and the use of videoconferencing technology for teaching his intercollegiate Statistical Mechanics course. He is a member of the London Low Temperature Laboratory under the direction of John Saunders at Royal Holloway. He was appointed Reader in Physics in 1992 and Professor of Physics in 1999. In 2003, he was elected a Fellow of the Institute of Physics. He has served two terms as Head of the Physics Department at Royal Holloway. He is a member of the Institute of Physics Degree Accreditation panel and has acted as External Examiner in Physics for various UK universities and as Ph.D. examiner both in the UK and abroad.

# CONTENTS

# CHAPTER 1

# THE METHODOLOGY OF STATISTICAL MECHANICS

This chapter provides an overview of statistical mechanics and thermodynamics. Although the discussion is reasonably self-contained, it is assumed that this is not the reader's first exposure to these subjects. More than many other branches of physics, these topics are treated in a variety of different ways in the textbooks in common usage. The aim of this chapter is thus to establish the perspective of the book: the standpoint from which the topics in the following chapters will be viewed. The books by Guénault [1] and Bowley [2] are recommended to the novice as highly accessible and clearly argued introductions to the subject; indeed, all readers will find many of the examples of this book treated by those books in complementary fashion.

## 1.1 Terminology and Methodology

### 1.1.1 *Approaches to the subject*

*Thermodynamics* is the study of the relationship between *macroscopic* properties of systems, such as temperature, volume, pressure, magnetisation, and compressibility. *Statistical Mechanics* is concerned with understanding how the various macroscopic properties arise as a consequence of the *microscopic* nature of the system. In essence, it makes macroscopic deductions from microscopic models.

The power of thermodynamics, as formulated in the traditional manner (Zemansky [3], for example), is that its deductions are quite general;

they do not rely, for their validity, on the microscopic nature of the system. Einstein expressed this quite impressively when he wrote [4]:

> *A theory is the more impressive the greater the simplicity of its premises is, the more different kinds of things it relates, and the more extended is its area of applicability. Therefore the deep impression which classical thermodynamics made upon me; it is the only physical theory of universal content concerning which I am convinced that, within the framework of the applicability of its basic concepts, will never be overthrown.*

On the other hand, statistical mechanics, as conventionally presented (Hill [5], for example) is system-specific. One starts from particular microscopic models, say the Debye model for a solid, and derives macroscopic properties such as the heat capacity. It is true that statistical mechanics will give relationships between the various macroscopic properties of a system, but they will only apply to the system/model under consideration. Results obtained from thermodynamics, on the other hand, are model-independent and general.

Traditionally, thermodynamics and statistical mechanics were developed as independent subjects, with "bridge equations" making the links between the two. Alternatively, the subjects can be developed together, where the Laws of Thermodynamics are justified by microscopic arguments. This reductionist view was adopted by Landau. In justification he wrote [6]:

> *Statistical physics and thermodynamics together form a unit. All the concepts and quantities of thermodynamics follow most naturally, simply and rigorously from the concepts of statistical physics. Although the general statements of thermodynamics can be formulated non-statistically, their application to specific cases always requires the use of statistical physics.*

The contrast between the views of Einstein and those of Landau is apparent. The paradox, however, is that so much of Einstein's work was reductionist and microscopic in nature whereas Landau was a master of the macroscopic description of phenomena. This book considers thermodynamics and statistical mechanics in a synthetic manner; in that respect, it follows more closely the Landau approach.

### 1.1.2 *Description of states*

The state of a system described at the macroscopic level is called a *macrostate*. Macrostates are described by relatively few variables, such as temperature, pressure, and volume.

The state of a system described at the microscopic level is called a *microstate*. Microstates are described by a very large number of variables. Classically, you would need to specify the position and momentum of each particle in the system. Using a quantum-mechanical description you would have to specify all the quantum numbers of the entire system. So, whereas a macrostate is likely to be described by under ten variables, a microstate will be described by over $10^{23}$ variables.

The fundamental methodology of statistical mechanics involves applying probabilistic arguments about microstates, regarding macrostates as statistical averages. It is because of the very large number of particles involved that the mean behaviour corresponds so well to the observed behaviour — that fluctuations are negligible. Recall (or see Section 1.4.6) that for $N$ particles the likely fractional deviation from the mean will be $1/\sqrt{N}$. So, if you calculate the mean pressure of one mole of air at one atmosphere, it is likely to be correct to within $\sim 10^{-12}$ of an atmosphere. It is also because of the large number of particles involved that the mean, observed behaviour corresponds to the most probable behaviour — the *mean* and the *mode* of the distribution differ negligibly.

Statistical Mechanics is slightly unusual in that its formalism is easier to understand using quantum mechanics for descriptions at the microscopic level rather than classical mechanics. This is because the idea of a quantum state is familiar, and often the quantum states are discrete. It is more difficult to enumerate the (micro)states of a classical system; it is best done by analogy with the quantum case. The classical case will be taken up in Section 1.6.

### 1.1.3 *Extensivity and the thermodynamic limit*

The thermodynamic variables describing macrostates fall into two classes. Quantities such as energy $E$, number of particles $N$ and volume $V$, which add when systems are combined, are known as *extensive* variables. And quantities such as pressure $p$ and temperature $T$, which remain independent when similar systems are combined, are known as *intensive* variables.

Extensive and intensive variables come in pairs, such as volume and pressure. Very generally, they are connected by a differential relation of the form

$$p = -\frac{\partial E}{\partial V}.$$  (1.1.1)

Thus, the product of an extensive variable and its intensive conjugate has the dimensions of energy.

As we have argued, thermodynamics concerns itself with the behaviour of "large" systems and usually, for these, finite-size effects are not of interest. Thus, for example, unless one is concerned specifically with surface phenomena, it will be sensible to focus on systems that are sufficiently large that the surface contribution to the energy will be negligible when compared to the volume contribution. This is possible because the energy of interaction between atoms is usually sufficiently short-ranged. In general, one will consider properties in the limit $N \to \infty$, $V \to \infty$, while $N/V$ remains constant. This is called the *thermodynamic limit*.

It should be clear that true extensivity of a quantity, such as energy, emerges only in the thermodynamic limit.

Gravity is a long-range force. It is apparent that gravitational energy is not truly extensive; this is examined in Problem 1.1. This non-extensivity makes for serious difficulties when trying to treat the statistical thermodynamics of gravitating systems.

## 1.2   The Fundamental Principles

### 1.2.1   *The laws of thermodynamics*

Thermodynamics, as a logical structure, is built on its four assumptions or laws (including the zeroth).

*Zeroth Law*: **If system *A* is in equilibrium with system *B* and with system *C*, then system *B* is in equilibrium with system *C*.** Equilibrium here is understood in the sense that when two systems are brought into contact, then there is no change. This law was formalised after the first three laws were formulated and numbered. Because it was believed to be more fundamental, it was thus called the *Zeroth* Law. The Zeroth Law recognises the existence of states of equilibrium and it points us to the concept of temperature, a non-mechanical quantity that can label (and order) equilibrium states.

*First Law*: **The internal energy of a body can change by the flow of heat or by doing work**

$$\Delta E = \Delta Q + \Delta W. \qquad (1.2.1)$$

Here, $\Delta Q$ is the energy increase as a consequence of heat flow and $\Delta W$ is the energy increase resulting from work done. We usually regard this as a statement about the conservation of energy. But in its historical context the law asserted that as well as the familiar mechanical form of energy, heat also was a form of energy. Today we understand this as the kinetic energy of the constituent particles of a system; in earlier times, the nature of heat was unclear.

Many (indeed most) books use the symbol $U$ to denote internal energy. I have chosen to use $E$ in order to signify the microscopic origin of this quantity. Note that some older books adopt the *opposite sign* for $\Delta W$; they consider $\Delta W$ to be the work done *by* the system rather than the work done *on* the system.

*Second Law* (this law has many formulations): **Heat flows from hot to cold,** or **It is not possible to convert** *all* **heat energy to work.** These statements have the great merit of being reflections of common experience. There are other formulations such as the Carathéodory statement (see the books by Adkins [7] or Pippard [8]): **In the neighbourhood of any equilibrium state of a thermally isolated system there are states which are inaccessible,** and the entropy statement (see Callen's book [9]): **There is an extensive quantity, which we call entropy, which never decreases in a physical process.** The claimed virtue of the Carathéodory statement is that it leads more rapidly to the important thermodynamic concepts of temperature and entropy: this, at the expense of common experience. Indeed, if that is believed to be a virtue, then one may as well go the "whole hog" and adopt the Callen statement.

A major exercise in classical thermodynamics is proving the equivalence of the various statements of the Second Law. In whatever form, the Second Law leads to the concept of entropy and the *quantification* of temperature (the Zeroth Law just gives an *ordering*: $A$ is hotter than $B$). And it tells us there is an absolute zero of temperature.

*Third Law*: **The entropy of a body tends to zero as the temperature tends to absolute zero.** The Third Law will be discussed in Section 1.7. We shall see that it arises as a consequence of the quantum behaviour of matter at the microscopic level. However, we see immediately that the Third Law is telling us there is an absolute zero of *entropy*.

An even more fundamental aspect of the Zeroth Law is the fact of the *existence* of equilibrium states. If systems did not exist in states of equilibrium, then there would be no macrostates and no hope of description in terms of small numbers of variables. Then there would be no discipline of thermodynamics and phenomena would have to be discussed solely in terms of their intractable microscopic description. Fortunately, this is not the case; the existence of states of equilibrium allows our simple minds to make some sense of a complex world.

### 1.2.2  *Probabilistic interpretation of the First Law*

The First Law discusses the way the energy of a system can change. From the statistical standpoint, we understand the energy of a macroscopic system as the *mean* value since the system can exist in a large number of different microstates. If the energy of the $j$th microstate is $E_j$ and the probability of occurrence of this microstate is $P_j$, then the (mean) energy of the system is given by

$$E = \sum_j P_j E_j. \qquad (1.2.2)$$

The differential of this expression is

$$dE = \sum_j P_j \, dE_j + \sum_j E_j \, dP_j, \qquad (1.2.3)$$

this indicates the energy of the system can change in two different ways: (a) the energy levels $E_j$ may change or (b) the probabilities $P_j$ may change.

The first term $\sum_j P_j \, dE_j$ relates to the change in the energy levels $dE_j$. This term is the mean energy change of the microstates and we shall show in what follows that this corresponds to the familiar "mechanical" energy: the work done on the system. The second term $\sum_j E_j \, dP_j$ is a consequence of the change in probabilities or occupation of the energy states. This is fundamentally probabilistic and we shall see that it corresponds to the heat flow into the system.

In order to understand that the first term corresponds to the work done, let us consider a $pV$ system. We shall see (Section 2.1.3) that the energy levels depend on the size (volume) of the system:

$$E_j = E_j(V),\tag{1.2.4}$$

so that the change in the energy levels when the volume changes is

$$dE_j = \frac{\partial E_j}{\partial V}dV.\tag{1.2.5}$$

Then

$$\sum_j P_j dE_j = \sum_j P_j \frac{\partial E_j}{\partial V}dV.\tag{1.2.6}$$

We are assuming that the change in volume occurs at constant $P_j$, then

$$\sum_j P_j dE_j = \sum_j \frac{\partial}{\partial V}P_j E_j\, dV$$

$$= \frac{\partial}{\partial V}E\, dV.\tag{1.2.7}$$

But we identify $\partial E/\partial V = -p$, so that

$$\sum_j P_j dE_j = -p\, dV.\tag{1.2.8}$$

And thus, we see that the term $\sum_j P_j dE_j$ corresponds to the work done on the system. Then the term $\sum_j E_j dP_j$ corresponds to the energy increase of the system that occurs when no work is done; this is what we understand as heat flow.

We have learned, in this section, that the idea of heat arises quite logically from the probabilistic point of view.

### 1.2.3 *Microscopic basis for entropy*

By contrast to macroscopic thermodynamics, statistical mechanics is built on a single assumption, which we will call the *Fundamental Postulate* of statistical mechanics. We shall see how the Laws of Thermodynamics may be understood in terms of this Fundamental Postulate. These ideas date back to Boltzmann. The Fundamental Postulate states: **All microstates of an isolated system are equally likely**. Note, in particular, that an isolated system will have fixed energy $E$, volume $V$ and number of particles $N$ (fixed extensive quantities). Conventionally, we denote by $\Omega(E, V, N)$ the number of microstates corresponding to a given macrostate $(E, V, N)$. Then from the Fundamental Postulate it follows that the probability of a given macrostate is proportional to the number of microstates corresponding to it: $\Omega(E, V, N)$

$$P \propto \Omega(E, V, N). \tag{1.2.9}$$

If we understand the observed equilibrium state of a system as the most probable macrostate, then it follows from the Fundamental Postulate that the equilibrium state corresponds to the macrostate with the largest number of microstates. We are saying that $\Omega$ is maximum for an equilibrium state.

Since $\Omega$ for two isolated systems is multiplicative, it follows that the *logarithm* of $\Omega$ is additive. In other words, $\ln \Omega$ is an *extensive* quantity.

Following Boltzmann, we define *entropy S* as

$$S = k \ln \Omega. \tag{1.2.10}$$

At this stage $k$ is simply a constant; later we will identify it as Boltzmann's constant. We should note that $\ln \Omega$ is dimensionless, so $S$ will have the same dimensions as $k$.

Since the logarithm is a monotonic function, it follows that the equilibrium state will have maximal entropy. So, we immediately obtain the Second Law. And we now understand the Second Law from the microscopic point of view; it is hardly more than the tautology "we are most likely to observe the most probable state"!

## 1.3 Interactions — The Conditions for Equilibrium

When systems interact, their states will often change, the composite system evolving to a state of equilibrium. We shall investigate what determines the final state. We will see that quantities such as temperature emerge in the characterisation of these equilibrium states.

### 1.3.1 *Thermal interaction — Temperature*

Let us allow two, otherwise isolated, systems to exchange energy without changing volume or numbers of particles. In other words, we allow thermal interaction only; the systems are separated by a *diathermal* wall, as in Fig. 1.1.

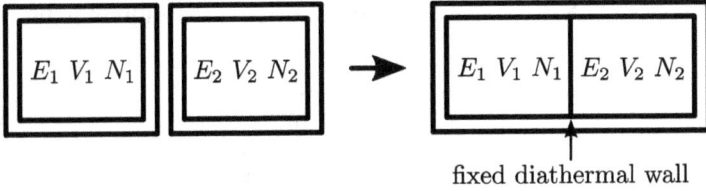

fixed diathermal wall

Fig. 1.1.   Thermal interaction.

Now,

$$\Omega_1 = \Omega_1(E_1, V_1, N_1)$$
$$\Omega_2 = \Omega_2(E_2, V_2, N_2)$$

(1.3.1)

and $V_1, N_1, V_2, N_2$ are all fixed.

The energies $E_1$ and $E_2$ can vary subject to the constraint $E_1 + E_2 = E_0$, a constant.

Our problem is this: after the two systems are brought together, what will be the equilibrium state? We know that the systems will exchange energy, and they will do this so as to maximise the total number of microstates for the composite system.

For different systems, the $\Omega$s multiply so that we have

$$\Omega = \Omega_1(E_1)\Omega_2(E_2),$$

(1.3.2)

we can ignore $V_1, N_1, V_2, N_2$ as they don't change.

The systems will exchange energy so as to maximise $\Omega$. Writing

$$\Omega = \Omega_1(E)\Omega_2(E_0 - E)$$

(1.3.3)

we allow the systems to vary $E$ so that $\Omega$ is a maximum:

$$\frac{\partial \Omega}{\partial E} = \frac{\partial \Omega_1}{\partial E}\Omega_2 - \Omega_1\frac{\partial \Omega_2}{\partial E} = 0$$

(1.3.4)

or

$$\frac{1}{\Omega_1}\frac{\partial \Omega_1}{\partial E} = \frac{1}{\Omega_2}\frac{\partial \Omega_2}{\partial E}.$$

(1.3.5)

In this form, the left-hand side is all about system 1 and the right-hand side is all about system 2. In other words, the equilibrium state is characterised by a certain property of system 1 being equal to a certain property of system 2.

We note that $\frac{1}{\Omega}\frac{\partial \Omega}{\partial E} = \frac{\partial \ln \Omega}{\partial E}$ so the equilibrium condition may be written as

$$\frac{\partial \ln \Omega_1}{\partial E} = \frac{\partial \ln \Omega_2}{\partial E}. \tag{1.3.6}$$

Observe the natural occurrence of the *logarithm* of $\Omega$.

And from the definition of entropy, $S = k \ln \Omega$, we see this means that the equilibrium state is specified by

$$\frac{\partial S_1}{\partial E} = \frac{\partial S_2}{\partial E}. \tag{1.3.7}$$

That is, when the systems have reached equilibrium, the quantity $\partial S/\partial E$ of system 1 is equal to $\partial S/\partial E$ of system 2. This is the condition for equilibrium when systems exchange only thermal energy.

Clearly, $\partial S/\partial E$ must be related to the temperature of the system. The Second Law requires that $\Delta S \geq 0$; that is

$$\Delta S = \left( \frac{\partial S_1}{\partial E} - \frac{\partial S_2}{\partial E} \right) \Delta E_1 \geq 0. \tag{1.3.8}$$

This means that

$$\begin{aligned}
E_1 \text{ increases (and } E_2 \text{ decreases)} \quad &\text{if} \quad \frac{\partial S_1}{\partial E} > \frac{\partial S_2}{\partial E} \\
E_1 \text{ decreases (and } E_2 \text{ increases)} \quad &\text{if} \quad \frac{\partial S_1}{\partial E} < \frac{\partial S_2}{\partial E}
\end{aligned} \tag{1.3.9}$$

so energy flows from systems with small $\partial S/\partial E$ to systems with large $\partial S/\partial E$.

Since we know that heat flows from hot systems to cold systems, we therefore identify

$$\begin{aligned}
\text{high } T &\equiv \text{low } \frac{\partial S}{\partial E} \\
\text{low } T &\equiv \text{high } \frac{\partial S}{\partial E}.
\end{aligned} \tag{1.3.10}$$

There is thus an *inverse* relation between $\partial S/\partial E$ and temperature.

We are led to define *statistical temperature* by

$$\frac{1}{T} = \frac{\partial S}{\partial E}. \tag{1.3.11}$$

When applied to the ideal gas, this will give us the result (Section 2.3.3)

$$pV = NkT,$$ (1.3.12)

and it is from *this* we conclude that the statistical temperature corresponds to the intuitive concept of temperature as measured by an ideal gas thermometer. Furthermore, the *scale* of temperatures will agree with the Kelvin scale (ice point at 273.18 K) when the constant $k$ in the definition of $S$ is identified with Boltzmann's constant.

When the derivative $\partial S / \partial E$ is evaluated, $N$ and $V$ are constant. So, the only energy flow is heat flow. Thus, the equation defining statistical temperature can also be written as

$$\Delta Q = T \Delta S.$$ (1.3.13)

We can now write the energy conservation expression for the First Law:

$$\Delta E = \Delta Q + \Delta W$$ (1.3.14)

as

$$\Delta E = T\Delta S - p\Delta V \quad \text{(for } pV \text{ systems).}$$ (1.3.15)

### 1.3.2 Volume change — Pressure

We now allow the volumes of the interacting systems to vary as well, subject to the total volume being fixed. Thus, we consider two systems separated by a *movable* diathermal wall, as in Fig. 1.2.

The constraints on this system are

$$E_1 + E_2 = E_0 = \text{const.}$$
$$V_1 + V_2 = V_0 = \text{const.}$$ (1.3.16)

while $N_1$ and $N_2$ are individually fixed.

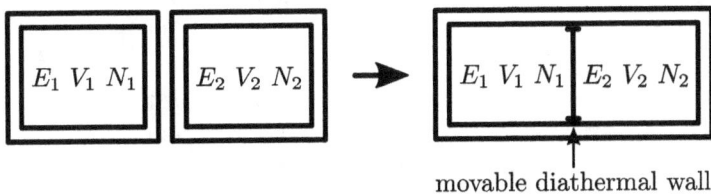

movable diathermal wall

Fig. 1.2. Mechanical and thermal interaction.

Maximising the entropy with respect to both energy flow and volume change then gives the two conditions

$$\frac{\partial S_1}{\partial E} = \frac{\partial S_2}{\partial E}$$
$$\frac{\partial S_1}{\partial V} = \frac{\partial S_2}{\partial V}. \tag{1.3.17}$$

The first of these gives, we know, the equality of temperature at equilibrium:

$$T_1 = T_2. \tag{1.3.18}$$

What does the second relation tell us? What is $\partial S / \partial V$? This may be found by rearranging the differential expression for the First Law:

$$dE = TdS - pdV. \tag{1.3.19}$$

This may be re-written as

$$dS = \frac{1}{T}dE + \frac{p}{T}dV \tag{1.3.20}$$

so just as we identified

$$\left.\frac{\partial S}{\partial E}\right|_V = \frac{1}{T}, \tag{1.3.21}$$

we now identify

$$\left.\frac{\partial S}{\partial V}\right|_E = \frac{p}{T}. \tag{1.3.22}$$

Thus, the condition that $\partial S / \partial V$ be the same for both systems means that $p/T$ must be the same. But we have already established that $T$ is the same so the new information is that at equilibrium the pressures are equalised:

$$p_1 = p_2. \tag{1.3.23}$$

A paradox arises if the movable wall is not diathermal, that is, if it is thermally isolating. Then one would conclude, from an analysis similar to that above, that while $p/T$ becomes equalised for the two sides, $T$ does not. On the other hand, a purely mechanical argument would say that the pressures $p$ should become equal. The paradox is resolved when one appreciates that without a flow of heat, thermodynamic equilibrium is not possible and so the entropy maximum principle is not applicable. Thus, $p/T$ will not be equalised. This issue is discussed in greater detail by Callen [9].

### 1.3.3  *Particle interchange — Chemical potential*

Let us keep the volumes of the two systems fixed, but allow particles to traverse the immobile diathermal wall, as in Fig. 1.3.

The constraints on this system are

$$E_1 + E_2 = E_0 = \text{const.}$$
$$N_1 + N_2 = N_0 = \text{const.} \tag{1.3.24}$$

while $V_1$ and $V_2$ are individually fixed.

Maximising the entropy with respect to both energy flow and particle flow then gives the two conditions

$$\frac{\partial S_1}{\partial E} = \frac{\partial S_2}{\partial E}$$
$$\frac{\partial S_1}{\partial N} = \frac{\partial S_2}{\partial N}. \tag{1.3.25}$$

The first of these gives, we know, the equality of temperature at equilibrium:

$$T_1 = T_2. \tag{1.3.26}$$

What does the second relation tell us? What is $\partial S/\partial N$? This may be found from the First Law in its extended form:

$$dE = TdS - pdV + \mu dN, \tag{1.3.27}$$

where $\mu$ is the chemical potential. This may be re-written as

$$dS = \frac{1}{T}dE + \frac{p}{T}dV - \frac{\mu}{T}dN, \tag{1.3.28}$$

so that we may identify

$$\left.\frac{\partial S}{\partial N}\right|_{E,V} = -\frac{\mu}{T}. \tag{1.3.29}$$

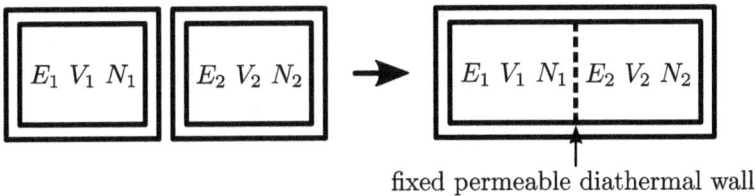

fixed permeable diathermal wall

Fig. 1.3.   Heat and particle exchange.

Thus, the condition that $\partial S / \partial N$ be the same for both systems means that $\mu / T$ must be the same. But we have already established that $T$ is the same so the new information is that at equilibrium the chemical potentials are equalised:

$$\mu_1 = \mu_2. \tag{1.3.30}$$

We see that just as pressure drives volume changes, chemical potential drives particle flow. And arguments similar to those of Section 1.3.1 indicate that particles flow from high values of $\mu$ to low values of $\mu$.

### 1.3.4   *Thermal interaction with the rest of the world — The Boltzmann factor*

For an isolated system, all microstates are equally likely; this is our Fundamental Postulate. It follows that the probability of the occurrence of a given *microstate* is given by

$$P_j = \frac{1}{\Omega}. \tag{1.3.31}$$

But what about a non-isolated system? What can we say about the occurrence of microstates of such a system? Here, the probability of a microstate will depend on properties of the surroundings.

In effect, we are seeking an extension of our Fundamental Postulate. We shall see how we can use the Fundamental Postulate itself to effect its own extension!

We consider a system interacting with its surroundings through a fixed diathermal wall; this non-isolated system can exchange thermal energy with its surroundings. We ask the question "what is the probability of this non-isolated system being in a given microstate"?

We shall idealise the surroundings by a "large" system, which we will call a *heat bath* or reservoir, as in Fig. 1.4. We shall regard the composite system of bath plus our system of interest as isolated — so to *this* we can apply the Fundamental Postulate. In this way, we shall be able to find the probability that the system of interest is in a particular microstate. This is the "wine bottle in the swimming pool" model of Reif [10].

The $\Omega$s multiply, thus

$$
\begin{array}{ccc}
\Omega_T & = & \Omega_B \times \Omega \\
\uparrow & \uparrow & \uparrow \\
\text{Total} & \text{Bath} & \text{System of interest}
\end{array}
\tag{1.3.32}
$$

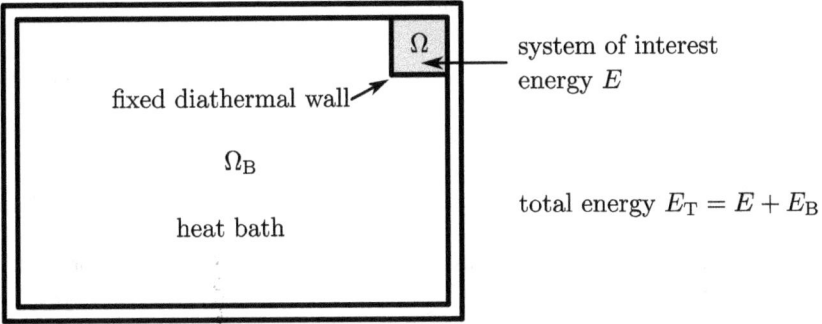

Fig. 1.4. Thermal interaction with the rest of the world.

or

$$\Omega_T = \Omega_B(E_T - E)\Omega(E). \tag{1.3.33}$$

Now, the Fundamental Postulate tells us that the probability the system of interest has energy $E$ is proportional to the number of microstates of the composite system that correspond to that energy partition

$$P(E) \propto \Omega_B(E_T - E)\Omega(E). \tag{1.3.34}$$

But here $\Omega(E) = 1$, since we are looking at a given *microstate* of energy $E$; there is one microstate. So,

$$P(E) \propto \Omega_B(E_T - E). \tag{1.3.35}$$

It depends solely on the bath. In terms of entropy, since $S = k \ln \Omega$,

$$P(E) \propto e^{S(E_T - E)/k} \tag{1.3.36}$$

where $S$ is the entropy of the *bath*. This type of expression, where probability is expressed in terms of entropy, is an inversion of the usual usage where entropy and other thermodynamic properties are found in terms of probabilities. This form was much used by Einstein in his treatment of fluctuations.

Now, the system of interest is very small compared with the bath; $E \ll E_T$. So, we can perform a Taylor expansion of $S$:

$$S(E_T - E) = S(E_T) - E\frac{\partial S}{\partial E} + \cdots \tag{1.3.37}$$

but

$$\frac{\partial S}{\partial E} = \frac{1}{T}, \tag{1.3.38}$$

the temperature of the bath, so that

$$S(E_T - E) = S(E_T) - \frac{E}{T},$$                    (1.3.39)

assuming we can ignore the higher terms. Then

$$P(E) \propto e^{S(E_T)/k} e^{-E/kT}.$$                    (1.3.40)

But the first term $e^{S(E_T)/k}$ is simply a constant, so we finally obtain the probability

$$P(E) \propto e^{-E/kT}.$$                    (1.3.41)

This is the probability that a system in equilibrium (with a bath) at a temperature $T$ will be found in a microstate of energy $E$. The exponential factor $e^{-E/kT}$ is known as the Boltzmann factor, the Boltzmann distribution function or the canonical distribution function.

The Boltzmann factor is a key result. Feynman says [11]:

> *This fundamental law is the summit of statistical mechanics, and the entire subject is either a slide-down from the summit, as the principle is applied to various cases, or the climb-up to where the fundamental law is derived and the concepts of thermal equilibrium and temperature are clarified.*

### 1.3.5   *Particle and energy exchange with the rest of the world —*
### *The Gibbs factor*

We now consider an extension of the Boltzmann factor to account for microstates where the number of particles may vary. Our system here can exchange both energy and particles with the rest of the world, as in Fig. 1.5. The microstate of our system of interest is now specified by a given energy and a given number of particles. We are asking: what is the probability the system of interest will be found in the microstate with energy $E$ and $N$ particles?

In this case, $\Omega_T$ is a function of both $E$ and $N$

$$\Omega_T = \Omega_B(E_T - E, N_T - N)\Omega(E, N).$$                    (1.3.42)

Now, the Fundamental Postulate tells us that the probability the system of interest has energy $E$ and $N$ particles is proportional to the number of

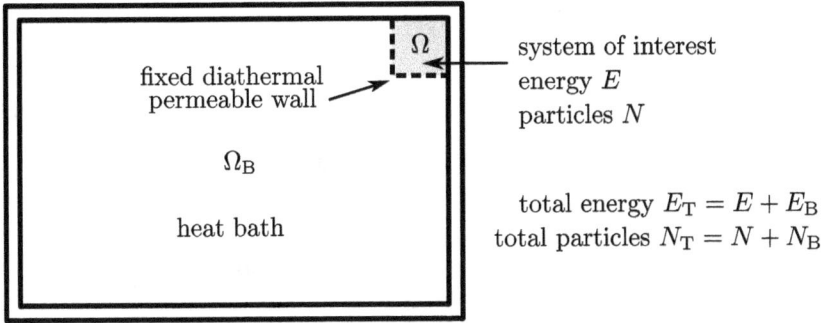

Fig. 1.5. Particle and energy exchange with the rest of the world.

microstates of the composite system that correspond to that energy and particle number partition

$$P(E,N) \propto \Omega_B(E_T - E, N_T - N)\Omega(E,N). \tag{1.3.43}$$

But as before, $\Omega(E,N) = 1$ since we are looking at a single microstate. So,

$$P(E,N) \propto \Omega_B(E_T - E, N_T - N). \tag{1.3.44}$$

It depends solely on the bath. In terms of entropy, since $S = k \ln \Omega$,

$$P(E,N) \propto e^{S(E_T - E, N_T - N)/k} \tag{1.3.45}$$

where $S$ is the entropy of the *bath*.

Now, the system of interest is very small compared with the bath; $E \ll E_T$ and $N \ll N_T$. So, as before, we can perform a Taylor expansion of $S$:

$$S(E_T - E, N_T - N) = S(E_T, N_T) - E\frac{\partial S}{\partial E} - N\frac{\partial S}{\partial N} + \cdots \tag{1.3.46}$$

but

$$\begin{aligned} \frac{\partial S}{\partial E} &= \frac{1}{T}, \\ \frac{\partial S}{\partial N} &= -\frac{\mu}{T}, \end{aligned} \tag{1.3.47}$$

so that

$$S(E_T - E, N_T - N) = S(E_T, N_T) - \frac{E}{T} + \frac{\mu N}{T} \tag{1.3.48}$$

assuming we can ignore the higher terms. Then

$$P(E,N) \propto e^{S(E_T, N_T)/k} e^{-(E - \mu N)/kT}. \tag{1.3.49}$$

But the first term $e^{S(E_T,N_T)/k}$ is simply a constant, so we finally obtain the probability

$$P(E,N) \propto e^{-(E-\mu N)/kT}. \tag{1.3.50}$$

This is the probability that a system in equilibrium (with a bath) at a temperature $T$ and chemical potential $\mu$ will be found in a microstate of energy $E$, with $N$ particles. The exponential factor $e^{-(E-\mu N)/kT}$ is sometimes known as the Gibbs factor, the Gibbs distribution function or the grand canonical distribution function.

## 1.4   Thermodynamic Averages

The importance of the previously derived probability distribution functions is that they may be used in calculating average (observed) values of various macroscopic properties of systems. In this way, the aims of Statistical Mechanics, as outlined in Section 1.1.1, are achieved.

### 1.4.1   *The partition function*

The probability that a system is in the $j$th microstate, of energy $E_j(N,V)$, is given by the Boltzmann factor, which we write as

$$P_j(N,V,T) = \frac{e^{-E_j(N,V)/kT}}{Z(N,V,T)} \tag{1.4.1}$$

where the normalisation quotient $Z$ is given by

$$Z(N,V,T) = \sum_i e^{-E_i(N,V)/kT}. \tag{1.4.2}$$

Here, we have been particular to indicate the functional dependencies. Energy eigenstates depend on the size of the system (standing waves) and the number of particles. And we are considering our system to be in thermal contact with a heat bath; thus, the temperature dependence. We do not, however, allow particle interchange.

The quantity $Z$ is called the (canonical) *partition function*. The letter $Z$ stands for the German word *Zustandssumme*, meaning "sum over states". Although $Z$ has been introduced simply as a normalisation factor, we shall see that it is a very useful quantity indeed.

### 1.4.2 *Gibbs expression for entropy*

For an isolated system, the micro–macro connection is given by the Boltzmann formula $S = k \ln \Omega$, where $\Omega$ is a function of the extensive variables of the system

$$\Omega = \Omega(E, V, N). \tag{1.4.3}$$

But now, at a specified temperature, the energy $E$ is not fixed; rather, it fluctuates about a mean value $\langle E \rangle$.

To make the micro–macro connection when $E$ is not fixed, we must generalise the Boltzmann expression for entropy. We shall consider a collection of (macroscopically) identical systems in thermal contact, Fig. 1.6.[1] The composite system may be regarded as being isolated, so to *that* we may apply the rule $S = k \ln \Omega$, and from that the mean entropy of a representative single system may be found.

Let us consider $M$ identical systems, and let there be $n_j$ of these systems in the $j$th microstate. We assume that this is a collection of a very large number of systems. Then the systems other than our one of particular interest may be regarded as a heat bath.

collection of identical systems

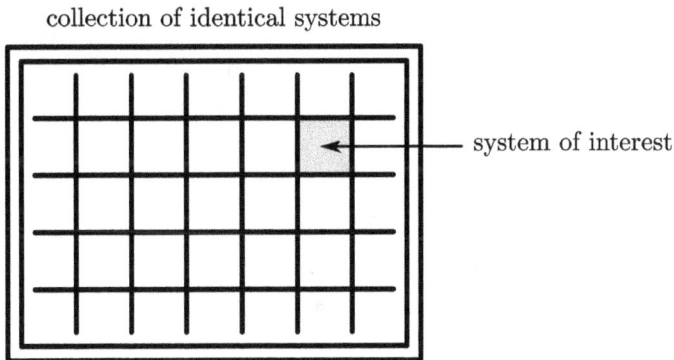
system of interest

Fig. 1.6.   Gibbs ensemble for evaluating generalised entropy.

---

[1] Such a collection is called a Gibbs ensemble; the designation will become clear when you have studied Section 1.6.2.

The number of possible microstates of the composite system corresponds to the number of ways of rearranging the subsystems:

$$\Omega = \frac{M!}{n_1! \, n_2! \, n_3! \ldots} \qquad (1.4.4)$$

and the total entropy of the composite system is then

$$S_{\text{tot}} = k \ln \left( \frac{M!}{n_1! \, n_2! \, n_3! \ldots} \right). \qquad (1.4.5)$$

Since all the numbers here are large, we may make use of Stirling's approximation for the logarithm of a factorial, $\ln n! \approx n \ln n - n$, so that

$$S_{\text{tot}} = k \left( M \ln M - M - \sum_j n_j \ln n_j + \sum_j n_j \right), \qquad (1.4.6)$$

the cancellation occurring because $M = \sum_j n_j$. Then if we express the first $M$ as $\sum_j n_j$, we have

$$
\begin{aligned}
S_{\text{tot}} &= k \left( \sum_j n_j \ln M - \sum_j n_j \ln n_j \right) \\
&= -k \sum_j n_j \ln \left( \frac{n_j}{M} \right).
\end{aligned}
\qquad (1.4.7)
$$

We are interested in the mean entropy of our particular system. We have been considering a composite of $M$ (macroscopically) similar systems, so the mean entropy of our system is simply the total entropy divided by $M$. Thus,

$$S = -k \sum_j \frac{n_j}{M} \ln \left( \frac{n_j}{M} \right). \qquad (1.4.8)$$

But $n_j/M$ is the fraction of systems in the $j$th state, or the *probability* of finding our representative system in the $j$th state:

$$P_j = \frac{n_j}{M}. \qquad (1.4.9)$$

So, we can now express the mean entropy of a non-isolated system in terms of the state probabilities as

$$S = -k \sum_j P_j \ln P_j, \qquad (1.4.10)$$

or

$$S = -k\langle \ln P \rangle, \tag{1.4.11}$$

the average value of the logarithm of the microstate probabilities. Equation (1.4.10) is known as the Gibbs expression for the entropy. For an isolated system, this reduces to the original Boltzmann expression.

### 1.4.3 Free energy

In the Gibbs expression for entropy, we actually know the values for the probabilities — they are given by the Boltzmann factor:

$$P_j(N, V, T) = \frac{e^{-E_j(N,V)/kT}}{Z(N, V, T)} \tag{1.4.12}$$

where we recall that the normalisation factor is given by the sum over states, the partition function $Z$

$$Z(N, V, T) = \sum_i e^{-E_i(N,V)/kT}. \tag{1.4.13}$$

We then have

$$\ln P_j = -\left(\frac{E_j}{kT} + \ln Z\right). \tag{1.4.14}$$

Thus,

$$S = k\left\langle \frac{E_j}{kT} + \ln Z \right\rangle \tag{1.4.15}$$

and since $Z$ is independent of $j$, we have

$$S = \frac{\langle E \rangle}{T} + k \ln Z. \tag{1.4.16}$$

Now, in the spirit of thermodynamics we don't distinguish between mean and actual values — since fluctuations will be of order $1/\sqrt{N}$. Thus, we write

$$E - TS = -kT \ln Z. \tag{1.4.17}$$

The quantity $E - TS$ is rather important and it is given a special name: *Helmholtz free energy*, or simply *free energy*. The symbol $F$ is used[2]:

$$F = E - TS, \tag{1.4.18}$$

so that we can write

$$F = -kT \ln Z. \tag{1.4.19}$$

It might be helpful to recall that $Z(N, V, T)$ is a *dimensionless* quantity.

### 1.4.4 *Thermodynamic variables*

A host of thermodynamic variables can be obtained from the partition function. This is seen from the differential of the free energy. Since

$$dE = TdS - pdV + \mu dN, \tag{1.4.20}$$

it follows that

$$dF = -SdT - pdV + \mu dN. \tag{1.4.21}$$

We can then immediately identify the various partial derivatives:

$$\left.\begin{aligned}
S &= -\left.\frac{\partial F}{\partial T}\right|_{V,N} = kT \left.\frac{\partial \ln Z}{\partial T}\right|_{V,N} + k \ln Z \\
p &= -\left.\frac{\partial F}{\partial V}\right|_{T,N} = kT \left.\frac{\partial \ln Z}{\partial V}\right|_{T,N} \\
\mu &= \left.\frac{\partial F}{\partial N}\right|_{T,V} = -kT \left.\frac{\partial \ln Z}{\partial N}\right|_{T,V}
\end{aligned}\right\} \tag{1.4.22}$$

Since $E = F + TS$, we can then express the internal energy as

$$E = kT^2 \left.\frac{\partial \ln Z}{\partial T}\right|_{V,N}. \tag{1.4.23}$$

Thus, we see that once the partition function is evaluated by summing over the states, all relevant thermodynamic variables can be obtained by differentiating $Z$.

---

[2]Some texts use $A$ for the Helmholtz free energy.

### 1.4.5 *The beta trick*

There is a mathematical trick that can be very useful in many applications in Statistical Mechanics. This relies on a change of variables to "inverse temperature" defined by

$$\beta = 1/kT. \tag{1.4.24}$$

Many calculations are more easily done in this way; moreover, the physical meaning of such calculations can be more apparent.

As an example, let's consider a direct calculation of the mean energy of a system $E = \sum_j E_j P_j$ (the internal energy). We now write this mean as

$$E = \frac{1}{Z} \sum_j E_j e^{-\beta E_j}. \tag{1.4.25}$$

We observe that the $E_j$ in the sum may be "brought down" by a differentiation with respect to $\beta$:

$$E_j e^{-\beta E_j} = -\frac{\partial}{\partial \beta} e^{-\beta E_j}. \tag{1.4.26}$$

So, the internal energy is simply

$$E = -\frac{1}{Z} \sum_j \frac{\partial}{\partial \beta} e^{-\beta E_j}. \tag{1.4.27}$$

Then by swapping the order of summation and differentiation, this reduces to

$$E = -\frac{1}{Z} \frac{\partial}{\partial \beta} \sum_j e^{-\beta E_j} \tag{1.4.28}$$

or

$$E = -\frac{1}{Z} \frac{\partial Z}{\partial \beta}. \tag{1.4.29}$$

Equivalently, we may write this in terms of $\ln Z$ as

$$E = -\frac{\partial \ln Z}{\partial \beta}. \tag{1.4.30}$$

Upon changing variables back to $T = 1/k\beta$, this is equivalent to

$$E = kT^2 \frac{\partial \ln Z}{\partial T}; \tag{1.4.31}$$

and we have recovered Eq. (1.4.23).

### 1.4.6  *Fluctuations*

An isolated system has a well-defined energy. But the energy of a system
in contact with a heat bath continually fluctuates about some mean value.
The equilibrium with the heat bath is characterised by a temperature $T$
and the system's mean energy is calculated from Eq. (1.4.23) or equivalently, from Eq. (1.4.29). In this section, we shall (in contrast to the usual
practice in Statistical Mechanics) distinguish between instantaneous and
mean values. So, here we write this equation as

$$\langle E \rangle = -\frac{1}{Z}\frac{\partial Z}{\partial \beta}. \tag{1.4.32}$$

What is the magnitude of the fluctuations about this mean?

We shall evaluate the RMS (root mean square) of the energy fluctuations $\sigma_E$, defined by

$$\sigma_E = \left\langle (E - \langle E \rangle)^2 \right\rangle^{1/2}. \tag{1.4.33}$$

By expanding out (the square), we obtain

$$\begin{aligned}\sigma_E^2 &= \langle E^2 \rangle - 2\langle E \rangle^2 + \langle E \rangle^2 \\ &= \langle E^2 \rangle - \langle E \rangle^2. \end{aligned} \tag{1.4.34}$$

In order to evaluate $\sigma_E$ using this equation, we have $\langle E \rangle^2$ by squaring
Eq. (1.4.32). We now apply the beta trick to the calculation of $\langle E^2 \rangle$. We
require to find

$$\langle E^2 \rangle = \frac{1}{Z}\sum_j E_j^2 e^{-\beta E_j} \tag{1.4.35}$$

so we need to bring down *two* lots of $E_j$. This can be done with *two* differ-
entiations

$$E_j^2 e^{-\beta E_j} = \frac{\partial^2}{\partial \beta^2} e^{-\beta E_j}. \tag{1.4.36}$$

So, the mean square energy is simply

$$\langle E^2 \rangle = -\frac{1}{Z}\sum_j \frac{\partial^2}{\partial \beta^2} e^{-\beta E_j}. \tag{1.4.37}$$

Then by swapping the order of summation and differentiation, this reduces to

$$\langle E^2 \rangle = \frac{1}{Z} \frac{\partial^2}{\partial \beta^2} \sum_j e^{-\beta E_j}$$

$$= \frac{1}{Z} \frac{\partial^2 Z}{\partial \beta^2}.$$

(1.4.38)

We need the second derivative of $Z$. The first derivative is given, from Eq. (1.4.32), by

$$\frac{\partial Z}{\partial \beta} = -Z \langle E \rangle.$$

(1.4.39)

Differentiating again gives

$$\frac{\partial^2 Z}{\partial \beta^2} = -\frac{\partial Z}{\partial \beta} \langle E \rangle - Z \frac{\partial \langle E \rangle}{\partial \beta}$$

$$= Z \langle E \rangle^2 - Z \frac{\partial \langle E \rangle}{\partial \beta},$$

(1.4.40)

so that

$$\langle E^2 \rangle = \langle E \rangle^2 - \frac{\partial \langle E \rangle}{\partial \beta}.$$

(1.4.41)

This gives us $\sigma_E^2$ as

$$\sigma_E^2 = -\frac{\partial \langle E \rangle}{\partial \beta}.$$

(1.4.42)

We now return to conventional notation, discarding the averaging brackets, so that

$$\sigma_E^2 = -\frac{\partial E}{\partial \beta}$$

$$= kT^2 \frac{\partial E}{\partial T}$$

(1.4.43)

upon changing back to $T$ as the independent variable. However, since we recognise the derivative of energy, $\partial E / \partial T$, as the heat capacity, the RMS

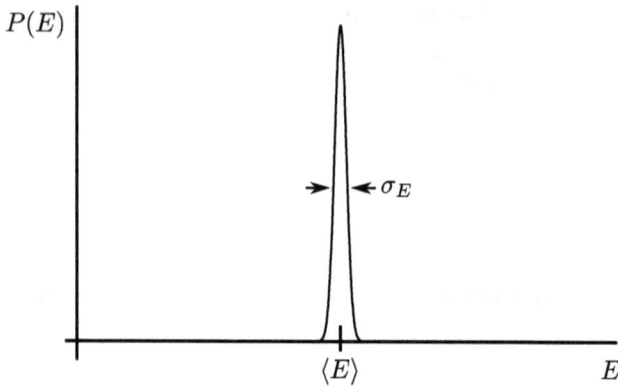

Fig. 1.7.   Fluctuations in energy for a system at fixed temperature.

variation in the energy may then be expressed, in terms of the heat capacity $C$, as

$$\sigma_E^2 = kT^2 C_V \tag{1.4.44}$$

(the derivatives, following Eq. (1.4.23) are taken at constant $V$) or

$$\sigma_E = \sqrt{kT^2 C_V}. \tag{1.4.45}$$

Since $C_V$ and $E$ are both proportional to the number of particles in the system (extensive variables), the *fractional* fluctuations in energy vary as

$$\frac{\sigma_E}{\langle E \rangle} \sim \frac{1}{\sqrt{N}} \tag{1.4.46}$$

which gets smaller and smaller as $N$ increases, as in Fig. 1.7.

We thus see that the significance of fluctuations vanishes in the $N \to \infty$ limit, while $N/V$ remains constant: the thermodynamic limit. And it is in this limit that statistical mechanics has its greatest applicability.

We might note that the usual use of Eq. (1.4.44) is the opposite: to show how the heat capacity is related to the energy fluctuations

$$C_V = \frac{1}{kT^2} \sigma_E^2. \tag{1.4.47}$$

### 1.4.7   *The grand partition function*

Here, we are concerned with systems of variable numbers of particles. The energy of a (many-body) state will depend on the number of particles in the system. As before, we label the (many-body) states of the system by $j$, but note that the $j$th state will be different for different $N$. In other words, we need the pair $\{N, j\}$ for specification of a state. The probability that a system has $N$ particles and is in the $j$th microstate, corresponding to a total energy $E_{N,j}$, is given by the Gibbs factor, which we write as

$$P_{N,j}(V, T, \mu) = \frac{e^{-[E_{N,j}(N,V) - \mu N]/kT}}{\Xi(V, T, \mu)} \tag{1.4.48}$$

where the normalisation constant $\Xi$ is given by

$$\Xi(V, T, \mu) = \sum_{N,j} e^{-[E_{N,j}(N,V) - \mu N]/kT}. \tag{1.4.49}$$

Here, both $T$ and $\mu$ are properties of the bath. The quantity $\Xi$ is called the *grand canonical partition function*. It may also be written in terms of the canonical partition function $Z$ for different $N$ as

$$\Xi(V, T, \mu) = \sum_{N} Z(N, V, T) e^{\mu N/kT}. \tag{1.4.50}$$

We will see that $\Xi$ also is a useful quantity.

### 1.4.8 *The grand potential*

The generalised expression for entropy in this case is

$$S = -k \left\langle \ln P_{N,j} \right\rangle. \tag{1.4.51}$$

Here, the probabilities are given by the Gibbs factor:

$$P_{N,j}(V, T, \mu) = \frac{e^{-[E_{N,j}(N,V) - \mu N]/kTkT}}{\Xi(V, T, \mu)} \tag{1.4.52}$$

where the normalisation factor, the sum over states, is the grand partition function $\Xi$. We then have

$$\ln P_{N,j} = -\left(\frac{E_{N,j}}{kT} - \frac{\mu N}{kT} + \ln \Xi\right). \tag{1.4.53}$$

Thus,

$$S = k \left\langle \frac{E_{N,j}}{kT} - \frac{\mu N}{kT} + \ln \Xi \right\rangle \tag{1.4.54}$$

which is given by

$$S = \frac{\langle E \rangle}{T} - \frac{\mu \langle N \rangle}{T} + k \ln \Xi. \tag{1.4.55}$$

Now, in the spirit of thermodynamics we don't distinguish between mean and actual values — since fluctuations will be of order $1/\sqrt{N}$. Thus, we write

$$E - TS - \mu N = -kT \ln \Xi. \tag{1.4.56}$$

The Euler relation (see Appendix A) is

$$E = TS - pV + \mu N. \tag{1.4.57}$$

Thus, the combination $E - TS - \mu N$ is equal to $-pV$, so that we can write

$$pV = kT \ln \Xi. \tag{1.4.58}$$

The quantity $pV$ is referred to as the *grand potential*.

### 1.4.9   *Thermodynamic variables*

Just as with the partition function, a host of thermodynamic variables can be obtained from the grand partition function. This is seen from the differential of the grand potential. It's easiest to differentiate $pV$ as $-E + TS + \mu N$; that is

$$\begin{aligned}
\mathrm{d}(pV) &= \mathrm{d}(-E + TS + \mu N) \\
&= -\mathrm{d}E + T\mathrm{d}S + S\mathrm{d}T + \mu \mathrm{d}N + N\mathrm{d}\mu.
\end{aligned} \tag{1.4.59}$$

But since $\mathrm{d}E = T\mathrm{d}S - p\mathrm{d}V + \mu \mathrm{d}N$, this gives

$$\mathrm{d}(pV) = S\mathrm{d}T + p\mathrm{d}V + N\mathrm{d}\mu. \tag{1.4.60}$$

We can then identify the various partial derivatives:

$$
\left.
\begin{aligned}
S &= \left.\frac{\partial(pV)}{\partial T}\right|_{V,\mu} = kT \left.\frac{\partial \ln \Xi}{\partial T}\right|_{V,\mu} + k \ln \Xi \\
p &= \left.\frac{\partial(pV)}{\partial V}\right|_{T,\mu} = kT \left.\frac{\partial \ln \Xi}{\partial V}\right|_{T,\mu} = \frac{kT}{V} \ln \Xi \\
N &= \left.\frac{\partial(pV)}{\partial \mu}\right|_{T,V} = kT \left.\frac{\partial \ln \Xi}{\partial \mu}\right|_{T,V}.
\end{aligned}
\right\}
\tag{1.4.61}
$$

Since $E - \mu N = -pV + TS$, we can then express the combination $E - \mu N$ as

$$
E - \mu N = kT^2 \left.\frac{\partial \ln \Xi}{\partial T}\right|_{V,\mu}.
\tag{1.4.62}
$$

Thus, we see that once the grand partition function is evaluated by summing over the states, all relevant thermodynamic variables can be obtained by differentiating $\Xi$.

## 1.5 Quantum Distributions

### 1.5.1 *Bosons and Fermions*

All particles in nature can be classified into one of two groups according to the behaviour of their wave function under the exchange of identical particles. For simplicity, let us consider just two identical particles. The wave function can then be represented as

$$
\Psi = \Psi(r_1, r_2)
\tag{1.5.1}
$$

where

$r_1$ is the position of the first particle

and

$r_2$ is the position of the second particle.

Let us interchange the particles. We denote the operator that effects this by $\mathscr{P}$ (the permutation operator). Then

$$
\mathscr{P}\Psi(r_1, r_2) = \Psi(r_2, r_1).
\tag{1.5.2}
$$

We are interested in the behaviour of the wave function under interchange of the particles. So far, we have not drawn much of a conclusion. Let us now perform the swapping operation again. Then we have

$$\mathscr{P}^2 \Psi(r_1, r_2) = \mathscr{P}\Psi(r_2, r_1)$$
$$= \Psi(r_1, r_2); \tag{1.5.3}$$

the effect is to return the particles to their original states. Thus, the operator $\mathscr{P}$ must obey

$$\mathscr{P}^2 = 1. \tag{1.5.4}$$

And taking the square root of this, we find for $\mathscr{P}$.

$$\mathscr{P} = \pm 1. \tag{1.5.5}$$

In other words, the effect of swapping two identical particles is either to leave the wave function unchanged or to change the sign of the wave function.

This property continues for all time since the permutation operator commutes with the Hamiltonian. Thus, all particles in nature belong to one class or the other. Particles for which

$$\mathscr{P} = +1 \text{ are called } \textit{bosons}$$

while those for which

$$\mathscr{P} = -1 \text{ are called } \textit{fermions}.$$

Fermions have the important property of not permitting multiple occupancy of quantum states. Consider two particles in the same state, at the same position $r$. The wave function is then

$$\Psi = \Psi(r, r). \tag{1.5.6}$$

Swapping over the particles, we have

$$\mathscr{P}\Psi = -\Psi. \tag{1.5.7}$$

But $\Psi = \Psi(r, r)$, so that $\mathscr{P}\Psi = +\Psi$ since both particles are in the same state. The conclusion is that

$$\Psi(r, r) = -\Psi(r, r) \tag{1.5.8}$$

and this can only be so if

$$\Psi(r, r) = 0. \tag{1.5.9}$$

Now, since $\Psi$ is related to the *probability* of finding particles in the given state, the result $\Psi = 0$ implies a state of zero probability — an impossible state. We conclude that it is impossible to have more than one fermion in a given quantum state.

This discussion was carried out using $r_1$ and $r_2$ to denote *position* states. However, that is not an important restriction. In fact, they could have designated any sort of quantum state and the same argument would follow.

This is the explanation of the Pauli exclusion principle obeyed by electrons.

We conclude:

- For bosons, we can have any number of particles in a quantum state.
- For fermions, we can have either 0 or 1 particle in a quantum state.

But what determines whether a given particle is a boson or a fermion? The answer is provided by quantum field theory. And it depends on the *spin* of the particle. Particles whose spin angular momentum is an integral multiple of $\hbar$ are bosons, while particles whose spin angular momentum is integer plus a half $\hbar$ are fermions. (In quantum theory, $\hbar/2$ is the smallest unit of spin angular momentum.) It is not straightforward to demonstrate this fundamental connection between spin and statistics. Feynman's heroic attempt is contained in his 1986 Dirac memorial lecture [12]. However, a slightly more accessible account is contained in Tomonaga's book *The Story of Spin* [13].

For some elementary particles, we have:

$$
\left.
\begin{array}{c}
\text{electrons} \\
\text{protons} \\
\text{neutrons}
\end{array}
\right\}
\left.
S = \frac{1}{2}
\right\}
\rightarrow \text{fermions}
$$

$$
\left.
\begin{array}{cc}
\text{photon} & S = 1 \\
\left.
\begin{array}{c}
\pi \text{ meson} \\
K \text{ meson}
\end{array}
\right\} & S = 0
\end{array}
\right\}
\rightarrow \text{bosons.}
$$

For composite particles (such as atoms), we simply add the spins of the constituent parts. And since protons, neutrons and electrons are all fermions, we can say:

- odd number of fermions → fermion;
- even number of fermions → boson.

The classic example of this is the two isotopes of helium. Thus,

- $^3$He is a fermion;
- $^4$He is a boson.

Although $^3$He and $^4$He are *chemically* equivalent, at low temperatures the isotopes have very different behaviour.

### 1.5.2 *Grand potential for identical particles*

The grand potential allows the treatment of systems of variable numbers of particles. We may exploit this in the study of systems of non-interacting (or weakly interacting) particles in the following way. We focus attention on a single-particle state, which we label by $k$. The state of the entire system is specified when we know how many particles are in each different (single-particle) quantum state.

$$\text{Many-particle state} \equiv \{n_1, n_2, \ldots, n_k, \ldots\}$$

$$\text{Energy of state} = \sum_k n_k \varepsilon_k$$

$$\text{No. of particles} = \sum_k n_k. \tag{1.5.10}$$

Here, $\varepsilon_k$ is the energy of the $k$th single-particle state. Note that the $\varepsilon_k$ are independent of $N$.

Now, since the formalism of the grand potential is appropriate for systems that exchange particles and energy with their surroundings, we may consider as our "system" the subsystem comprising the particles in a given state $k$. For this subsystem,

$$E = n_k \varepsilon_k,$$
$$N = n_k, \tag{1.5.11}$$

so that the probability of observing this, i.e. the probability of finding $n_k$ particles in the $k$th state (provided this is allowed by the statistics), is

$$P_{n_k}(V, T, \mu) = \frac{e^{-(n_k \varepsilon_k - n_k \mu)/kT}}{\Xi_k} \tag{1.5.12}$$

where the grand partition function for the subsystem can be written as

$$\Xi_k = \sum_{n_k} \left\{ e^{-(\varepsilon_k - \mu)/kT} \right\}^{n_k}. \qquad (1.5.13)$$

Here, $n_k$ takes only values 0 and 1 for fermions and $0, 1, 2, \ldots, \infty$ for bosons. The grand potential for the "system" is

$$
\begin{aligned}
(pV)_k &= kT \ln \Xi_k \\
&= kT \ln \sum_{n_k} \left\{ e^{-(\varepsilon_k - \mu)/kT} \right\}^{n_k}.
\end{aligned} \qquad (1.5.14)
$$

The grand partition function for the entire system is the product

$$\Xi = \prod_k \Xi_k, \qquad (1.5.15)$$

so that the grand potential (and any other extensive quantity) for the entire system is found by summing over all single-particle state contributions:

$$pV = \sum_k (pV)_k. \qquad (1.5.16)$$

### 1.5.3 The Fermi–Dirac distribution

For fermions, the grand potential for the single state is

$$
\begin{aligned}
(pV)_k &= kT \ln \sum_{n_k = 0,1} \left\{ e^{-(\varepsilon_k - \mu)/kT} \right\}^{n_k} \\
&= kT \ln \left\{ 1 + e^{-(\varepsilon_k - \mu)/kT} \right\}.
\end{aligned} \qquad (1.5.17)
$$

From this we can find the mean number of particles in the state using

$$
\begin{aligned}
\bar{n}_k &= \left. \frac{\partial (pV)_k}{\partial \mu} \right|_{T,V} = \frac{e^{-(\varepsilon_k - \mu)/kT}}{1 + e^{-(\varepsilon_k - \mu)/kT}} \\
&= \frac{1}{e^{(\varepsilon_k - \mu)/kT} + 1}.
\end{aligned} \qquad (1.5.18)
$$

This is known as the Fermi–Dirac distribution function.

The grand potential for the entire system of fermions is found by summing the single-state grand potentials

$$pV = kT \sum_k \ln \left\{ 1 + e^{-(\varepsilon_k - \mu)/kT} \right\}. \qquad (1.5.19)$$

### 1.5.4  *The Bose–Einstein distribution*

For bosons the grand potential for the single state is

$$(pV)_k = kT \ln \sum_{n_k=0}^{\infty} \left\{ e^{-(\varepsilon_k - \mu)/kT} \right\}^{n_k}$$

$$= kT \ln \left\{ \frac{1}{1 - e^{-(\varepsilon_k - \mu)/kT}} \right\} \qquad (1.5.20)$$

$$= -kT \ln \left\{ 1 - e^{-(\varepsilon_k - \mu)/kT} \right\}$$

assuming the geometric progression is convergent. From this we can find the mean number of particles in the state using

$$\bar{n}_k = \left. \frac{\partial (pV)_k}{\partial \mu} \right|_{T,V} = \frac{e^{-(\varepsilon_k - \mu)/kT}}{1 - e^{-(\varepsilon_k - \mu)/kT}}$$

$$= \frac{1}{e^{(\varepsilon_k - \mu)/kT} - 1}. \qquad (1.5.21)$$

This is known as the Bose–Einstein distribution function.

The grand potential for the entire system of bosons is found by summing the single-state grand potentials

$$pV = -kT \sum_k \ln \left\{ 1 - e^{-(\varepsilon_k - \mu)/kT} \right\}. \qquad (1.5.22)$$

(An elegant derivation of the Bose and the Fermi distributions which indicates how the + and − signs in the denominators arise directly from the eigenvalue of the $\mathscr{P}$ operator is given in the Quantum Mechanics text book by Merzbacher [14]. Beware, however — it uses the method of Second Quantisation.) See also the simple derivation of the Bose and Fermi distributions given in Appendix E, inspired by Feynman, based on the canonical distribution function.

### 1.5.5  *The classical limit — The Maxwell–Boltzmann distribution*

The Bose–Einstein and Fermi–Dirac distributions give the mean numbers of particles in the microstate of energy $\varepsilon_j$ as a function of $(\varepsilon_j - \mu)/kT$. When this quantity is large, we observe two things. Firstly, the denominator of the distributions will be very much larger than one, so the +1 or −1 distinguishing fermions from bosons may be neglected. And secondly, the large value for the denominator means that the $\bar{n}_j$, the mean occupation of the state, will be very much less than unity.

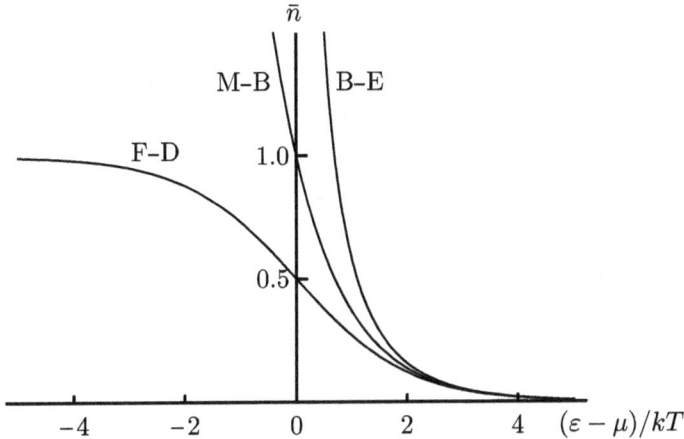

Fig. 1.8. Fermi–Dirac, Maxwell–Boltzmann and Bose–Einstein distribution functions.

This condition will apply to all states, down to the ground state of $\varepsilon_j = 0$, if $\mu/kT$ is large and negative. This is the classical limit where the issue of multiple state occupancy does not arise and the distinction between fermions and bosons becomes unimportant. We refer to such (hypothetical) particles as maxwellons, obeying Maxwell–Boltzmann statistics. Thus, for these particles, the mean number of particles in the state is given by

$$\bar{n}_k = e^{-(\varepsilon_k - \mu)/kT}. \qquad (1.5.23)$$

This is essentially the Boltzmann distribution function.

The three distribution functions are shown in Fig. 1.8. Observe, in particular, that when $\mu = \varepsilon$, the Fermi occupation is one half, the Maxwell occupation is unity and the Bose occupation is infinite.

## 1.6 Classical Statistical Mechanics

### 1.6.1 *Phase space and classical states*

The formalism of statistical mechanics developed thus far relies very much, at the microscopic level, on the use of (micro)states. We count the number of states, we sum over states, etc. This is all very convenient to do within the framework of a quantum-mechanical description of systems where states of a (finite or bound) system are discrete, but what about classical systems. How is the formalism of *classical* statistical mechanics developed — what is a "classical state"?

To specify a state in classical mechanics, we must know the position and velocity of all particles in the system. (Position *and* velocity, since the equations of motion — Newton's laws — are second-order differential equations.) For reasons which become clear in the Lagrangian and the Hamiltonian formulations of mechanics, it proves convenient to use the position and the *momentum* rather than velocity of the particles in the system. This is because it is then possible to work in terms of *generalised* coordinates and momenta — such as angles and angular momenta — in a completely general way; one is not constrained to a particular coordinate system. Thus, we will say that a classical state is specified by the coordinates and the momenta of all the constituent particles. A single-particle has three coordinates, $x$, $y$, $z$, and three momentum components, $p_x$, $p_y$, $p_z$, so it needs *six* components to specify its state. The generalised coordinates are conventionally denoted by $q$ and the momenta by $p$. This $p, q$ space is called *phase space*.

The classical state of a particle is denoted by a *point* in phase space. Its evolution in time is represented by a *curve* in phase space, as in Fig. 1.9. It should be evident that during its evolution, the phase curve of a point cannot intersect itself since there is a unique path proceeding from each location in phase space.

There is a difficulty in counting these states since the $p$ and $q$ vary continuously; there would be an *infinite* number of states in any region of phase space. In classical statistical mechanics, it is expedient to erect a *grid* in phase space with cells $\Delta q_x \, \Delta q_y \, \Delta q_z \, \Delta p_x \, \Delta p_y \, \Delta p_z$. Then the classical analogue of a quantum microstate is a cell in phase space. In other words,

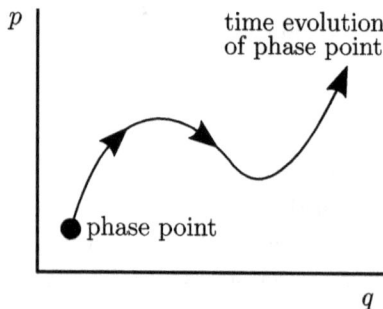

Fig. 1.9.   Trajectory in phase space.

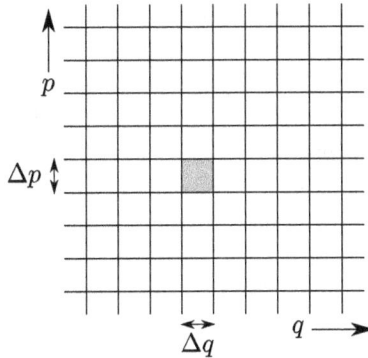

Fig. 1.10.  Classical microstate in phase space.

a system is in a given microstate if it is in a specified cell in phase space, as in Fig. 1.10.

What should be the size of these cells in phase space? From the historical perspective, we would say that the "area" of each product $\Delta p \, \Delta q$ will be a constant $h$ and then we will observe:

- how that constant appears in calculated physical quantities and
- how such calculated physical quantities compare with those calculated quantum-mechanically.

Looking ahead we will see that $h$ is identified with Planck's constant. This might be surprising. But we note that the Uncertainty Principle of quantum mechanics tells us we cannot discern a state to better than a $\Delta p \, \Delta q$ cell smaller than Planck's constant.

The general rule is that the sums over states in the quantum case correspond to integrals over phase space, with appropriate normalisation, in the classical case:

$$\sum_{\substack{\text{single-}\\\text{particle}\\\text{states}}} \cdots \quad \rightarrow \quad \frac{1}{h^3} \int \cdots \mathrm{d}^3 p \, \mathrm{d}^3 q. \tag{1.6.1}$$

### 1.6.2  Boltzmann and Gibbs phase spaces

The microstate of a system is represented by a point in phase space. A different microstate will be represented by a different point in the phase space. In developing the statistical approach to mechanics, we must talk about different microstates — so we are considering different points in phase space and the probabilities associated with them. Boltzmann and Gibbs looked at this in different ways. Boltzmann's idea was that a gas of $N$ particles would be represented by $N$ points in the 6-dimensional phase space. The evolution of the state of the system with time is then described by the "flow" of the "gas" of points in the phase space. This sort of argument only works for weakly interacting particles, since:

- only then can you talk about the state of an individual particle, and
- later arguments are based on the movement of the points in phase space being independent.

Gibbs adopted a rather more general approach. He regarded the state of a system of $N$ particles as being specified by a single point in a $6N$-dimensional phase space.

In both cases, one applies probabilistic arguments to the collection of points in phase space. This collection is called an *ensemble*. So, in Boltzmann's view a single-particle is the system and the $N$ particles comprise the ensemble, while in Gibbs's view the assembly of particles is the system and many imaginary copies of the system comprise the ensemble. In the Boltzmann case, one performs averages over the possible states of a single particle, while in the Gibbs case, one is considering possible states of the entire system and applying probabilistic arguments to those.

Both views are useful. The Boltzmann approach is easier to picture, but it can only be applied to weakly interacting particles. The Gibbs approach is more powerful as it can be applied to strongly interacting systems where the particles cannot be regarded as being even *approximately* independent.

### 1.6.3 *The Fundamental Postulate in the classical case*

If we say that all (quantum) states are equally likely, then the classical analogue will be that all points in phase space are equally likely. The quantum version of the Fundamental Postulate refers to an *isolated system*. This is a system for which $E$, $V$ and $N$ are fixed. Classically, when the energy is fixed, this restricts the accessible region of phase space to a constant-energy *hypersurface*. Similarly, fixing $V$ and $N$ determines which regions of phase space are available. The classical version of the Fundamental

Postulate then states that **for an isolated system all available regions of phase space on the constant energy hypersurface are equally likely.**

The probability of a macrostate corresponding to a region of phase space will then be proportional to the number of phase points in the region. So, correctly normalised, it will be given by the *density* of points $\rho(p,q)$ where by $p$ and $q$ we mean the set of all momentum coordinates and all position coordinates.

### 1.6.4 The classical partition function

The classical analogue of the quantum partition function, Eq. (1.4.2), is given by

$$Z = \frac{1}{h^{3N}} \int e^{-H(p_i, q_i)/kT} d^{3N}p \, d^{3N}q. \tag{1.6.2}$$

The function $H(p_i, q_i)$ is the energy of the system expressed as a function of the position and momentum coordinates $q_i$ and $p_i$ (called the hamiltonian).

### 1.6.5 The equipartition theorem

The equipartition theorem is concerned with the internal energy associated with individual degrees of freedom of a system. It has important consequences for the behaviour of the heat capacity of classical systems.

We ask the question "What is the internal energy associated with a given degree of freedom — say $p_i$"? That is easy to write down:

$$\langle E_i \rangle = \frac{1}{Z} \int E_i e^{-E(q_1 \ldots q_i \ldots q_N, \, p_1 \ldots p_i \ldots p_N)/kT} d^{3N}q \, d^{3N}p. \tag{1.6.3}$$

Assuming the energy $E_i$ depends only on the $p_i$ and not on the other $p$ s and $q$ s, we can factorise that bit out of the exponential and write the integral as

$$\downarrow \cdots \text{no } p_i \cdots \downarrow$$

$$\langle E_i \rangle = \frac{\int E_i e^{-E_i/kT} dp_i \int e^{-E(q_1 \ldots q_i \ldots q_N, p_1 \ldots p_n)/kT} d^{3N}q \, d^{3N-1}p}{\int e^{-E_i/kT} dp_i \int [\text{same integral as above} - \text{no } p_i]}. \tag{1.6.4}$$

So, the second integral in numerator and denominator cancel, leaving the simple expression

$$\langle E_i \rangle = \frac{\int E_i e^{-E_i/kT} dp_i}{\int e^{-E_i/kT} dp_i}. \tag{1.6.5}$$

This may be simplified by using the beta trick discussed in Section 1.4.5. We write $\beta = 1/kT$ and use

$$E_i e^{-\beta E_i} = -\frac{\partial}{\partial \beta} e^{-\beta E_i}, \qquad (1.6.6)$$

so that

$$\langle E_i \rangle = -\frac{\frac{\partial}{\partial \beta} \int e^{-\beta E_i} dp_i}{\int e^{-\beta E_i} dp_i} \qquad (1.6.7)$$

or

$$\langle E_i \rangle = -\frac{\partial}{\partial \beta} \ln \int e^{-\beta E_i} dp_i. \qquad (1.6.8)$$

At this stage, we must be more specific about the functional form of $E_i(p_i)$. Since the $p_i$ is a momentum, then for a classical particle $E_i(p_i) = p_i^2/2m$: a quadratic dependence. For simplicity, let us write

$$E_i = bp_i^2 \qquad (1.6.9)$$

for some positive constant $b$. The integral is then

$$\int e^{-\beta E_i} dp_i = \int e^{-\beta b p_i^2} dp_i. \qquad (1.6.10)$$

We don't actually need to evaluate the integral! Remember that we are going to differentiate the logarithm of the integral with respect to $\beta$; so all we want is the $\beta$-dependence. Let us make a change of variable and put

$$\beta p_i^2 = y^2. \qquad (1.6.11)$$

The integral then becomes

$$\beta^{-1/2} \int e^{-by^2} dy, \qquad (1.6.12)$$

so that

$$\begin{aligned} \langle E_i \rangle &= -\frac{\partial}{\partial \beta} \ln \left( \beta^{-1/2} \int e^{-by^2} dy \right) \\ &= -\frac{\partial}{\partial \beta} \left\{ -\frac{1}{2} \ln \beta + \ln \int e^{-by^2} dy \right\}. \end{aligned} \qquad (1.6.13)$$

The second term is independent of $\beta$ so upon differentiation, it vanishes. Thus, by differentiating we obtain

$$\langle E_i \rangle = \frac{1}{2\beta} \qquad (1.6.14)$$

or, in terms of $T$:

$$\langle E_i \rangle = \frac{1}{2}kT. \tag{1.6.15}$$

So, you see we didn't have to evaluate the integral — and the $b$ has vanished; the physics came out of the integral upon change of variables.

The general conclusion here may be stated as the **Equipartition theorem**: For a classical (non-quantum) system, each degree of freedom with a quadratic dependence on coordinate or momentum gives a contribution to the internal energy of $kT/2$.

(Incidentally, if $E_i \propto q_i^n$ or $p_i^n$ (for even $n$), then the corresponding equipartition energy is given by $\langle E_i \rangle = kT/n$.)

### 1.6.6   Consequences of equipartition

We consider two examples — lattice vibrations and a gas of particles.

#### Lattice vibrations

For the case of lattice vibrations, each atom is essentially three harmonic oscillators, one in the $x$, $y$ and $z$ directions. Thus, for $N$ atoms, we have $3N$ harmonic oscillators. Now, in this case *both* the position and momentum coordinates contribute a quadratic term to the energy. The internal energy is then

$$E = 3NkT \tag{1.6.16}$$

in the non-quantum (high temperature) limit. Differentiation with respect to temperature gives the isochoric heat capacity

$$\begin{aligned} C_V &= 3Nk \\ &= 3R \text{ per mol.} \end{aligned} \tag{1.6.17}$$

#### Gas of particles

Considering now a gas of non-interacting particles, there is no contribution to the energy from the position coordinates. Only the momentum coordinates contribute a quadratic term to the energy and the internal energy is then

$$E = \frac{3}{2}NkT \tag{1.6.18}$$

in the non-quantum (high temperature) limit. Differentiation with respect to temperature gives the (isochoric) heat capacity

$$C_V = \frac{3}{2}Nk$$
$$= \frac{3}{2}R \text{ per mol.} \tag{1.6.19}$$

The heat capacity of the solid is double that of the fluid because in the solid the position coordinates also contribute to the internal energy.

(In fact, the walls of a box of gas can be modelled as an oscillator with a power law potential $V \propto x^n$ where $n \to \infty$.)

### Breakdown of equipartition

Equipartition breaks down when quantum effects become important. In Section 2.3.3, we shall see that the internal energy of a *single* quantum free particle corresponds to $\frac{3}{2}kT$: the equipartition value for the three spatial degrees of freedom. However, once we have a collection of $N$ identical particles comprising a quantum gas, the internal energy is given, from Section 2.7.4, by

$$E = \frac{3}{2}NkT \left\{ 1 + a\frac{N}{V}\frac{1}{\alpha}\left(\frac{\pi\hbar^2}{mkT}\right)^{3/2} + a^2 \ldots \right\} \tag{1.6.20}$$

where $a = +1$ for fermions, zero for "classical" particles and $-1$ for bosons and $\alpha$ is the spin degeneracy factor. The equipartition result occurs at high temperatures; as the gas cools, quantum effects become important. For fermions the internal energy increases above the equipartition value, while for bosons the internal energy decreases below the equipartition value.

In Problem 1.19, you will see that the internal energy of a quantum harmonic oscillator may be written as

$$E = kT + \frac{\hbar^2\omega^2}{12kT} + \cdots . \tag{1.6.21}$$

The first term represents the high-temperature equipartition value. The second (and higher) terms indicate the deviations, showing the internal energy increasing above its high temperature value as the temperature is lowered.

### 1.6.7 Liouville's theorem

We ask the question "How does a macroscopic system evolve in time"? The answer is that it will develop in accordance with the Second Law of thermodynamics; the system evolves to the state of maximum entropy consistent with the constraints imposed. Can this be understood from microscopic first principles? In other words, can the law of entropy increase be derived from Newton's laws? Both Boltzmann and Gibbs agonised over this.

We need a definition of entropy which will be suitable for use in the classical case. The problem is that there are no *discrete* states now, since the $p$ and the $q$ can vary continuously. By analogy with the Gibbs expression for entropy:

$$S = -k \sum_j P_j \ln P_j, \qquad (1.6.22)$$

since the probability (density) is given by the density of points in phase space, we now have

$$S = -k \int \rho \ln \rho \, dp \, dq. \qquad (1.6.23)$$

It is essentially (minus) the average of the logarithm of the density of points in phase space.

If one calculates the way points move around phase space under the influence of the laws of mechanics, one finds that the "flow" is incompressible. Thus, the density remains constant. This result is known as Liouville's theorem. We need the machinery of Hamiltonian mechanics to show this. If you are happy with Hamiltonian mechanics, the proof is sketched in what follows. But the implication is that since $\rho$ remains constant, then the entropy remains constant, so the Second Law of thermodynamics seems to be inconsistent with the laws of mechanics at the microscopic level.

To demonstrate Liouville's theorem, we first note that the flow of points in phase space must obey the equation of continuity, since the number of points is conserved:

$$\frac{\partial \rho}{\partial t} + \operatorname{div} \mathbf{v} \rho = 0. \qquad (1.6.24)$$

However, in this case $\rho$ depends on the position and momentum coordinates, $q$ and $p$. Thus, the divergence contains all the $\partial/\partial q$ derivatives and all the $\partial/\partial p$ derivatives. And the "velocity" $\mathbf{v}$ has components $dp/dt$ as well as the usual $dq/dt$. Thus, the divergence term is actually

$$\operatorname{div}\mathbf{v}\rho = \frac{\partial}{\partial p}\left(\frac{dp}{dt}\rho\right) + \frac{\partial}{\partial q}\left(\frac{dq}{dt}\rho\right) \tag{1.6.25}$$

(these equations really contain all the $q$ and $p$ coordinates; the above, as elsewhere, is a shorthand simplification). We expand the $p$ and $q$ derivatives to give

$$\operatorname{div}\mathbf{v}\rho = \left(\frac{\partial}{\partial p}\frac{dp}{dt} + \frac{\partial}{\partial q}\frac{dq}{dt}\right)\rho + \frac{\partial\rho}{\partial p}\frac{dp}{dt} + \frac{\partial\rho}{\partial q}\frac{dq}{dt} \tag{1.6.26}$$

and then we use Hamilton's equations

$$\frac{dp}{dt} = -\frac{\partial H}{\partial q}, \quad \frac{dq}{dt} = \frac{\partial H}{\partial p} \tag{1.6.27}$$

in the first bracket. Then

$$\frac{\partial}{\partial p}\frac{dp}{dt} + \frac{\partial}{\partial q}\frac{dq}{dt} = -\frac{\partial^2 H}{\partial p\partial q} + \frac{\partial^2 H}{\partial q\partial p} = 0. \tag{1.6.28}$$

Thus, we find that

$$\frac{\partial\rho}{\partial t} + \frac{\partial\rho}{\partial p}\frac{dp}{dt} + \frac{\partial\rho}{\partial q}\frac{dq}{dt} = 0. \tag{1.6.29}$$

But we recognise this as the *total* derivative of $\rho$; it is the derivative of the density when moving with the flow in phase space. This is zero. Thus, as the representative points evolve and flow in phase space, the local density remains constant. This is the content of Liouville's theorem, expressed in its usual form as

$$\frac{d\rho}{dt} = 0. \tag{1.6.30}$$

We have a paradox: since $\rho$ remains constant during evolution, then *the entropy remains constant*; the Second Law of thermodynamics seems to be inconsistent with the laws of mechanics at the microscopic level.

### 1.6.8 *Boltzmann's H theorem*

The resolution of the paradox of the incompatibility between Liouville's theorem and the Second Law may be understood from the *nature* of the flow of points in phase space. Boltzmann defined a quantity $H$, which was essentially the integral of $\rho \ln \rho$ over phase space and he obtained an equation of motion for $H$ as a probabilistic differential equation for the flow of points into and out of regions of phase space. We shall adopt a variant of the approach of Gibbs to study the evolution of Boltzmann's $H$. Please note this $H$ is neither the hamiltonian of the previous section nor the enthalpy function; this is *Boltzmann's H*.

The flow of points in phase space is complicated. Since we have a given number of elements in our ensemble, the number of points in phase space is fixed. And Liouville's theorem says that the multi-dimensional volume occupied by the points is constant. But the flow can be "dendritic" with fingers spreading and splitting in all directions, as in Fig. 1.11.

And as this happens, there will come a time when it is difficult to distinguish between what is an occupied region and what is an unoccupied region of phase space; they will be continually folded into each other. Gibbs argued that there was a scale in phase space, beyond which it was not possible (or at least reasonable) to discern. If the details of the state "3" in Fig. 1.11 are too fine to discern, then it will simply appear as a region of greater volume and, therefore, lesser density. This is shown in Fig. 1.12.

The procedure of taking an average over small regions of phase space in this way is known as "coarse graining". Gibbs showed rigorously (rather than by just using diagrams as here) that the coarse grained density of points in phase space decreased as time proceeded [15].

Thus, the conclusion is that the coarse-grained $H$ decreases; so the coarse-grained entropy increases.

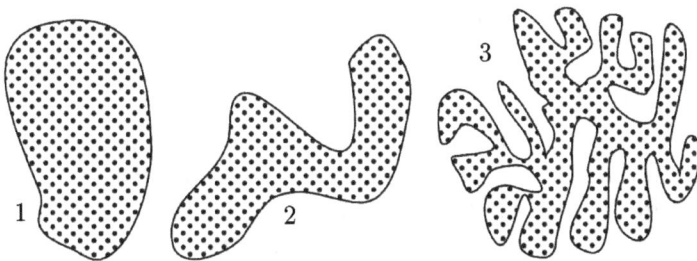

Fig. 1.11. Evolution of a region of phase space.

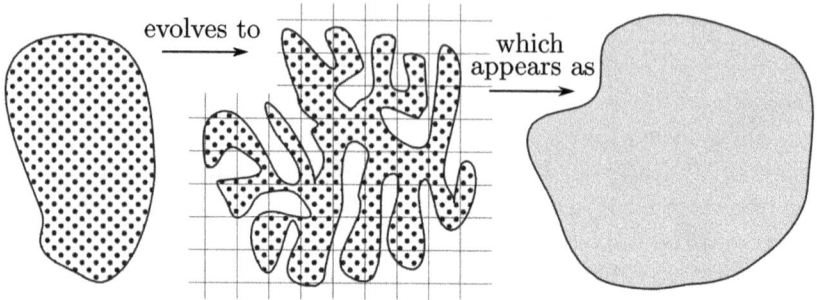

Fig. 1.12.   Apparent reduction in density in phase space.

There is a connection with quantum mechanics, which may be invoked in the question of coarse-graining. The volume of a "cell" in phase space is a product of $p, q$ pairs. Now, the Uncertainty Principle tells us that we cannot locate a point within a $pq$ area to within better than Planck's constant. This gives the ultimate resolution that is achievable in specifying the state of a system — so at the fundamental level, there is indeed a firm justification for coarse-graining.

Quantum mechanics has a habit of popping up in the most unexpected areas of statistical thermodynamics. This theme continues into the following section.

## 1.7   The Third Law of Thermodynamics

### 1.7.1   *History of the Third Law*

The Third Law of thermodynamics arose as the result of experimental work in chemistry, principally by the famous physical chemist Walther Nernst. He published what he called his "heat theorem" in 1906. A readable account of the history of the Third Law and the controversies surrounding its acceptance is given by Dugdale [16].

Nernst measured the change in Gibbs free energy and the change in enthalpy for chemical reactions which started and finished at the same temperature. At lower and lower temperatures, he found that the changes in $G$ and the changes in $H$ became closer and closer. This is shown schematically, in Fig. 1.13.

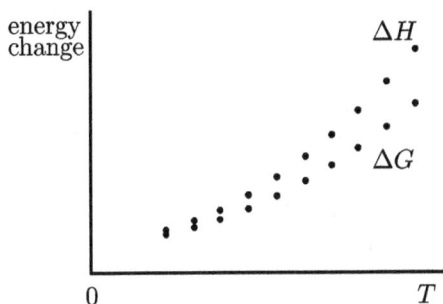

Fig. 1.13. Nernst's observations.

Nernst was led to conclude that at $T = 0$ the changes in $G$ and $H$ were the same. And from some elementary thermodynamic arguments he was able to infer the behaviour of the entropy at low temperatures.

Changes in $H$ and $G$ are given by

$$\Delta H = \ \ T\Delta S + V\Delta p,$$
$$\Delta G = -S\Delta T + V\Delta p. \tag{1.7.1}$$

Thus, $\Delta G$ and $\Delta H$ are related by

$$\Delta G = \Delta H - T\Delta S - S\Delta T, \tag{1.7.2}$$

and if the temperature is the same before and after, $\Delta T = 0$, so then

$$\Delta G = \Delta H - T\Delta S. \tag{1.7.3}$$

This is a very important equation for chemists.

Now, Nernst's observation may be stated as

$$\Delta H - \Delta G \to 0 \quad \text{as} \quad T \to 0, \tag{1.7.4}$$

which he realised implied that

$$T\Delta S \to 0 \quad \text{as} \quad T \to 0. \tag{1.7.5}$$

### 1.7.2 *Entropy*

On the face of it, this result is no surprise since the factor $T$ will ensure the product $T\Delta S$ goes to zero. But Nernst took the result further. He studied *how fast* $\Delta H - \Delta G$ tended to zero. And his observation was that it always went faster than linearly. In other words, he concluded that

$$\frac{\Delta H - \Delta G}{T} \to 0 \quad \text{as} \quad T \to 0. \tag{1.7.6}$$

So, even though $1/T$ was getting bigger and bigger, the quotient $(\Delta H - \Delta G)/T$ still tended to zero. But we know that

$$\frac{\Delta H - \Delta G}{T} = \Delta S. \tag{1.7.7}$$

So, from this Nernst drew the conclusion

$$\Delta S \to 0 \quad \text{as} \quad T \to 0. \tag{1.7.8}$$

The entropy change in a process tends to zero at $T = 0$. The entropy thus remains a constant in any process at absolute zero. We conclude:

- The entropy of a body at zero temperature is a constant, independent of all other external parameters.

This was the conclusion of Nernst; it is sometimes called Nernst's heat theorem. It was subsequently to be developed into the Third Law of thermodynamics.

### 1.7.3 *Quantum viewpoint*

From the purely macroscopic perspective, the third law is as stated above: at $T = 0$ the entropy of a body is a constant. And many conclusions can be drawn from this. One might ask the question "what is the constant"? However, we do know that thermodynamic conclusions about measurable quantities are not influenced by any such additive constants since one usually differentiates to find observables.

If we want to ask about the constant, then we must look into the microscopic model for the system under investigation. Recall the Boltzmann expression for entropy:

$$S = k \ln \Omega$$

where $\Omega$ is the number of microstates in the macrostate. Now, consider the situation at $T = 0$. Then we know the system will be in its ground state, the lowest energy state. But this is a *unique* quantum state. Thus, for the ground state

$$\Omega = 1 \qquad (1.7.9)$$

and so

$$S = 0. \qquad (1.7.10)$$

Nernst's constant is thus zero and we then have the expression for the Third Law:

- As the absolute zero of temperature is approached, the entropy of all bodies tends to zero.

We note that this applies specifically to bodies that are in *thermal equilibrium*. The Third Law can be summarised as

$$\frac{\partial S}{\partial\, \text{anything}} \to 0 \quad \text{as} \quad T \to 0. \qquad (1.7.11)$$

The above discussion is actually an over-simplification. In reality, there may be degeneracy in the ground state of the system; then the above argument appears to break down. However, recall that entropy is an extensive quantity and that the entropy of the system should be considered in the thermodynamic limit. In other words, strictly, we should examine how the intensive quantity $S/V$ or $S/N$ behaves in the limit $V \to \infty, N \to \infty$ while $V/N$ remains constant.

If the degeneracy of the ground state is $g$, then we must look at the behaviour of $\ln(g)/N$. This will tend to zero in the thermodynamic limit so long as $g$ increases with $N$ no faster than exponentially. This is the fundamental quantum-mechanical principle behind the Third Law. The interested reader should consult the paper by Leggett [17] for a deeper discussion of these points.

To complete this discussion, it is instructive to see how the Third Law would fail if classical mechanics were to apply down to $T = 0$. We saw, in Section 1.6.7, that the Gibbs expression for entropy:

$$S = -k \sum_j P_j \ln P_j \qquad (1.7.12)$$

must be replaced, in the classical case by

$$S = -k \int \rho \ln \rho \, dp \, dq, \tag{1.7.13}$$

where $\rho$ is the density of points in phase space. This is necessary because in the classical case there are no discrete states and the momenta and coordinates, the $p$ and $q$, can vary continuously.

As the temperature is lowered, the mean energy of the system will decrease. And corresponding to this, the "volume" of phase space occupied will decrease. In particular, the momentum coordinates $q$ will vary over a smaller and smaller range. In the $T \to 0$ limit, the momentum range will become localised closer and closer to $q = 0$. The volume of occupied phase space shrinks to zero and the entropy thus tends to $-\infty$. This indeed is the limiting value indicated by the classical treatment of the Ideal Gas, as we shall see in Section 2.3.3.

The Uncertainly Principle of quantum mechanics limits the low temperature position–momentum specification of a system; you cannot localise points in phase space to a volume smaller than the appropriate power of Planck's constant. This fundamental limitation of the density of phase points recovers the Third Law. Thus, again, we see the intimate connection between quantum mechanics and the Third Law.

The Second Law tells us that there is an absolute zero of temperature. Now, we see that the Third Law tells us there is an absolute zero of entropy.

### 1.7.4  *Unattainability of absolute zero*

The Third Law has important implications concerning the possibility of cooling a body to absolute zero. Let us consider a sequence of adiabatic and isothermal operations on two systems, one obeying the Third Law and one not, as in Fig. 1.14.

Taking a sequence of adiabatics and isothermals between two values of some external parameter, we see that the existence of the Third Law implies that you cannot get to $T = 0$ in a finite number of steps. This is, in fact, another possible statement of the Third Law.

Although one cannot get all the way to $T = 0$, it is possible to get closer and closer. Figure 1.15, adapted and extended from Pobel's book [18], indicates the success in this venture.

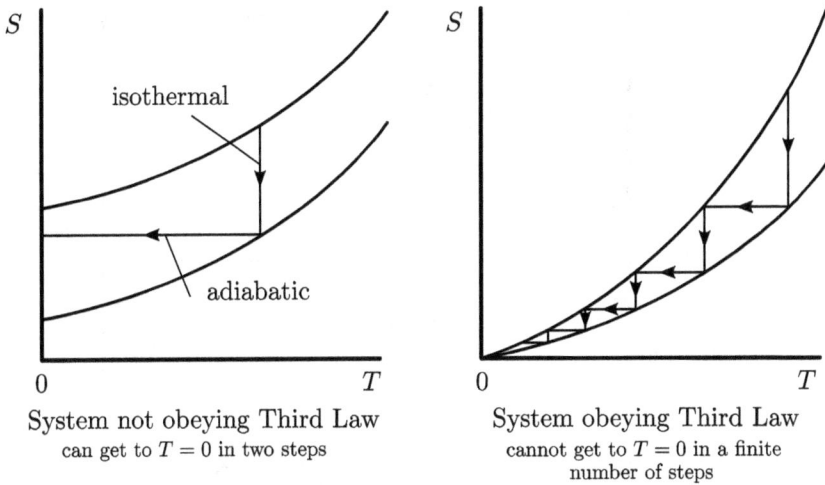

Fig. 1.14.   Approaching absolute zero.

### 1.7.5  *Heat capacity at low temperatures*

The Third Law has important consequences for the heat capacity of bodies at low temperatures. Since

$$C = \frac{\partial Q}{\partial T}$$
$$= T\frac{\partial S}{\partial T},$$

(1.7.14)

and the Third Law tells us that

$$\frac{\partial S}{\partial T} \to 0 \quad \text{as} \quad T \to 0$$

(1.7.15)

we then have

$$C \to 0 \quad \text{as} \quad T \to 0.$$

(1.7.16)

Classical models often give a constant heat capacity — from equipartition. For an ideal gas (Section 2.3.3)

$$C_V = \frac{3}{2}Nk$$

(1.7.17)

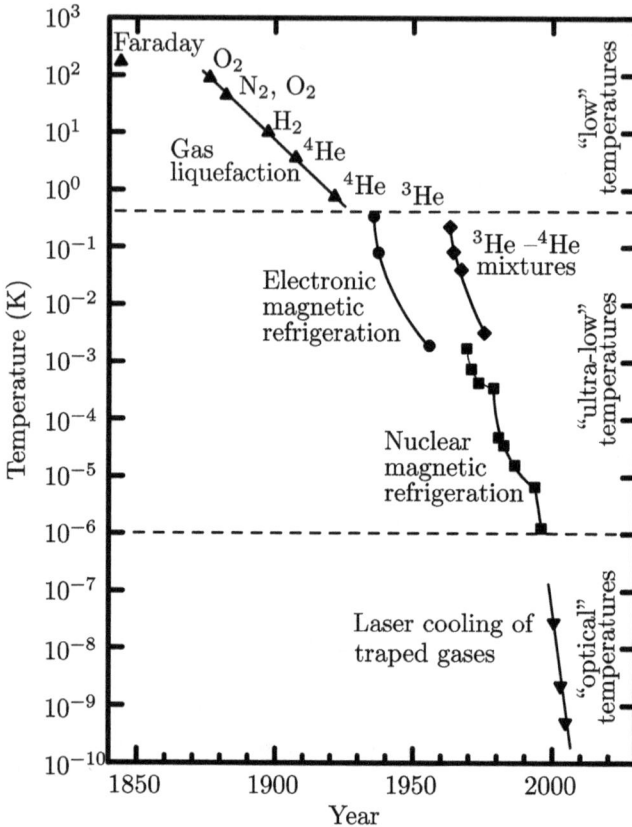

Fig. 1.15. The road to absolute zero (adapted, with grateful thanks, from Pobel [18]).

independent of temperature. The Third Law tells us that this cannot hold at low temperatures. And indeed we shall see that for both Fermi and Bose gases, $C_V$ does go to zero as $T \to 0$.

### 1.7.6 *Other consequences of the Third Law*

Most "response functions" or susceptibilities — generalised spring constants — go to zero or to a constant as $T \to 0$ as a consequence of the Third Law. This is best seen by examining the relevant Maxwell relation (Appendix B.6). For example, consider the thermal expansion coefficient.

The Maxwell relation here is

$$\left.\frac{\partial V}{\partial T}\right|_p = -\left.\frac{\partial S}{\partial p}\right|_T. \tag{1.7.18}$$

The right-hand side is zero by virtue of the Third Law. Thus, we conclude that

$$\left.\frac{\partial V}{\partial T}\right|_p \to 0 \quad \text{as} \quad T \to 0; \tag{1.7.19}$$

the expansion coefficient goes to zero.

An interesting example is the magnetic susceptibility of a paramagnet, which we will treat in Section 2.10. The parallel with the ubiquitous model $pV$ system is made by noting that the magnetic work increment is

$$\Delta W_{\mathrm{m}} = -M\Delta B, \tag{1.7.20}$$

corresponding to the mechanical work increment

$$\Delta W = -p\,\Delta V. \tag{1.7.21}$$

Thus, we can take over $pV$ results through the identification

$$\begin{aligned} M &\to p \\ B &\to V \end{aligned} \tag{1.7.22}$$

where $M$ is the total magnetic moment.

The magnetic susceptibility is

$$\chi = \frac{\mu_0}{V}\frac{\partial M}{\partial B}, \tag{1.7.23}$$

so that

$$\chi V/\mu_0 \to \frac{\partial p}{\partial V}. \tag{1.7.24}$$

There is no Maxwell relation for this, but consider the variation of the susceptibility with temperature:

$$\frac{\partial(\chi V/\mu_0)}{\partial T} = \frac{\partial^2 M}{\partial T \partial B}$$
$$\to \frac{\partial^2 p}{\partial T \partial V}. \tag{1.7.25}$$

The order of differentiation can be reversed here. In other words,

$$\frac{\partial}{\partial B}\frac{\partial M}{\partial T} \rightarrow \frac{\partial}{\partial V}\frac{\partial p}{\partial T}. \qquad (1.7.26)$$

And now we do have a Maxwell relation:

$$\left.\frac{\partial p}{\partial T}\right|_V = \left.\frac{\partial S}{\partial V}\right|_T \quad \text{gives} \quad \left.\frac{\partial M}{\partial T}\right|_V = \left.\frac{\partial S}{\partial B}\right|_T. \qquad (1.7.27)$$

The Third Law tells us that the right-hand side of these equations goes to zero as $T \rightarrow 0$. We conclude then that

$$\frac{\partial \chi}{\partial T} \rightarrow 0 \quad \text{as} \quad T \rightarrow 0 \qquad (1.7.28)$$

or

$$\chi \rightarrow \text{const} \quad \text{as} \quad T \rightarrow 0. \qquad (1.7.29)$$

The Third Law tells us that the magnetic susceptibility becomes constant as $T \rightarrow 0$. But what does Curie's law, Eq. (2.10.12), say? This states

$$\chi = \frac{C}{T} \qquad (1.7.30)$$

where $C$ is the Curie constant. From this, we conclude

$$\chi \rightarrow \infty \quad \text{as} \quad T \rightarrow 0 \,!! \qquad (1.7.31)$$

This is *completely incompatible* with the Third Law.

However, Curie's law is a specifically high-temperature result (strictly, it applies to the small $B/T$ limit). The general expression for the magnetisation of an ideal paramagnet of $N$ spin ½ moments $\mu$ is, Eq. (2.10.9):

$$M = N\mu \tanh\left(\frac{\mu B}{kT}\right) \qquad (1.7.32)$$

and corresponding to this, the (differential) susceptibility is

$$\chi = \frac{N\mu^2}{VkT}\text{sech}^2\left(\frac{\mu B}{kT}\right). \qquad (1.7.33)$$

Now, we see that

$$\chi \rightarrow 0 \quad \text{as} \quad T \rightarrow 0 \qquad (1.7.34)$$

in conformity with the Third Law, so long as the magnetic field $B$ is finite. Of course, you *have* to use a magnetic field, however small, to measure the susceptibility. Nevertheless, even in the absence of an externally applied magnetic field, there will be an internal field present: the dipole fields of the magnetic moments themselves. Thus, the Third Law is not under threat.

There is a further consideration in the case of fluid magnetic systems. In a fluid, where the particles must be treated as delocalised, the statistics will also have an effect. Consider the behaviour of fermions at low temperatures, to be treated in Section 2.5. Very roughly, only a fraction $T/T_F$ of the particles are free and available to participate in "normal" behaviour. We then expect that the Curie law behaviour will be modified to

$$\chi \sim \left(\frac{T}{T_F}\right) \times \frac{C}{T} \tag{1.7.35}$$

or

$$\chi \sim \frac{C}{T_F} \tag{1.7.36}$$

which is indeed a constant, in conformity with the Third Law. This result is correct, but a more detailed calculation must be done to determine the numerical constants involved.

### 1.7.7  Pessimist's statement of the laws of thermodynamics

As we have now covered all the laws of thermodynamics, we can present statements on them in terms of what they prohibit in the operation of Nature.

- **First Law:** You cannot convert heat to work at greater than 100% efficiency.
- **Second Law:** You cannot even achieve 100% efficiency — except at $T = 0$.
- **Third Law:** You cannot get to $T = 0$.

This is a simplification, but it encapsulates the underlying truths, and it is easy to remember.

## Problems

1.1  Demonstrate that gravitational energy is not extensive: show that the gravitational energy of a sphere of radius $r$ and uniform density varies with volume as $V^n$ and find the exponent $n$.

1.2   Demonstrate that *entropy*, as given by the Boltzmann expression $S = k \ln \Omega$, is an *extensive* property. The best way to do this is to argue *clearly* that $\Omega$ is multiplicative.

1.3   In investigating the conditions for the establishment of equilibrium through the transfer of thermal energy, the fundamental requirement is that the entropy of the equilibrium state should be a maximum. Equality of temperature was established from the vanishing of the first derivative of $S$. What follows from a consideration of the *second derivative*? **Hint:** Consider the heat capacity.

1.4   Do particles flow from high $\mu$ to low $\mu$ or *vice versa*? Explain your reasoning.

1.5   In the derivation of the Boltzmann factor, the entropy of the bath was expanded in powers of the energy of the "system of interest". The higher order terms of the expansion were neglected. Discuss the validity of this.

1.6   The Boltzmann factor might have been derived by expanding $\Omega$ rather than by expanding $S$. In that case, however, the expansion cannot be terminated. Why not?

1.7   Show that $\ln N! = \sum_{n=1}^{N} \ln n$. By approximating this sum by an integral, obtain *Stirling's approximation*: $\ln N! \approx N \ln N - N = N \ln (N/e)$.

1.8   Show that the Gibbs expression for entropy: $S = -k \sum_j P_j \ln P_j$ reduces to the Boltzmann expression $S = k \ln \Omega$ in the case of an isolated system.

1.9   This problem considers the probability distribution for the energy fluctuations in the canonical ensemble. The *moments* of the energy fluctuations are defined by

$$\sigma_n = \frac{1}{Z} \sum_j \left( E_j - \varepsilon \right)^n e^{-\beta E_j}$$

where $\beta = 1/kT$ and $\varepsilon$ is an arbitrary (at this stage) energy.

Show that

$$\sigma_n = (-1)^n \frac{1}{Ze^{\beta\varepsilon}} \frac{\partial^n \{Ze^{\beta\varepsilon}\}}{\partial \beta^n}$$

and use this to prove that the energy fluctuations in an ideal gas, in the thermodynamic limit, follow a *normal distribution*. It will prove to be convenient to take $\varepsilon$ as the mean energy. (This is difficult. You really need to use a computer algebra system to do this problem.)

1.10 Starting from the expression for the Gibbs factor for a many-particle system, write down the grand partition function $\Xi$ and show how it may be expressed as the product of $\Xi_k$, the grand partition function for the subsystem comprising particles in the $k$th single-particle state.

1.11 What is the condition for the geometric progression in the derivation of the Bose–Einstein distribution to be convergent?

1.12 Why can't the evolutionary curve in phase space intersect? You need to demonstrate that the evolution from a point is unique.

1.13 Show that the trajectory of a 1d harmonic oscillator is an ellipse in phase space. What would the trajectory be if the oscillator were *weakly* damped.

1.14 The Fundamental Postulate of classical statistical mechanics states that for an isolated system all available regions of phase space on the constant energy hypersurface are equally likely.

In terms of this, discuss the properties of the phase space of a Boltzmann ensemble of simple harmonic oscillators of identical energy.

1.15 A *quartic* oscillator has a potential energy that varies with its displacement as $V(x) = gx^4$. What would be the equipartition thermal energy corresponding to the displacement degree of freedom?

1.16 Consider a particle subject to a hypothetical confining potential $V(x) = gx^n$ (where $n$ is even and positive).

(a) Calculate the heat capacity of a collection of such particles as a function of $n$.

(b) Show that in the limit $n \to \infty$ the heat capacity tends to that for a free particle.

(c) Comment on this limit — in the context of a gas of free particles.

1.17 For a single-component system with a variable number of particles, the Gibbs free energy is a function of temperature, pressure and number of particles: $G = G(T, p, N)$. Since $N$ is the only extensive variable upon which $G$ depends, show that the chemical potential for this system is equal to the Gibbs free energy per particle: $G = N\mu$.

1.18 Use the definition of the Gibbs free energy together with the result of the previous question, Problem 1.17, to obtain the Euler relation of Appendix A.2:

$$E = TS - pV + \mu N.$$

1.19 The energy of a harmonic oscillator may be written as $m\omega^2 x^2 / 2 + p^2 / 2m$ so it is quadratic in both position and momentum — thus equipartition will give a classical internal energy of $kT$.

The energy levels of the quantum harmonic oscillator are given by $\varepsilon_n = (\frac{1}{2} + n)\hbar\omega$.

(a) Show that the partition function of this system is given by

$$Z = \frac{1}{2}\operatorname{cosech}\frac{\hbar\omega}{2kT} \qquad (1.7.37)$$

and that the internal energy is given by

$$E = \frac{1}{2}\hbar\omega \coth\frac{\hbar\omega}{2kT} = \frac{\hbar\omega}{e^{\hbar\omega/kT} - 1} + \frac{\hbar\omega}{2}. \qquad (1.7.38)$$

(b) Show that at high temperatures $E$ may be expanded as

$$E = kT + \frac{\hbar^2\omega^2}{12kT} + \cdots. \qquad (1.7.39)$$

(c) Identify the terms in this expansion.

# References

[1] A. M. Guénault, *Statistical Physics* (Springer, 2007).
[2] R. Bowley and M. Sànchez, *Introductory Statistical Mechanics*, 2nd ed. (Oxford University Press, 1999).

[3]  M. W. Zemansky and R. H. Dittman, *Heat and Thermodynamics* (McGraw-Hill, 1968).

[4]  A. Einstein, *Albert Einstein: Philosopher-Scientist*, P. A. Schlipp (ed.) (Open Court Publishing Co., 1949).

[5]  T. L. Hill, *Introduction to Statistical Thermodynamics* (Addison Wesley, 1960).

[6]  L. D. Landau and E. M. Lifshitz, *Statistical Physics* (Pergamon Press, 1970).

[7]  C. Adkins, *Equilibrium Thermodynamics* (Cambridge University Press, 1983).

[8]  A. B. Pippard, *Elements of Classical Thermodynamics* (Cambridge University Press, 1966).

[9]  H. B. Callen, *Thermodynamics and an Introduction to Thermostatistics* (John Wiley, 1985).

[10]  F. Reif, *Fundamentals of Statistical and Thermal Physics* (McGraw-Hill, 1965).

[11]  R. P. Feynman, *Statistical Mechanics* (Benjamin, 1972).

[12]  R. P. Feynman, The reason for antiparticles. In *Elementary Particles and the Laws of Physics — The 1986 Dirac Memorial Lectures* (Cambridge University Press, 1987).

[13]  S. Tomonaga, *The Story of Spin* (University of Chicago Press, 1997).

[14]  E. Merzbacher, *Quantum Mechanics* (John Wiley, 1970).

[15]  R. C. Tolman, *The Principles of Statistical Mechanics* (Oxford University Press, 1938).

[16]  J. S. Dugdale, *Entropy and its Physical Meaning* (Taylor and Francis, 1996).

[17]  A. J. Leggett, On the minimum entropy of a large system at low temperatures, *Ann. Phys. N.Y.*, **72** (1972) 80–106.

[18]  F. Pobel, *Matter and Methods at Low Temperatures* (Springer-Verlag, 1992).

# CHAPTER 2

# PRACTICAL CALCULATIONS WITH IDEAL SYSTEMS

We understand an *ideal system* to be one where there are no interactions between the microscopic constituents. But the concept of a non-interacting system is hypothetical since a system of truly non-interacting components could not achieve thermal equilibrium. Thus, we are really interested in assemblies of *very weakly* interacting systems. To be precise, the interactions must be sufficient to lead to thermal equilibrium, but weak enough that these interactions have negligible effect on the energy of the individual particles. We note, parenthetically, that the *rate* at which an equilibrium state is established will depend on the strength of the interactions; this is the subject of *non-equilibrium statistical mechanics*. Some aspects will be touched upon in Chapter 5.

## 2.1 The Density of States

### 2.1.1 *Non-interacting systems*

Since, for a non-interacting system, the single-particle states have well-defined energies, it follows that one can obtain full thermodynamic information about such a system once the energies of the single-particle states and the mean number of particles in each state are known. Then all thermodynamic properties are found by performing sums over states and distribution functions. And for infinite or very large systems, where the spacings between the energy levels become very small, such sums may usually be converted to integrals.

### 2.1.2 *Converting sums to integrals*

In quantum statistical mechanics, there are many sums over states to be evaluated. An example is the partition function we encountered in the previous chapter:

$$Z(N, V, T) = \sum_j e^{-E_j(N,V)/kT}. \tag{2.1.1}$$

It is often convenient to approximate the sums by integrals. And since the individual states are densely packed, negligible error is introduced in so doing. Now, if $g(\varepsilon)d\varepsilon$ is the number of states with energy between $\varepsilon$ and $\varepsilon + d\varepsilon$, then the sum may be approximated by

$$\sum_j e^{-E_j(N,V)/kT} \rightarrow \int_0^\infty g(\varepsilon)e^{-\varepsilon/kT}d\varepsilon. \tag{2.1.2}$$

Here, $g(\varepsilon)$ is referred to as the *(energy) density of states*. If we are studying the properties of a gas, then the microstates to be considered are the quantum states of a "particle in a box". And the density of states for a particle in a box may be evaluated in the following way.

### 2.1.3 *Enumeration of states*

We consider a cubic box of volume $V$. Each side has length $V^{1/3}$. Elementary quantum mechanics tells us that the wave function of a particle in the box must go to zero at the boundary walls; only standing waves are allowed, Fig. 2.1. In the general case, the allowed wavelengths $\lambda$ satisfy

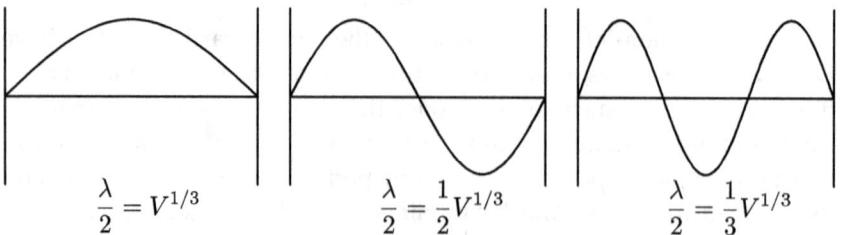

$$\frac{\lambda}{2} = V^{1/3} \qquad \frac{\lambda}{2} = \frac{1}{2}V^{1/3} \qquad \frac{\lambda}{2} = \frac{1}{3}V^{1/3}$$

Fig. 2.1.	Standing waves in a box.

$\lambda/2 = V^{1/3}/n$:

$$\lambda_n = \frac{2}{n} V^{1/3}, \quad n = 1, 2, 3, 4, \ldots, \infty. \tag{2.1.3}$$

In three dimensions, there will be a $\lambda$ for the $x$, $y$, and $z$ directions:

$$\lambda_{n_x} = 2\frac{V^{1/3}}{n_x}, \quad \lambda_{n_y} = 2\frac{V^{1/3}}{n_y}, \quad \lambda_{n_z} = 2\frac{V^{1/3}}{n_z}. \tag{2.1.4}$$

Or, since this corresponds to the components of the wave vector $k_x = 2\pi/\lambda_{n_x}$, etc.,

$$k_x = \frac{\pi}{V^{1/3}} n_x, \quad k_y = \frac{\pi}{V^{1/3}} n_y, \quad k_z = \frac{\pi}{V^{1/3}} n_z. \tag{2.1.5}$$

We can now use the de Broglie relation $\mathbf{p} = \hbar\mathbf{k}$ to obtain the momentum and hence the energy:

$$p_x = \frac{\pi\hbar}{V^{1/3}} n_x, \quad p_y = \frac{\pi\hbar}{V^{1/3}} n_y, \quad p_z = \frac{\pi\hbar}{V^{1/3}} n_z. \tag{2.1.6}$$

And so for a free particle, the energy is then

$$\varepsilon = \frac{p^2}{2m} = \frac{p_x^2 + p_y^2 + p_z^2}{2m}, \tag{2.1.7}$$

which is

$$\varepsilon = \frac{\pi^2\hbar^2}{2mV^{2/3}}(n_x^2 + n_y^2 + n_z^2). \tag{2.1.8}$$

In this expression, it is the triple of quantum numbers $\{n_x, n_y, n_z\}$ which specifies the quantum state. Now, each triple defines a point on a cubic grid. If we put

$$R^2 = n_x^2 + n_y^2 + n_z^2, \tag{2.1.9}$$

then the energy is given by

$$\varepsilon = \frac{\pi^2\hbar^2}{2mV^{2/3}} R^2. \tag{2.1.10}$$

Observe that the energy levels depend on the size of the container. We used this fact in the considerations of Section 1.2.2 on the probabilistic interpretation of the First Law.

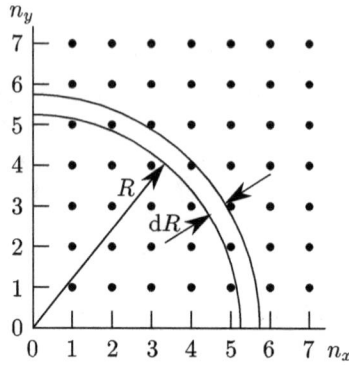

Fig. 2.2.   Counting of quantum states.

### 2.1.4   *Counting states*

The number of states of energy up to $\varepsilon$, denoted by $\mathcal{N}(\varepsilon)$, is given by the number of points in the octant up to $\varepsilon(R)$, Fig. 2.2. (An octant is used since $n_x$, $n_y$ and $n_z$ are restricted to being positive.) And the number of points in the octant is approximately equal to the volume of the octant:

$$\mathcal{N}(\varepsilon) = \frac{1}{8}\frac{4}{3}\pi R^3. \tag{2.1.11}$$

But since, from Eq. (2.1.10),

$$R = \left(\frac{2mV^{2/3}}{\pi^2\hbar^2}\right)^{1/2}\varepsilon^{1/2}, \tag{2.1.12}$$

we then obtain

$$\mathcal{N}(\varepsilon) = \frac{1}{6}\frac{V}{\pi^2\hbar^3}(2m\varepsilon)^{3/2}. \tag{2.1.13}$$

Recall that the density of states $g(\varepsilon)$ is defined by saying that the number of states with energy between $\varepsilon$ and $\varepsilon + d\varepsilon$ is $g(\varepsilon)d\varepsilon$. In other words,

$$g(\varepsilon)d\varepsilon = \mathcal{N}(\varepsilon + d\varepsilon) - \mathcal{N}(\varepsilon) \tag{2.1.14}$$

or, simply,

$$g(\varepsilon) = \frac{d\mathcal{N}(\varepsilon)}{d\varepsilon}. \tag{2.1.15}$$

So, differentiating $\mathcal{N}(\varepsilon)$, we obtain

$$g(\varepsilon) = \frac{1}{4}\frac{V}{\pi^2\hbar^3}(2m)^{3/2}\varepsilon^{1/2}, \tag{2.1.16}$$

which is the required expression for the density of states.

For completeness, we shall also give the expressions for the density of states in two dimensions and in one dimension; these are found by arguments similar to those above. They will be discussed in Problem 2.1. In a two-dimensional system of area $A$, the density of states is

$$g(\varepsilon) = \frac{mA}{2\pi\hbar^2} \quad \text{(two dimensions)}, \tag{2.1.17}$$

observe this is *independent of energy*. And in a one-dimensional system of length $L$, the density of states is

$$g(\varepsilon) = \frac{L}{\pi\hbar} \left(\frac{m}{2}\right)^{1/2} \varepsilon^{-1/2} \quad \text{(one dimension)}. \tag{2.1.18}$$

### 2.1.5 General expression for the density of states

The result of the previous section gives the (energy) density of states for free particles confined to a box of specified volume. This, we argued, was appropriate for the consideration of ideal gases. And it will indeed be quite adequate for the consideration of the ideal gas in both the classical and quantum mechanical cases.

In order to be able to treat systems comprising elements other than free particles, we need to adopt a more general approach than that of the previous section. Central to that approach was the energy–momentum relation (really the energy–wave vector relation) of the particles: $\varepsilon = p^2/2m = \hbar^2 k^2/2m$. But in the general case, this expression may not be appropriate. Thus, at relativistic speeds, free particles would obey $\varepsilon = \sqrt{c^2 p^2 + m^2 c^4}$ and extreme-relativistic/massless particles, $\varepsilon = cp$. In these cases as well as those with other energy–momentum relations, a more general approach is needed for the density of states.

The key to the general treatment is to consider the density of states in $k$-space. The point is that the quantisation condition discussed above applies fundamentally to the $k$-states: the wave function for any elements confined to a box must vanish at the walls. This means that the standing wave condition, Eq. (2.1.5), applies generally:

$$k_x = \frac{\pi}{V^{1/3}} n_x, \quad k_y = \frac{\pi}{V^{1/3}} n_y, \quad k_z = \frac{\pi}{V^{1/3}} n_z.$$

The allowed states correspond to triples of integers $\{n_x, n_y, n_z\}$; in other words, there is a uniform density of states in $k$-space.

The number of states up to a maximum $k$ value is then the volume of the octant

$$\mathcal{N}(k) = \frac{1}{8}\frac{4}{3}\pi R^3 = \frac{1}{6}\pi(n_x^2 + n_y^2 + n_z^2)^{3/2}. \qquad (2.1.19)$$

And since $n_x = k_x V^{1/3}/\pi$, etc., it follows that

$$\mathcal{N}(k) = \frac{V}{6\pi^2}k^3. \qquad (2.1.20)$$

The density of states in $k$-space, which we shall denote by $g(k)$, is then the derivative of $\mathcal{N}(k)$, so that

$$g(k) = \frac{V}{2\pi^2}k^2. \qquad (2.1.21)$$

Recall that the energy density of states is defined as $g(\varepsilon) = \mathrm{d}\mathcal{N}/\mathrm{d}\varepsilon$. And this can be expressed in terms of $g(k)$ by using the chain rule for differentiation:

$$g(\varepsilon) = \frac{\mathrm{d}\mathcal{N}}{\mathrm{d}\varepsilon} = \frac{\mathrm{d}\mathcal{N}}{\mathrm{d}k}\frac{\mathrm{d}k}{\mathrm{d}\varepsilon} = g(k)\left/\frac{\mathrm{d}\varepsilon}{\mathrm{d}k}\right.. \qquad (2.1.22)$$

Thus, we have the general result

$$g(\varepsilon) = \frac{V}{2\pi^2}k^2\left/\frac{\mathrm{d}\varepsilon}{\mathrm{d}k}\right., \qquad (2.1.23)$$

where we must eliminate $k$ in favour of $\varepsilon$. Thus, to find the energy density of states in this general case, we need to know the energy–momentum (energy–wave vector) relation for the particles or excitations.

For completeness, we shall also quote the corresponding expressions for the density of states in two dimensions and in one dimension. In a two-dimensional system of area $A$, the general density of states is

$$g(\varepsilon) = \frac{A}{2\pi}k\left/\frac{\mathrm{d}\varepsilon}{\mathrm{d}k}\right. \quad \text{(two dimensions)}. \qquad (2.1.24)$$

And in a one-dimensional system of length $L$, the general density of states is

$$g(\varepsilon) = \frac{L}{\pi}\left/\frac{\mathrm{d}\varepsilon}{\mathrm{d}k}\right. \quad \text{(one dimension)}. \qquad (2.1.25)$$

### 2.1.6 *Relation between pressure and energy*

It is possible to obtain a relation between the internal energy and the pressure of a gas of (non-interacting) particles. Central to the argument is the understanding, discussed in Section 2.1.3, that the energy levels of a particle in a box depend on the volume of the box, Eq. (2.1.10).

From the differential expression for the First Law, it follows that

$$p = -\left.\frac{\partial E}{\partial V}\right|_S . \tag{2.1.26}$$

We may write the internal energy in terms of the energy levels $\varepsilon_i$ and the occupation numbers of these states $n_i$:

$$E = \sum_i n_i \varepsilon_i . \tag{2.1.27}$$

The differential of this expression is

$$dE = \sum_i n_i d\varepsilon_i + \sum_i \varepsilon_i dn_i . \tag{2.1.28}$$

These expressions should be compared with those in the discussion of the probabilistic interpretation of the First Law in Section 1.2.2. There we used (what we now understand as) the Gibbs approach to the calculation of mean values; here we use the Boltzmann approach. Apart from that they are equivalent.

The second term of the differential expression vanishes at constant entropy. Thus, we identify

$$p = -\left.\frac{\partial E}{\partial V}\right|_S = -\sum_i n_i \frac{d\varepsilon_i}{dV} . \tag{2.1.29}$$

Now, we take the energy levels to depend on $V$ as a simple power. We shall write

$$\varepsilon_i = AV^\alpha , \tag{2.1.30}$$

so that

$$\frac{d\varepsilon_i}{dV} = A\alpha V^{\alpha-1} , \tag{2.1.31}$$

which is convenient to write as

$$\frac{d\varepsilon_i}{dV} = \frac{A\alpha V^\alpha}{V} = \alpha\frac{\varepsilon_i}{V} . \tag{2.1.32}$$

Thus, the expression for the pressure becomes

$$p = -\sum_i n_i \frac{d\varepsilon_i}{dV} = -\frac{\alpha}{V} \sum_i n_i \varepsilon_i = -\alpha \frac{E}{V}. \tag{2.1.33}$$

For (non-relativistic) particles of mass $m$ (where the kinetic energy is $p^2/2m$), we saw in Section 2.1.3, Eq. (2.1.10), that $\alpha = -2/3$. So, in this case,

$$p = \frac{2}{3} \frac{E}{V} \tag{2.1.34}$$

or

$$pV = \frac{2}{3} E. \tag{2.1.35}$$

It is important to appreciate the generality of this result. It relies only on the power-law dependence of the energy levels upon volume. It applies to quantum particles irrespective of statistics and thus it applies to classical particles as well. It only holds, however, for non-interacting particles; with interactions, the energy levels will depend on volume in a much more complicated way.

As a further example, we can consider a gas of ultra-relativistic/massless particles. In this case, the energy–momentum relation $\varepsilon = cp$ leads to the energy levels

$$\varepsilon = \frac{c\pi\hbar}{V^{1/3}} (n_x^2 + n_y^2 + n_z^2)^{1/2}, \tag{2.1.36}$$

so now the index $\alpha$ is $-1/3$ and then

$$pV = \frac{1}{3} E. \tag{2.1.37}$$

This result applies to a gas of photons, and in that case, the expression gives the radiation pressure as one-third of the energy density. The identical expression for radiation pressure is conventionally derived in electromagnetism [1] without the need for quantum arguments.

## 2.2   Identical Particles

### 2.2.1   *Indistinguishability*

Since the partition function is proportional to probabilities, it follows that for composite systems the partition function is essentially a product of the partition functions for the individual subsystems. The free energy is proportional to the logarithm of the partition function and this leads to the extensive variables of composite systems being additive.

In this section, we shall examine how the (canonical) partition function of a many-particle system is related to the partition function of a single-particle.

If we had an assembly of $N$ identical but *distinguishable* particles, the resultant partition function would be the product of the $N$ (same) partition functions of a single-particle, $z$:

$$Z = z^N. \tag{2.2.1}$$

The key question here is that of *indistinguishability* of the atoms or molecules of a many-body system. When two indistinguishable molecules are interchanged, the system is still in the same microstate, so the distinguishable particle result *overcounts* the states in this case. Now, the number of ways of redistributing $N$ particles when there are $n_1$ particles in the first state, $n_2$ particles in the second state, etc. is

$$\frac{N!}{n_1!\, n_2!\, n_3!\, \ldots}, \tag{2.2.2}$$

so that for a given distribution $\{n_i\}$ the partition function for identical indistinguishable particles is

$$Z = \frac{n_1!\, n_2!\, n_3!\, \ldots}{N!} z^N. \tag{2.2.3}$$

### 2.2.2 *Classical approximation*

The problem here is the occupation numbers $\{n_i\}$; we do not know these in advance. However, at high temperatures, the probability of occupancy of any state is small; the probability of multiple occupancy is then negligible. This is the classical régime, where the thermal de Broglie wavelength is very much less than the inter-particle spacing. The thermal de Broglie wavelength is the wavelength, corresponding to the momentum, corresponding to the kinetic energy corresponding to a temperature $T$. This will appear in Section 2.3.1. Under these circumstances, the factors $n_1!, n_2!, n_3!, \ldots$ can be ignored and we have a soluble problem.

In the classical case, we have then

$$Z = \frac{1}{N!}z^N.$$

(2.2.4)

The Helmholtz free energy

$$F = -kT \ln Z$$

(2.2.5)

is thus

$$F = -NkT \ln z + kT \ln N!.$$

(2.2.6)

This is $N$ times the Helmholtz free energy for a single-particle plus an extra term depending on $T$ and $N$. So, the second term can be ignored so long as we differentiate with respect to something other than $T$ or $N$. Specifically, when differentiating with respect to volume to find the pressure, the result is $N$ times that for a single-particle.

The structure of Eq. (2.2.6) and in particular the logical necessity for the $1/N!$ term in the classical gas partition function will be explored further in Section 2.3.4.

## 2.3   The Ideal Gas

### 2.3.1   *Quantum approach*

The partition function for a single-particle is

$$z(V, T) = \sum_i e^{-\varepsilon_i(V)/kT},$$

(2.3.1)

where $\varepsilon_i$ is the energy of the $i$th single-particle state; these states were enumerated in Section 2.1.3. As explained previously, the energy states are closely packed and this allows us to replace the sum over states by an integral using the density of states $g(\varepsilon) = V(2m)^{3/2}\varepsilon^{1/2}/4\pi^2\hbar^3$:

$$z = \int_0^\infty g(\varepsilon)e^{-\varepsilon/kT}d\varepsilon$$

$$= \frac{1}{4}\frac{V}{\pi^2\hbar^3}(2m)^{3/2}\int_0^\infty \varepsilon^{1/2}e^{-\varepsilon/kT}d\varepsilon.$$

(2.3.2)

Following a change of variable, this may be expressed, in terms of a standard integral, as [1]

$$z = V \left( \frac{mkT}{2\pi\hbar^2} \right)^{3/2} \frac{2}{\sqrt{\pi}} \int_0^\infty w^{1/2} e^{-w} \, dw. \tag{2.3.3}$$

The integral is $\sqrt{\pi}/2$, so that

$$z = \left( \frac{mkT}{2\pi\hbar^2} \right)^{3/2} V = \frac{V}{\Lambda^3}, \tag{2.3.4}$$

where

$$\Lambda = \sqrt{\frac{2\pi\hbar^2}{mkT}}. \tag{2.3.5}$$

The parameter $\Lambda$ is known as the *thermal de Broglie wavelength*; within a numerical factor, it is the wavelength corresponding to the momentum corresponding to the thermal kinetic energy of the particle. Thus, in a sense, it represents the quantum "size" of the particle.

For a gas of $N$ particles, we then have

$$Z = \frac{1}{N!} z^N = \frac{1}{N!} \left( \frac{V}{\Lambda^3} \right)^N. \tag{2.3.6}$$

We shall use Stirling's approximation, $\ln N! = N \ln N - N = N \ln(N/e)$, when evaluating the logarithm so that

$$\ln Z = N \ln(ze/N) = N \ln \left[ \left( \frac{mkT}{2\pi\hbar^2} \right)^{3/2} \frac{Ve}{N} \right], \tag{2.3.7}$$

from which all thermodynamic properties can be found, by the formulae of Section 1.4.4.

Since Stirling's approximation becomes true in the thermodynamic limit, it follows that by invoking Stirling's approximation we are implicitly taking the thermodynamic limit.

---

[1] Observe: The change of variable has brought the "physics" out of the integral.

### 2.3.2  Classical approach

It is instructive to consider the calculation of the single-particle partition function from the classical point of view. The classical partition function is given by the integral

$$z = \frac{1}{h^3} \int e^{-\varepsilon/kT} d^3p\, d^3q$$

$$= \frac{1}{8\pi\hbar^3} \int e^{-\varepsilon/kT} d^3p\, d^3q, \qquad (2.3.8)$$

where for the ideal gas $\varepsilon = p^2/2m$. The $q$ integrals are trivial, giving a factor $V$, and we have

$$z = \frac{V}{8\pi^3\hbar^3} \left[ \int_{-\infty}^{\infty} e^{-p^2/2mkT} dp \right]^3. \qquad (2.3.9)$$

The integral is transformed to a pure number by changing variables: $p = x\sqrt{2mkT}$ so that

$$z = V \left( \frac{mkT}{2\pi\hbar^2} \right)^{3/2} \frac{1}{\pi^{3/2}} \left[ \int_{-\infty}^{\infty} e^{-x^2} dx \right]^3. \qquad (2.3.10)$$

As in the quantum calculation, the physics is all outside the integral and the integral is just a pure number. The value of this integral is $\sqrt{\pi}$ so that

$$z = \left( \frac{mkT}{2\pi\hbar^2} \right)^{3/2} V = \frac{V}{\Lambda^3}, \qquad (2.3.11)$$

just as in the "quantum" calculation Eq. (2.3.4); we obtain the identical result. *This* justifies the use of Planck's constant $h$ in the normalisation factor for the classical state element of phase space.

### 2.3.3  Thermodynamic properties

In order to investigate the thermodynamic properties of the classical ideal gas, we start from the Helmholtz free energy, Eq. (2.3.7):

$$F = -kT \ln Z = NkT \ln \left[ \left( \frac{mkT}{2\pi\hbar^2} \right)^{3/2} \frac{Ve}{N} \right]. \qquad (2.3.12)$$

Then upon differentiation, we obtain

$$p = kT \left. \frac{\partial \ln Z}{\partial V} \right|_{T,N} = NkT \left. \frac{\partial \ln z}{\partial V} \right|_{T} = \frac{NkT}{V}. \qquad (2.3.13)$$

This is the ideal gas equation of state, and from this, we identify $k$ as Boltzmann's constant. Furthermore, we see that the ideal gas temperature thus corresponds to the "statistical" temperature introduced in the previous chapter.

The internal energy is

$$E = kT^2 \left. \frac{\partial \ln Z}{\partial T} \right|_{V,N} = NkT^2 \frac{d \ln T^{3/2}}{dT} = \frac{3}{2} NkT. \qquad (2.3.14)$$

This is the result we obtained previously from equipartition. This gives another important property of an ideal gas: the internal energy depends *only* on temperature (not pressure or density). This is known as Joule's law. From the energy expression, we obtain the heat capacity

$$C_V = \left. \frac{\partial E}{\partial T} \right|_{V,N} = \frac{3}{2} Nk. \qquad (2.3.15)$$

This is a constant, independent of temperature, in violation of the Third Law. This is because of the classical approximation, ignoring multiple state occupancy, etc. We also find the entropy and chemical potential:

$$S = Nk \ln \left[ \left( \frac{mkT}{2\pi\hbar^2} \right)^{3/2} \frac{V}{N} e^{5/2} \right],$$

$$\mu = -kT \ln \left[ \left( \frac{mkT}{2\pi\hbar^2} \right)^{3/2} \frac{V}{N} \right]. \qquad (2.3.16)$$

The formula for the entropy is often expressed as

$$S = Nk \ln V - Nk \ln N + \frac{3}{2} Nk \ln T + Nks_0, \qquad (2.3.17)$$

where the reduced "entropy constant" $s_0$ is given by

$$s_0 = \frac{3}{2} \ln \left( \frac{mk}{2\pi\hbar^2} \right) + \frac{5}{2}. \qquad (2.3.18)$$

The above expression for the entropy is known as the Sackur–Tetrode equation. It is often interpreted as indicating different contributions to the entropy: the volume contribution in the first term, the number contribution in the second term and the temperature contribution in the third term.

Such an identification is entirely *incorrect*; this matter is explored in Problem 2.4.

**Measuring Planck's constant from thermodynamics alone:**    The entropy constant may be determined experimentally, as explained by Guénault [2]. One measures the heat capacity of a substance as a function of temperature from very low temperatures (essentially $T = 0$) up to a temperature sufficiently high so that the substance behaves as an ideal gas. One also needs the latent heats of melting $L_m$ and boiling $L_b$. Then the entropy at temperature $T$ is given by

$$S(T) = \int_0^T \frac{C_V(T')}{T'} dT' + \frac{L_m}{T_m} + \frac{L_b}{T_b},\qquad(2.3.19)$$

where $T_m$ and $T_b$ are the melting and boiling temperatures. This gives the entropy of the ideal gas. By comparing this entropy with that given by Eq. (2.3.17) and subtracting off the first three terms, one obtains $s_0$. When the Avogadro constant is known, then Boltzmann's constant and the atomic mass are also known. Then it follows that Planck's constant may be found by "straightforward" calorimetric measurements. This relies on using the Third Law to fix the zero-temperature entropy at zero.

### 2.3.4   *The 1/N! term in the partition function*

In Section 2.2.2, we used quantum arguments to justify the factor $1/N!$ in the partition function for a gas of particles. Gibbs worked before quantum mechanics, so he had a problem. He *thought* the many-particle partition function was just the product of the individual particle partition functions, Eq. (2.2.1). This would give

$$\tilde{Z} = z^N$$
$$= \left(\frac{mkT}{2\pi\hbar^2}\right)^{3N/2} V^N.\qquad(2.3.20)$$

I have used $\tilde{Z}$ to indicate the erroneous partition function and I will use $\tilde{F}$ to indicate the corresponding (erroneous) Helmholtz free energy:

$$\tilde{F} = -kT \ln \tilde{Z}$$
$$= -NkT \ln \left\{ \left(\frac{mkT}{2\pi\hbar^2}\right)^{3/2} V \right\}.\qquad(2.3.21)$$

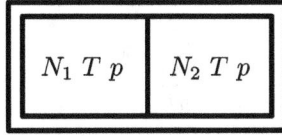

Fig. 2.3.   Mixing of gases.

You should note that this free energy will give, upon differentiation with respect to $V$, the correct equation of state. But, as Gibbs realised, the free energy expression is nevertheless problematic. While the argument of the logarithm is dimensionless, the $V$ factor makes it *extensive*. In other words, *the logarithm will diverge in the thermodynamic limit*. One would have expected the argument of the logarithm to be size-independent, since then the extensivity of $F$ is ensured by the $N$ prefactor.

This was the problem confronted by Gibbs. He *knew* the argument of the logarithm should be independent of size and he achieved this by inserting a factor of $1/N$ "by hand".

We know that the $1/N!$ multiplier of the partition function, required by quantum mechanics, results in a factor $e/N$ in the logarithm upon the use of Stirling's approximation. So, Gibbs's paradox is resolved using quantum mechanics and the thermodynamic limit.

### 2.3.5   *Entropy of mixing*

Let us consider the mixing of two gases. Figure 2.3 shows a container with a wall separating it into two regions. In the left-hand region, there are $N_1$ particles and in the right, there are $N_2$ particles. We shall assume that both regions are at the same temperature and pressure. Furthermore, we will assume that the number density $N/V$ is the same on both sides.

We shall investigate what happens when the wall is removed. In particular, we shall consider what happens when the gases on either side are (a) different and (b) when they are the same.

When the wall is in place, the partition function is simply the product of the partition functions of the two systems. We use the partition function expression of Eq. (2.1.16) so that

$$Z = Z_1 Z_2$$

$$= \frac{1}{N_1!} \left( \frac{mkT}{2\pi\hbar^2} \right)^{3N_1/2} V_1^{N_1} \frac{1}{N_2!} \left( \frac{mkT}{2\pi\hbar^2} \right)^{3N_2/2} V_2^{N_2}. \qquad (2.3.22)$$

This may be simplified to

$$Z = \frac{1}{N_1! \, N_2!} \left(\frac{mkT}{2\pi\hbar^2}\right)^{3(N_1+N_2)/2} V_1^{N_1} V_2^{N_2}. \tag{2.3.23}$$

This is the same whether the particles are identical or different.

Now, we shall remove the separating wall and allow the gases to mix. There are two cases to consider: identical and different particles.

(a) *Identical particles*

If the particles are identical, then the new partition function is that corresponding to $N = N_1 + N_2$, $V = V_1 + V_2$. Thus, the "after" partition function will be

$$Z = \frac{1}{(N_1 + N_2)!} \left(\frac{mkT}{2\pi\hbar^2}\right)^{3(N_1+N_2)/2} (V_1 + V_2)^{(N_1+N_2)}. \tag{2.3.24}$$

The ratio of the after to the before partition function is

$$\frac{Z_{\text{after}}}{Z_{\text{before}}} = \frac{N_1! N_2!}{(N_1 + N_2)!} \frac{(V_1 + V_2)^{(N_1+N_2)}}{V_1^{N_1} V_2^{N_2}}. \tag{2.3.25}$$

This may be simplified. Since $V_1/N_1 = V_2/N_2 = V/N$, it follows that $V_1 = VN_1/N$ and $V_2 = VN_2/N$ so that

$$\frac{Z_{\text{after}}}{Z_{\text{before}}} = \frac{N_1! N_2!}{N!} \frac{N^N}{N_1^{N_1} N_2^{N_2}}. \tag{2.3.26}$$

Upon using Stirling's approximation, that is, upon taking the thermodynamic limit, this expression becomes unity. In other words, in the case of identical particles,

$$Z_{\text{after}} = Z_{\text{before}}. \tag{2.3.27}$$

So, for identical particles, there is no effect on removing the separating wall and the original (macro-)state will be recovered if the wall is replaced. This is as expected on physical grounds, but note that the result would not have been obtained if the $1/N!$ term in the partition function had not been used.

(b) *Different particles*

If the particles on the two sides were different, then on removing the wall we have a composite system with $N_1$ type-1 particles in the volume $V = V_1 + V_2$ and $N_2$ type-2 particles in the same volume. Thus, we have

$$Z = \frac{1}{N_1!}\left(\frac{mkT}{2\pi\hbar^2}\right)^{3N_1/2} V^{N_1} \frac{1}{N_2!}\left(\frac{mkT}{2\pi\hbar^2}\right)^{3N_2/2} V^{N_2}$$

$$= \frac{1}{N_1!N_2!}\left(\frac{mkT}{2\pi\hbar^2}\right)^{3N/2} V^{N}.$$

(2.3.28)

The ratio of the after to the before partition function is then

$$\frac{Z_{\text{after}}}{Z_{\text{before}}} = \frac{V^N}{V_1^{N_1} V_2^{N_2}}.$$

(2.3.29)

In this case, the "after" partition function is different from the "before". Then, for unlike particles, the free energy and some of the other thermodynamic parameters will change when the wall is removed.

The change in the free energy is given by

$$F_{\text{after}} - F_{\text{before}} = -kT \ln \frac{Z_{\text{after}}}{Z_{\text{before}}}$$

$$= -kT \ln \left\{ \frac{V^N}{V_1^{N_1} V_2^{N_2}} \right\}.$$

(2.3.30)

This will lead to a change of entropy. Since

$$S = - \left. \frac{\partial F}{\partial T} \right|_{V,N},$$

(2.3.31)

it follows that the entropy change upon removal of the wall is

$$\Delta S = k \ln \left\{ \frac{V^N}{V_1^{N_1} V_2^{N_2}} \right\}$$

$$= Nk \ln V - N_1 k \ln V_1 - N_2 k \ln V_2.$$

(2.3.32)

This may be written as

$$\Delta S = N_1 k \ln \frac{V}{V_1} + N_2 k \ln \frac{V}{V_2},$$

(2.3.33)

which shows that the entropy increases when the wall is removed. This increase of entropy when two different gases are mixed is known as the *entropy of mixing*. The expression may be understood from the Boltzmann entropy expression, using simple counting arguments: the first term gives the entropy increase of the $N_1$ type-1 particles expanding from volume $V_1$ into volume $V$ while the second term gives the entropy increase of the $N_2$ type-2 particles expanding from volume $V_2$ into volume $V$.

## 2.4   The Quantum Gas

### 2.4.1   *Methodology for quantum gases*

The Bose–Einstein and the Fermi–Dirac distribution functions give the mean number of particles in a given single-particle quantum state in terms of the temperature $T$ and the chemical potential $\mu$. The temperature and chemical potential are the *intensive* variables that determine the equilibrium distribution $\bar{n}\,(\varepsilon)$. We have a good intuitive feel for the temperature of a system. But the chemical potential is different. This determines the number of particles, in a system that can exchange particles with its surrounding. In practice, however, it might be more intuitive to speak of a system containing a given number of particles. In that case, it is the number of particles in the system that determines the chemical potential.

Now, the number of particles in the system is given by

$$N = \alpha \sum_i \bar{n}(\varepsilon_i), \tag{2.4.1}$$

which converts to the integral

$$N = \alpha \int_0^\infty \bar{n}(\varepsilon)g(\varepsilon)\,\mathrm{d}\varepsilon, \tag{2.4.2}$$

where $g(\varepsilon)$ is the energy density of states and $\alpha$ is the factor which accounts for the degeneracy of the particles' spin states. This is 2 for electrons since there are two spin states for a spin $S = 1/2$; more generally, it will be $2S + 1$, (but see Section 2.9.3 dealing with photons; massless particles are different).

The expression for $N$ must be inverted to give $\mu$, which can then be used in the distribution function to find the other properties of the system. For instance, the internal energy of the system would be found from

$$E = \alpha \sum_i \varepsilon_i \bar{n}(\varepsilon_i) \tag{2.4.3}$$

or, in integral form,

$$E = \alpha \int_0^\infty \varepsilon \, \bar{n}(\varepsilon) g(\varepsilon) d\varepsilon. \tag{2.4.4}$$

More generally, the average value of a function of energy $f(\varepsilon)$ is given by

$$\bar{f} = \alpha \sum_i f(\varepsilon_i) \bar{n}(\varepsilon_i), \tag{2.4.5}$$

where the sum is taken over the single-particle energy states. In integral form, this is

$$\bar{f} = \alpha \int_0^\infty f(\varepsilon) g(\varepsilon) \bar{n}(\varepsilon) d\varepsilon. \tag{2.4.6}$$

Thus, the methodology for treating a quantum gas starts with the expression for the number of particles $N$, which is inverted to find the chemical potential $\mu$. This is substituted into the distribution function $\bar{n}(\varepsilon)$, from which all other properties may be found.

## 2.5 Fermi Gas at Low Temperatures

### 2.5.1 *Ideal Fermi gas at zero temperature*

The Fermi–Dirac distribution function is given by

$$n(\varepsilon) = \frac{1}{e^{(\varepsilon-\mu)/kT} + 1}. \tag{2.5.1}$$

Henceforth, for notational convenience, we shall drop the bar over $n$. We have plotted $n(\varepsilon)$ in Fig. 2.4, where the dashed line shows the zero-temperature case. At $T = 0$, the distribution becomes a box function

$$n(\varepsilon) = \begin{cases} 1 & \varepsilon < \mu, \\ 0 & \varepsilon > \mu. \end{cases} \tag{2.5.2}$$

Note that in general the chemical potential depends on temperature. Its zero-temperature value is called the Fermi energy:

$$\varepsilon_F = \mu(T = 0). \tag{2.5.3}$$

And the temperature defined by $kT_F = \varepsilon_F$ is known as the Fermi temperature.

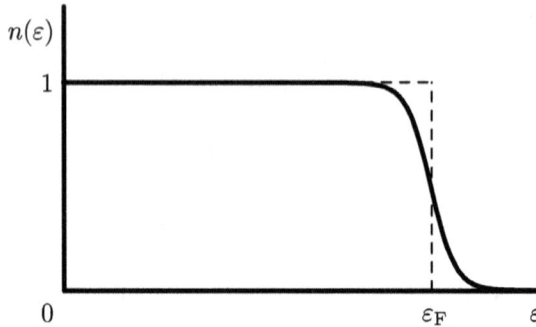

Fig. 2.4.   Fermi–Dirac distribution, with $T = 0$ limit.

In accordance with the methodology described in the previous section, the first thing to do is to evaluate the total number of particles in the system in order to find the chemical potential — the Fermi energy in the $T = 0$ case.

The density of states is given by Eq. (2.1.16):

$$g(\varepsilon) = \frac{1}{4} \frac{V}{\pi^2 \hbar^3} (2m)^{3/2} \varepsilon^{1/2}. \tag{2.5.4}$$

The total number of particles in the system is

$$N = \frac{\alpha V}{4\pi^2 \hbar^3} (2m)^{3/2} \int_0^{\varepsilon_F} \varepsilon^{1/2} d\varepsilon$$

$$= \frac{\alpha V}{6\pi^2 \hbar^3} (2m\varepsilon_F)^{3/2}. \tag{2.5.5}$$

And this may be inverted to obtain the Fermi energy:

$$\varepsilon_F = \frac{\hbar^2}{2m} \left( \frac{6\pi^2}{\alpha} \frac{N}{V} \right)^{2/3}. \tag{2.5.6}$$

This gives the chemical potential at zero temperature. Observe that it depends on the *density* of particles in the system $N/V$, so the Fermi energy is, as expected, an intensive variable.

*Fermi momentum and Fermi wave vector*

There is a host of Fermi-related quantities. We have introduced the Fermi energy and the Fermi temperature. It will prove convenient also to mention the Fermi velocity $v_F$, the Fermi momentum $p_F$, and the Fermi wave vector $k_F$. Since we are considering non-relativistic particles, we define $v_F$ and $k_F$ through

$$\varepsilon_F = \frac{1}{2}mv_F^2 = \frac{p_F^2}{2m}, \tag{2.5.7}$$

so that

$$v_F = \frac{\hbar}{m}\left(\frac{6\pi^2}{\alpha}\frac{N}{V}\right)^{1/3} \tag{2.5.8}$$

and

$$p_F = \hbar\left(\frac{6\pi^2}{\alpha}\frac{N}{V}\right)^{1/3}. \tag{2.5.9}$$

From this, the Fermi wave vector follows naturally as

$$k_F = \left(\frac{6\pi^2}{\alpha}\frac{N}{V}\right)^{1/3}. \tag{2.5.10}$$

Observe that $k_F$ depends only on the number density (and the spin degeneracy factor). In terms of the mean particle spacing $d = (V/N)^{1/3}$, we have

$$k_F = \left(\frac{6\pi^2}{\alpha}\right)^{1/3}\frac{1}{d}, \tag{2.5.11}$$

which for spin $\frac{1}{2}$ is

$$k_F \approx \frac{3.09}{d}. \tag{2.5.12}$$

So, the Fermi wave vector is essentially a measure of the inverse particle separation.

### Re-expression of the density of states

We see that the Fermi energy/velocity/temperature/momentum/wave vector provide natural and useful scales for the consideration of the properties of Fermi systems. Thus, it proves convenient and expedient to express the density of states in terms of the Fermi energy

$$g(\varepsilon) = \frac{3N}{2\alpha\varepsilon_F^{3/2}} \varepsilon^{1/2}$$

$$= \frac{3}{2\alpha} \frac{N}{\varepsilon_F} \left(\frac{\varepsilon}{\varepsilon_F}\right)^{1/2}.$$

(2.5.13)

Having obtained the zero-temperature chemical potential, the Fermi–Dirac function is now completely specified at $T = 0$, and we can proceed to find the zero-temperature internal energy of the system. This is given by

$$E(T = 0) = \frac{3N}{2\varepsilon_F^{3/2}} \int_0^{\varepsilon_F} \varepsilon^{3/2} d\varepsilon,$$

(2.5.14)

which evaluates to

$$E = \frac{3}{5} N\varepsilon_F.$$

(2.5.15)

The internal energy is proportional to the number of particles in the system and so it is, as expected, an extensive quantity.

Pressure is related to the internal energy. For non-relativistic particles (Eq. (2.1.35)), we have simply $pV = \frac{2}{3}E$. So, the zero-temperature pressure of a Fermi gas is

$$p(T = 0) = \frac{2}{5} \frac{N}{V} \varepsilon_F.$$

(2.5.16)

The zero-temperature pressure of a classical gas is zero. By contrast, the pressure of a Fermi gas at $T = 0$ is finite; it corresponds to the pressure of a classical gas at a temperature $T = \frac{2}{3} T_F$.

The considerable ground state energy and zero-temperature pressure of a Fermi gas are a consequence of the Pauli exclusion principle.

### 2.5.2 *Fermi gas at low temperatures — Simple model*

The effect of a small finite temperature may be modelled very simply by approximating the Fermi–Dirac distribution by a piecewise linear function, Fig. 2.5. This must match the slope of the curve at $\varepsilon = \mu$, and the derivative is found to be

$$\left.\frac{dn(\varepsilon)}{d\varepsilon}\right|_{\varepsilon=\mu} = -\frac{1}{4kT}.$$

(2.5.17)

This indicates that the energy width of the transition region is $\Delta\varepsilon \sim kT$ so that only a fraction $kT/\varepsilon_F$ of the particles in the vicinity of the Fermi

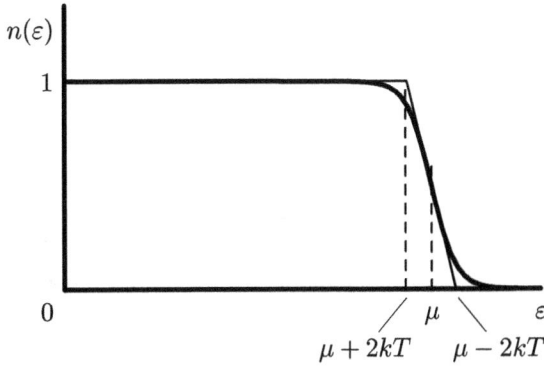

Fig. 2.5.   Simple modelling of Fermi–Dirac distribution.

energy are "excitable". In other words, in many respects, a degenerate gas of $N$ fermions at a temperature $T$ is "like" a classical gas of $N\frac{T}{T_F}$ particles.

The distribution function, in this approximation, is then

$$n(\varepsilon) = \begin{cases} 1 & 0 < \varepsilon < \mu - 2kT, \\ 1 - (\varepsilon - \mu + 2kT)/4kT & \mu - 2kT < \varepsilon < \mu + 2kT, \\ 0 & \mu + 2kT < \varepsilon, \end{cases} \qquad (2.5.18)$$

from which the thermodynamic properties may be calculated.

The chemical potential is found by considering the total number of particles in the system:

$$\begin{aligned} N &= \alpha \int_0^\infty n(\varepsilon) g(\varepsilon) \mathrm{d}\varepsilon \\ &= \frac{3N}{2\varepsilon_F^{3/2}} \int_0^\infty n(\varepsilon)\varepsilon^{1/2}\mathrm{d}\varepsilon. \end{aligned} \qquad (2.5.19)$$

Observe that here the number $N$ appears on both sides and so it cancels. The reason for this is that we have chosen to express the density of states in terms of the Fermi energy. Thus, while the $N$ cancels from both sides, the number (density) is contained in the Fermi energy. So, in this approach, we are really expressing the Fermi energy in terms of the chemical potential:

$$\varepsilon_F^{3/2} = \frac{3}{2} \int_0^\infty n(\varepsilon)\varepsilon^{1/2}\mathrm{d}\varepsilon, \qquad (2.5.20)$$

and we wish to invert this relation to give the chemical potential in terms of the Fermi energy (and temperature).

We use the approximate form for $n(\varepsilon)$. Upon integration, this gives

$$\varepsilon_F^{3/2} = \mu^{3/2} \frac{1}{10} \left\{ \sqrt{1 + 2\frac{kT}{\mu}} \left[ 4\frac{kT}{\mu} + 4 + \frac{\mu}{kT} \right] \right.$$

$$\left. - \sqrt{1 - 2\frac{kT}{\mu}} \left[ 4\frac{kT}{\mu} - 4 + \frac{\mu}{kT} \right] \right\}. \tag{2.5.21}$$

As we are interested specifically in the low-temperature behaviour, we shall expand as a series in powers of $kT/\mu$

$$\varepsilon_F^{3/2} = \mu^{3/2} \left\{ 1 + \frac{1}{2} \left( \frac{kT}{\mu} \right)^2 + \frac{3}{40} \left( \frac{kT}{\mu} \right)^4 \cdots \right\}. \tag{2.5.22}$$

The Fermi energy is then the 2/3 power of this, which, by the binomial theorem, is

$$\varepsilon_F = \mu \left\{ 1 + \frac{1}{3} \left( \frac{kT}{\mu} \right)^2 + \frac{1}{45} \left( \frac{kT}{\mu} \right)^4 + \cdots \right\}$$

$$= \mu + \frac{1}{3} \frac{(kT)^2}{\mu} + \frac{1}{45} \frac{(kT)^4}{\mu^3} + \cdots. \tag{2.5.23}$$

This series may be inverted to give $\mu$ in terms of $\varepsilon_F$:

$$\mu = \varepsilon_F \left\{ 1 - \frac{1}{3} \left( \frac{kT}{\varepsilon_F} \right)^2 - \frac{2}{15} \left( \frac{kT}{\varepsilon_F} \right)^4 + \cdots \right\}. \tag{2.5.24}$$

This shows that as the temperature is increased from $T = 0$ the chemical potential decreases from its zero temperature value, and the first term is proportional to $T^2$.

In a similar way, the internal energy is found to be

$$E = E_0 \left\{ 1 + \frac{5}{3} \left( \frac{kT}{\varepsilon_F} \right)^2 - \frac{2}{3} \left( \frac{kT}{\varepsilon_F} \right)^4 + \cdots \right\} \tag{2.5.25}$$

up to the term in $T^4$.

We should note that the approximation of the Fermi–Dirac distribution used in this section is only a model that allows the simple calculation of properties of the Fermi gas, indicating the general behaviour at low temperatures. But the coefficients of the powers of $kT/\varepsilon_F$ are incorrect; this is simply a model that permits demonstration of the *qualitative* low-temperature behaviour.

### 2.5.3 *Fermi gas at low temperatures — Series expansion*

We have formal expressions for the thermodynamic behaviour of a gas of fermions: expressions like Eqs. (2.4.2) and (2.4.4). In the previous section, a model was described, which treated the deviations from $T = 0$ behaviour in a simplistic and qualitative way. What is required is a systematic procedure for calculating finite (but low) temperature behaviour. We shall see that such a procedure is possible based on the special shape of the Fermi–Dirac distribution function; this approach was pioneered by Arnold Sommerfeld [3].

Very generally, one requires to evaluate integrals of the form

$$I = \int_0^\infty n(\varepsilon)\varphi(\varepsilon)d\varepsilon, \tag{2.5.26}$$

where $n(\varepsilon)$ is the Fermi–Dirac distribution function and $\varphi(\varepsilon)$ is the function to be integrated over, including the density of states. Equations (2.4.2) and (2.4.4) for $N$ and $E$ are of this form. If we define $\psi(\varepsilon)$, the integral of $\varphi(\varepsilon)$, by

$$\psi(\varepsilon) = \int_0^\varepsilon \varphi(\varepsilon')d\varepsilon', \tag{2.5.27}$$

then the expression for $I$ may be integrated by parts to give

$$I = n(\varepsilon)\psi(\varepsilon)\Big|_0^\infty - \int_0^\infty n'(\varepsilon)\psi(\varepsilon)d\varepsilon. \tag{2.5.28}$$

Now, since $n(\infty) = 0$ from the form of the Fermi–Dirac distribution function, and $\psi(0) = 0$ from its definition, the first term vanishes and one is left with

$$I = -\int_0^\infty n'(\varepsilon)\psi(\varepsilon)d\varepsilon, \tag{2.5.29}$$

an integral over $n'(\varepsilon)$. But $n'(\varepsilon)$ has an important shape (Fig. 2.6): it is sharply peaked at $\varepsilon = \mu$, particularly at the lowest of temperatures; this was Sommerfeld's key point. It means that in the integral for $I$ the behaviour of $\psi(\varepsilon)$ only in the vicinity of $\varepsilon = \mu$ is important.

It then follows that a convenient series for $I$ may be obtained by expanding $\psi(\varepsilon)$ as a series about $\varepsilon = \mu$ and integrating Eq. (2.5.29) term

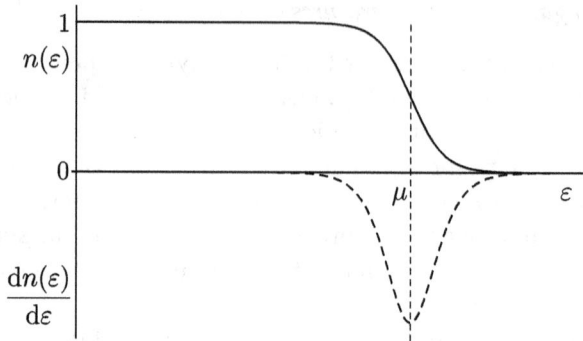

Fig. 2.6.   Fermi–Dirac distribution and its derivative.

by term. The Taylor expansion is

$$\psi(\varepsilon) = \sum_{n=0}^{\infty} \frac{1}{n!} \frac{d^n \psi}{d\varepsilon^n}\bigg|_{\varepsilon=\mu} (\varepsilon - \mu)^n, \tag{2.5.30}$$

so that

$$I = -\sum_{n=0}^{\infty} \frac{1}{n!} \frac{d^n \psi}{d\varepsilon^n}\bigg|_{\varepsilon=\mu} \int_0^{\infty} n'(\varepsilon)(\varepsilon - \mu)^n d\varepsilon. \tag{2.5.31}$$

The integral is simplified through the substitution $x = (\varepsilon - \mu)/kT$, and since $n'(\varepsilon)$ goes to zero away from $\varepsilon = \mu$, the lower limit of the integral can be extended to $-\infty$ without introducing significant error. Then

$$\int_0^{\infty} n'(\varepsilon)(\varepsilon - \mu)^n d\varepsilon = -(kT)^n \mathscr{F}_n, \tag{2.5.32}$$

where the $\mathscr{F}_n$ are just numbers:

$$\mathscr{F}_n = \int_{-\infty}^{\infty} \frac{e^x}{(e^x + 1)^2} x^n dx$$
$$= \int_{-\infty}^{\infty} \frac{x^n}{4\cosh^2(x/2)} dx. \tag{2.5.33}$$

Since $\cosh^2(x/2)$ is an even function, the $\mathscr{F}_n$ for odd $n$ vanish. We then find

$$I = \sum_{\substack{n=0 \\ (\text{even } n)}}^{\infty} \mathscr{F}_n \frac{(kT)^n}{n!} \frac{d^n \psi}{d\varepsilon^n}\bigg|_{\varepsilon=\mu}, \tag{2.5.34}$$

which is a series expansion for the quantity $I$ in powers of temperature. This thus provides our *low-temperature* expansion. By contrast, the high-temperature behaviour will be treated in Section 2.7.1.

The general expression for the integrals is (even $n$)

$$\mathscr{F}_n = 2(1 - 2^{1-n})\zeta(n)n!, \tag{2.5.35}$$

where $\zeta(n)$ is Riemann's zeta function.[2] So, we may write $I$ as

$$I = \sum_{\substack{n=0 \\ (\text{even } n)}}^{\infty} 2(1 - 2^{1-n})\zeta(n)(kT)^n \left.\frac{d^n \psi}{d\varepsilon^n}\right|_{\varepsilon=\mu}. \tag{2.5.36}$$

The first few $\mathscr{F}_n$ values are

$$\begin{aligned}
\mathscr{F}_0 &= 1, \\
\mathscr{F}_2 &= \pi^2/3, \\
\mathscr{F}_4 &= 7\pi^4/15, \\
\mathscr{F}_6 &= 31\pi^6/21, \\
\mathscr{F}_8 &= 127\pi^8/15, \\
\mathscr{F}_{10} &= 2555\pi^{10}/33,
\end{aligned} \tag{2.5.37}$$

so that

$$I = \psi(\mu) + \frac{\pi^2}{6}(kT)^2 \left.\frac{d^2\psi}{d\varepsilon^2}\right|_{\varepsilon=\mu} + \frac{7\pi^4}{360}(kT)^4 \left.\frac{d^4\psi}{d\varepsilon^4}\right|_{\varepsilon=\mu} + \cdots \tag{2.5.38}$$

or, in terms of the original function $\varphi(\varepsilon)$,

$$I = \int_0^\mu \varphi(\varepsilon)d\varepsilon + \frac{\pi^2}{6}(kT)^2 \left.\frac{d\varphi}{d\varepsilon}\right|_{\varepsilon=\mu} + \frac{7\pi^4}{360}(kT)^4 \left.\frac{d^3\varphi}{d\varepsilon^3}\right|_{\varepsilon=\mu} + \cdots. \tag{2.5.39}$$

Note this is not *quite* the simple power series it appears since the chemical potential $\mu$ also depends upon temperature. Thus, chemical potential must always be found before proceeding to the other thermodynamic quantities.

We should also point out that in extending the lower limit of the integral in Eq. (2.5.33) to minus infinity, we are neglecting exponentially

---

[2]The Riemann zeta function $\zeta(n)$ is defined as the infinite series $\zeta(n) = \frac{1}{1^n} + \frac{1}{2^n} + \frac{1}{3^n} + \cdots$. *Mathematica* provides this function as Zeta[n].

small terms. Landau and Lifshitz [4] point out that the resultant expansion is thus an *asymptotic* and not a convergent series. This is dramatically illustrated in the two-dimensional case, considered in Problem 2.8.

These series expansions are conveniently evaluated using a computer symbolic mathematics system such as *Mathematica*. It is then expedient to recast the expression for $I$ of Eq. (2.5.29) as

$$I = -\int_0^\infty n'(\varepsilon)\psi(\varepsilon)d\varepsilon$$

$$= \frac{1}{kT}\int_0^\infty \frac{e^{(\varepsilon-\mu)/kT}}{(e^{(\varepsilon-\mu)/kT}+1)^2}\psi(\varepsilon)d\varepsilon.$$

(2.5.40)

Then making the substitution $x = (\varepsilon - \mu)/kT$, and extending the lower limit of the integral to $-\infty$ as explained above,

$$I = \int_{-\infty}^\infty \frac{e^x}{(e^x+1)^2}\psi(kTx+\mu)dx.$$

(2.5.41)

We shall call this a Sommerfeld integral. The procedure for evaluating the low-temperature series for $I$ then comprises expanding $\psi$ in powers of $x$ before integrating term by term.

*Chemical potential*

The chemical potential is found by considering the total number of particles in the system:

$$N = \alpha \int_0^\infty n(\varepsilon)g(\varepsilon)d\varepsilon$$

$$= \frac{3N}{2\varepsilon_F^{3/2}}\int_0^\infty n(\varepsilon)\varepsilon^{1/2}d\varepsilon.$$

(2.5.42)

Observe that here the number $N$ appears on both sides and so it cancels. The reason for this is that we have chosen to express the density of states in terms of the Fermi energy. Thus, while the $N$ cancels from both sides, the number (density) is contained in the Fermi energy. So, in this approach, we are really expressing the Fermi energy in terms of the chemical potential

$$\varepsilon_F^{3/2} = \frac{3}{2}\int_0^\infty n(\varepsilon)\varepsilon^{1/2}d\varepsilon,$$

(2.5.43)

and we wish to invert this relation to give the chemical potential in terms of the Fermi energy (and temperature).

In this case,

$$\varphi(\varepsilon) = \frac{3N}{2\varepsilon_F^{3/2}} \varepsilon^{1/2}, \tag{2.5.44}$$

so that

$$\psi(\varepsilon) = \frac{N}{\varepsilon_F^{3/2}} \varepsilon^{3/2}. \tag{2.5.45}$$

Then using the above result in Eq. (2.5.39) or using the *Mathematica* procedure in the Appendix C.1, we find

$$\varepsilon_F^{3/2} = \mu^{3/2} \left\{ 1 + \frac{\pi^2}{6} \frac{3}{4} \left( \frac{kT}{\mu} \right)^2 + \frac{7\pi^4}{360} \frac{9}{16} \left( \frac{kT}{\mu} \right)^4 + \cdots \right\}, \tag{2.5.46}$$

or, upon simplification,

$$\varepsilon_F^{3/2} = \mu^{3/2} \left\{ 1 + \frac{1}{8} \pi^2 \left( \frac{kT}{\mu} \right)^2 + \frac{7}{640} \pi^4 \left( \frac{kT}{\mu} \right)^4 + \cdots \right\}. \tag{2.5.47}$$

The Fermi energy is then the 2/3 power of this which, by the binomial theorem, gives $\varepsilon_F$ as

$$\varepsilon_F = \mu \left\{ 1 + \frac{2}{3} \frac{1}{8} \pi^2 \left( \frac{kT}{\mu} \right)^2 + \frac{1}{180} \pi^4 \left( \frac{kT}{\mu} \right)^4 + \cdots \right\}$$

$$= \mu + \frac{\pi^2}{12} \frac{(kT)^2}{\mu} + \frac{\pi^4}{180} \frac{(kT)^4}{\mu^3} \cdots \tag{2.5.48}$$

and to the same order this may be inverted to give

$$\mu = \varepsilon_F \left\{ 1 - \frac{\pi^2}{12} \left( \frac{kT}{\varepsilon_F} \right)^2 - \frac{\pi^4}{80} \left( \frac{kT}{\varepsilon_F} \right)^4 + \cdots \right\}. \tag{2.5.49}$$

This shows that as the temperature is increased from $T = 0$ the chemical potential decreases from its zero temperature value, and the first term is in $T^2$. The result is similar to the qualitative approximation of the previous section, Eq. (2.5.24), but now the coefficients of the series are correct.

The procedure can be extended easily to find higher powers of $T$ in the expansion; indeed the *Mathematica* code in Appendix C can do this automatically. The starting function in this case is the Sommerfeld integral:

$$\varepsilon_F^{3/2} = \int_{-\infty}^{\infty} \frac{e^x}{(e^x + 1)^2} [\mu + kTx]^{3/2} dx. \tag{2.5.50}$$

The series for $\varepsilon_F$ in powers of $\mu$ is inverted to give a series for $\mu$ in powers of $\varepsilon_F$.

*Internal energy*

The internal energy for the gas of Fermions is given by

$$\begin{aligned} E &= \alpha \int_0^{\infty} \varepsilon n(\varepsilon) g(\varepsilon) d\varepsilon \\ &= \frac{3N}{2\varepsilon_F^{3/2}} \int_0^{\infty} n(\varepsilon) \varepsilon^{3/2} d\varepsilon, \end{aligned} \tag{2.5.51}$$

so that in this case

$$\varphi(\varepsilon) = \frac{3N}{2\varepsilon_F^{3/2}} \varepsilon^{3/2} \tag{2.5.52}$$

and

$$\psi(\varepsilon) = \frac{3}{5} \frac{N}{\varepsilon_F^{3/2}} \varepsilon^{5/2}. \tag{2.5.53}$$

Then using the above result in Eq. (2.5.39) or using the *Mathematica* procedure in Appendix C, we find

$$E = \frac{3}{5} \frac{N\mu^{5/2}}{\varepsilon_F^{3/2}} + \frac{\pi^2}{6} (kT)^2 \frac{9}{4} \frac{N\mu^{1/2}}{\varepsilon_F^{3/2}} - \frac{7\pi^4}{360} (kT)^4 \frac{9}{16} \frac{N\mu^{-3/2}}{\varepsilon_F^{3/2}} + \cdots . \tag{2.5.54}$$

This, however, gives the energy in terms of $\mu$, which has a temperature dependence of its own. So, we must substitute for $\mu$ from Eq. (2.5.49). This will then give the temperature dependence of the internal energy. Up to the term in $T^4$ this is

$$E = \frac{3}{5}N\varepsilon_F + \frac{\pi^2}{4}N\varepsilon_F \left(\frac{kT}{\varepsilon_F}\right)^2 - \frac{3\pi^4}{80}N\varepsilon_F \left(\frac{kT}{\varepsilon_F}\right)^4 + \cdots$$

$$= E_0 \left\{1 + \frac{5\pi^2}{12}\left(\frac{kT}{\varepsilon_F}\right)^2 - \frac{5\pi^4}{80}\left(\frac{kT}{\varepsilon_F}\right)^4 + \cdots\right\}.$$

(2.5.55)

This procedure can be extended easily to find higher powers of $T$ in the expansion; indeed the *Mathematica* code in Appendix C can do this automatically. The starting function in this case is the Sommerfeld integral:

$$E = \frac{3}{5}\frac{N}{\varepsilon_F^{3/2}} \int_{-\infty}^{\infty} \frac{e^x}{(e^x + 1)^2} (kTx + \mu)^{5/2} dx.$$

(2.5.56)

*Equation of state*

Since we know that $pV = \frac{2}{3}E$ for classical (non-relativistic) non-interacting particles, we obtain the low-temperature equation of state for fermions

$$pV = \frac{2}{5}N\varepsilon_F + \frac{\pi^2}{6}N\varepsilon_F \left(\frac{kT}{\varepsilon_F}\right)^2 - \frac{\pi^4}{40}N\varepsilon_F \left(\frac{kT}{\varepsilon_F}\right)^4 + \cdots.$$

(2.5.57)

*Heat capacity*

The heat capacity is found by differentiating the internal energy with respect to temperature:

$$C_V = \left.\frac{\partial E}{\partial T}\right|_V,$$

(2.5.58)

so that

$$C_V = Nk\left\{\frac{\pi^2}{2}\frac{kT}{\varepsilon_F} - \frac{3\pi^4}{20}\left(\frac{kT}{\varepsilon_F}\right)^3 + \cdots\right\}.$$

(2.5.59)

We see that the heat capacity goes to zero with temperature in agreement with the Third Law.

At low temperatures, it decreases linearly

$$C_V = Nk\frac{\pi^2}{2}\frac{kT}{\varepsilon_F},$$

(2.5.60)

and we observe that in this limit $C_V$ is $\frac{\pi^2}{3}\frac{kT}{\varepsilon_F}$ times the classical value. This is in accordance with the idea that in a degenerate gas of fermions, only a fraction $kT/\varepsilon_F$ of the particles in the vicinity of the Fermi energy are "excitable".

We may substitute for $\varepsilon_F$ to write the linear heat capacity as

$$\begin{aligned}
C_V &= N\frac{m\pi^2}{\hbar^2 k_F^2}k^2 T \\
&= N\frac{m}{\hbar^2}\left(\frac{\pi\alpha V}{6N}\right)^{2/3}k^2 T.
\end{aligned}$$

(2.5.61)

### 2.5.4 *More general treatment of low-temperature heat capacity*

The low-temperature heat capacity of real fermions, such as electrons or liquid $^3$He, is observed to follow the linear temperature behaviour above, but the coefficient can be different from that of the free particle expression, Eq. (2.5.61). The linearity of the heat capacity is interpreted as indicating that while the free-particle picture is inappropriate, the excitations in the system are nevertheless "particle-like". The dispersion (energy–momentum) may not be that of free particles and, correspondingly, the density of states may not correspond to that for free particles.

We shall approach this issue by considering the heat capacity of a system of fermions where the form for the density of states function is not assumed. This means we are no longer considering non-interacting fermions particles; instead we are considering a gas of non-interacting (weakly interacting) Fermi *quasi-particles*. In this sense, the discussion is appropriate for this chapter. We shall utilise a generalisation of the Sommerfeld expansion method. Our treatment was inspired by the discussion of Reif [5].

In accordance with our general procedures, the first thing to do is to obtain a suitable expression for the chemical potential for the system. Since we are considering low temperatures, we know that $\mu$ will be close to the Fermi energy, but we will require the small deviation from this value.

The number of particles in the system is given by

$$N = \alpha \int_0^\infty n(\varepsilon)g(\varepsilon)\,d\varepsilon,$$

(2.5.62)

where $g(\varepsilon)$ is our general expression for the density of states. In terms of the Sommerfeld expansion, Eq. (2.5.39), here

$$\varphi(\varepsilon) = \alpha g(\varepsilon) \tag{2.5.63}$$

so that up to second order in $T$

$$N = \alpha \int_0^\mu g(\varepsilon)d\varepsilon + \alpha \frac{\pi^2}{6}(kT)^2 g'(\varepsilon). \tag{2.5.64}$$

In this expression, we shall split the integral in the first term into the range from 0 to $\varepsilon_F$ and from $\varepsilon_F$ to $\mu$:

$$N = \alpha \int_0^{\varepsilon_F} g(\varepsilon)d\varepsilon + \alpha \int_{\varepsilon_F}^\mu g(\varepsilon)d\varepsilon + \alpha \frac{\pi^2}{6}(kT)^2 g'(\mu). \tag{2.5.65}$$

Here, the first integral is simply the total number of particles $N$. In the second integral, we note that $\varepsilon_F$ is close to $\mu$ and that over this small range the argument of the integral, $g(\varepsilon)$ will hardly change. Thus, the second term may be approximated by

$$\alpha(\mu - \varepsilon_F)g(\varepsilon_F). \tag{2.5.66}$$

Since the third term already has a $T^2$ factor, we introduce negligible error here by replacing $\mu$ by $\varepsilon_F$. Then we have

$$(\mu - \varepsilon_F)g(\varepsilon_F) + \frac{\pi^2}{6}(kT)^2 g'(\varepsilon_F) = 0, \tag{2.5.67}$$

so that to this order of approximation

$$\mu = \varepsilon_F - \frac{\pi^2}{6}(kT)^2 \frac{g'(\varepsilon_F)}{g(\varepsilon_F)}. \tag{2.5.68}$$

This gives the leading order deviation of the chemical potential from the zero temperature value of the Fermi energy. We see the leading order deviation goes as $T^2$, as we found for the free particle case, Eq. (2.5.49).

Now, moving to the calculation of the internal energy, in this case we have

$$\varphi(\varepsilon) = \alpha\varepsilon g(\varepsilon). \tag{2.5.69}$$

Then the series for the energy may be written in a similar way as

$$E = \alpha \int_0^{\varepsilon_F} \varepsilon g(\varepsilon)d\varepsilon + \alpha \int_{\varepsilon_F}^\mu \varepsilon g(\varepsilon)d\varepsilon + \alpha \frac{\pi^2}{6}(kT)^2\{g(\mu) + \mu g'(\mu)\}. \tag{2.5.70}$$

Here, the first term is the ground state energy $E_0$. In the second term, $\varepsilon_F$ is again close to $\mu$ and over this small range the argument of the integral, $\varepsilon g(\varepsilon)$, will hardly change. Thus, the second term may be approximated by

$$\alpha(\mu - \varepsilon_F)\varepsilon_F g(\varepsilon_F). \qquad (2.5.71)$$

But now we have an expression for $\mu$ in Eq. (2.5.68). Thus, the second term may be written as

$$-\alpha \frac{\pi^2}{6}(kT)^2 g'(\varepsilon_F)\varepsilon_F. \qquad (2.5.72)$$

And since the third term already has a $T^2$ factor, we introduce negligible error by replacing $\mu$ by $\varepsilon_F$. Then we have

$$E = E_0 - \alpha\frac{\pi^2}{6}(kT)^2 g'(\varepsilon_F)\varepsilon_F + \alpha\frac{\pi^2}{6}(kT)^2\{g(\varepsilon_F) + \varepsilon_F g'(\varepsilon_F)\}, \qquad (2.5.73)$$

which simplifies to

$$E = E_0 + \alpha\frac{\pi^2}{6}(kT)^2 g(\varepsilon_F) \qquad (2.5.74)$$

as the derivative terms cancel. This result shows that the internal energy depends on the density of states only at the Fermi surface.

The heat capacity is found by differentiating the internal energy. So, in the low-temperature limit, the heat capacity of an assembly of Fermions with arbitrary dispersion relation and thus arbitrary density of states is

$$C_V = \frac{1}{3}\alpha\pi^2 g(\varepsilon_F)k^2 T. \qquad (2.5.75)$$

We see that the low-temperature heat capacity remains linear in temperature, and we see that the coefficient is determined by the density of states at the Fermi surface. The general form for $g(\varepsilon)$ is unimportant; only its value at the Fermi level is required.

### Effective mass

The quasi-particle excitations giving rise to the heat capacity above may be regarded as having an effective mass $m^*$ such that the free fermion low-temperature heat capacity, Eq. (2.5.61),

$$C_V = N\frac{m\pi^2}{\hbar^2 k_F^2}k^2 T$$

is rehabilitated by the replacement $m \rightarrow m^*$ where

$$m^* = \frac{\alpha}{3} \hbar^2 k_F^2 \frac{g(\varepsilon_F)}{N}$$

$$= 2\pi^2 \frac{\hbar^2}{k_F} \frac{g(\varepsilon_F)}{V}. \qquad (2.5.76)$$

It is possible that the effective mass enhancement ratio $m^*/m$ can be as high as 1000 in some compounds. These are referred to as *heavy fermion* compounds.

## 2.6 Bose Gas at Low Temperatures

There is no restriction on the number of bosons occupying a single state. This means that at low temperatures there will be a substantial occupation of the low-energy states. This is in contrast to the Fermi case where the Pauli exclusion principle restricts the state occupation. Thus, the behaviour of the Bose gas and the Fermi gas will be different, particularly at low temperatures. Indeed the zero-temperature states are completely different. The Fermi gas has all states up to the Fermi level occupied, with a resultant significant internal energy. Since there is no restriction on occupation, the ground state of the Bose gas will have all particles occupying the same lowest energy single-particle state. This macroscopic occupation of the lowest energy state is known as Bose–Einstein condensation; its possibility was predicted by Einstein in 1925. The early treatment of this phenomenon by London [6] is still one of the clearest.

### 2.6.1 *General procedure for treating the Bose gas*

The Bose–Einstein distribution is

$$n(\varepsilon) = \frac{1}{e^{(\varepsilon - \mu)/kT} - 1}, \qquad (2.6.1)$$

which gives the mean number of bosons in the state of energy $\varepsilon$ in a system at temperature $T$ and chemical potential $\mu$. As with the Fermi case, the average value of a general function of energy $f(\varepsilon)$ is given by

$$\bar{f} = \alpha \sum_i f(\varepsilon_i) n(\varepsilon_i), \qquad (2.6.2)$$

where the sum is taken over the single-particle energy states. Convention-ally, this sum is transformed to an integral over energy states using the density of states function $g(\varepsilon)$ as

$$\bar{f} = \alpha \int_0^\infty f(\varepsilon)g(\varepsilon)n(\varepsilon)\mathrm{d}\varepsilon, \tag{2.6.3}$$

where $g(\varepsilon)$ is given by Eq. (2.1.16):

$$g(\varepsilon) = \frac{V}{4\pi^2\hbar^3}(2m)^{3/2}\varepsilon^{1/2}. \tag{2.6.4}$$

The procedure, as described, will give the expression for $\bar{f}$ in terms of the system's characterising intensive parameters $T$ and $\mu$. But, as we have seen already, the chemical potential is best eliminated in terms of the total num-ber of particles in the system. Thus, the first task must be to consider the expression for the number of particles. But since we know that the number of particles in the ground state can be significant at low temperatures, let us start by looking specifically at the ground state occupation.

### 2.6.2 Ground state occupation — Chemical potential

The number of particles in the ground state $N_0$ is given by

$$N_0 = \frac{\alpha}{e^{(\varepsilon_0-\mu)/kT} - 1}, \tag{2.6.5}$$

where $\alpha$ is the spin degeneracy factor; this will be unity for $S = 0$ particles. Here, $\varepsilon_0$ is the ground state energy. In the thermodynamic limit ($V \to \infty$), the ground state energy will vanish: $\varepsilon_0 \to 0$. In this case, we have

$$N_0 = \frac{1}{e^{-\mu/kT} - 1}. \tag{2.6.6}$$

If the ground state occupation is significant: above a few hundred, for instance, then the denominator will be small, which means that the expo-nential will differ only slightly from unity. Then $\mu$ must be very small (and negative) so that

$$N_0 = \frac{1}{1 - \mu/kT - \cdots - 1} \sim -kT/\mu \tag{2.6.7}$$

or

$$\mu \sim -kT/N_0. \tag{2.6.8}$$

We conclude that a macroscopic occupation of the ground state is associated with the vanishing of the chemical potential:

$$\mu = 0 \quad \text{when} \quad N_0 \text{ is macroscopic.} \quad (2.6.9)$$

It must also be noted that the chemical potential cannot be positive as this would give a negative number of bosons in the ground state; for bosons, $\mu$ is always negative.

### 2.6.3 Number of particles

One is inclined to write the expression for the number of particles in the system by following the recipe of Eq. (2.6.3) by integrating over the number of particles in each quantum state:

$$N = \sum_i n_i \rightarrow \int_0^\infty g(\varepsilon)n(\varepsilon)d\varepsilon. \quad (2.6.10)$$

However, there is a problem with this. We know that at low enough temperatures, for bosons, there will be a large number of particles in the ground state of energy $\varepsilon = 0$. But the density of states $g(\varepsilon)$ is proportional to $\varepsilon^{1/2}$ (in three dimensions). This means that it gives *zero weight* to the ground state. There is an error introduced in transforming from a sum over states to an integral using the density of states. Ordinarily, there will be no problem with the neglect of a single state. But if there is an appreciable occupation of this state, then the error becomes serious. Since the $\varepsilon^{1/2}$ factor neglects completely the ground state, we should add this "by hand" to the calculation for $N$. Thus, we write (using a spin degeneracy factor $\alpha$ of unity)

$$N = N_0 + \frac{V}{4\pi^2\hbar^3}(2m)^{3/2} \int_0^\infty \frac{\varepsilon^{1/2}}{e^{(\varepsilon-\mu)/kT} - 1}d\varepsilon, \quad (2.6.11)$$

where $N_0$ is the number of particles in the ground state. We should note that while $N_0$ may be a large number (macroscopic), the *fraction* of particles in the ground state $N_0/N$ may be small.

### 2.6.4 Low-temperature behaviour of Bose gas

At low temperatures, when $N_0$ is appreciable, the chemical potential is very small and it can be ignored. Then, within this approximation, the expression for $N$ becomes

$$N = N_0 + \frac{V}{4\pi^2\hbar^3}(2m)^{3/2}\int_0^\infty \frac{\varepsilon^{1/2}}{e^{\varepsilon/kT}-1}d\varepsilon. \qquad (2.6.12)$$

The integral may be "tidied" by the substitution $x = \varepsilon/kT$ whereupon[3]

$$N = N_0 + V\left(\frac{mkT}{2\pi\hbar^2}\right)^{3/2}\frac{2}{\sqrt{\pi}}\int_0^\infty \frac{x^{1/2}}{e^x-1}dx. \qquad (2.6.13)$$

The integral in this expression is a pure number. We denote it by $\mathscr{B}_{1/2}$ and we see it is a particular case of the integral

$$\mathscr{B}_n = \int_0^\infty \frac{x^n}{e^x-1}dx. \qquad (2.6.14)$$

These integrals may be expressed as

$$\mathscr{B}_n = \Gamma(n+1)\zeta(n+1), \qquad (2.6.15)$$

where $\Gamma()$ is Euler's gamma function[4] and $\zeta()$ is Riemann's zeta function (see footnote 2 on p. 87). So, in our present case,

$$\mathscr{B}_{1/2} = \Gamma(3/2)\,\zeta(3/2), \qquad (2.6.16)$$

and these have values

$$\begin{aligned}\Gamma(3/2) &= \sqrt{\pi}/2,\\ \zeta(3/2) &= 2.612\ldots,\end{aligned} \qquad (2.6.17)$$

so that $\mathscr{B}_{1/2} = \frac{\sqrt{\pi}}{2}\zeta\left(\frac{3}{2}\right)$.

---

[3]Here and in the following, I have used the bracket $(mkT/2\pi\hbar^2)$ since this is $\Lambda^{-2}$, where $\Lambda$ is the thermal de Broglie wavelength, Eq. (2.3.5).

[4]The gamma function $\Gamma(z)$ was introduced by Euler in order to extend the factorial function to non-integer arguments. It may be specified by an integral: $\Gamma(z) = \int_0^\infty t^{z-1}e^{-t}dt$ since when $z$ is a positive integer then $\Gamma(z) = (n-1)!$. However, the recursion relation $\Gamma(z+1) = z\Gamma(z)$ holds for non-integer $z$ as well. We note that $\Gamma\left(\frac{1}{2}\right) = \sqrt{\pi}$ and then by recursion $\Gamma\left(\frac{3}{2}\right) = \sqrt{\pi}/2$, $\Gamma\left(\frac{5}{2}\right) = 3\sqrt{\pi}/4$, etc. Another useful property is the reflection relation $\Gamma(z)\Gamma(1-z) = \pi/\sin(\pi z)$. The gamma function has the *Mathematica* symbol *Gamma* [z].

Then the number of excited particles in the system is given by

$$N_{\text{ex}} = V \left(\frac{mkT}{2\pi\hbar^2}\right)^{3/2} \frac{2}{\sqrt{\pi}} \mathscr{B}_{1/2}$$

$$= V \left(\frac{mkT}{2\pi\hbar^2}\right)^{3/2} \zeta\left(\frac{3}{2}\right). \tag{2.6.18}$$

As the temperature increases, the number of excited particles increases. There will be a certain temperature when $N_{\text{ex}}$ will be equal to the total number of particles in the system $N$ — at least according to this equation. In reality, the expression will become invalid when $N_0$ becomes *too* small (since then $\mu$ cannot be ignored), but let us just use this as a mathematical definition of a characteristic temperature $T_B$, that is, let us define $T_B$ by

$$N = V \left(\frac{mkT_B}{2\pi\hbar^2}\right)^{3/2} \zeta\left(\frac{3}{2}\right), \tag{2.6.19}$$

that is,

$$T_B = \frac{2\pi\hbar^2}{mk} \left\{\frac{1}{\zeta\left(\frac{3}{2}\right)} \frac{N}{V}\right\}^{2/3}. \tag{2.6.20}$$

Then the expression for $N$ becomes

$$N = N_0 + N \left(\frac{T}{T_B}\right)^{3/2}, \tag{2.6.21}$$

which may be inverted to give the number of particles in the ground state as

$$N_0 = N \left\{1 - \left(\frac{T}{T_B}\right)^{3/2}\right\}. \tag{2.6.22}$$

This is shown in Fig. 2.7. The expression is valid for very low temperatures; clearly we must have $T < T_c$, but how close to $T_B$ can one go? In fact, one can go very close indeed since we require the *absolute* value of $N_0$ to be large. And as one has $N \sim 10^{23}$, it follows that the *fractional* value $N_0/N$ may be very small while $N_0$ will still be *enormous*. So, we can use the above expression for $N_0$ for temperatures $T < T_B$ to within a whisker of $T_B$; indeed in the thermodynamic limit, we can go right up to $T_B$ (these issues are explored in Problem 2.14). Then we can identify the temperature $T_B$ as the transition temperature, below which there will be a *macroscopic* occupation of the ground state. This is known as *Bose–Einstein condensation*

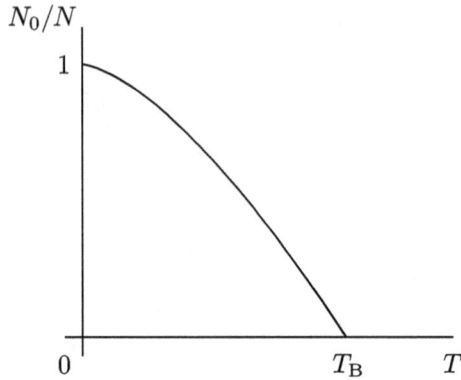

Fig. 2.7. Ground state occupation for Bose gas.

and it happens at the temperature $T_B$. It is the only phase transition in nature that happens in the absence of any interactions.

The transition temperature is given by Eq. (2.6.20):

$$T_B = \frac{2\pi\hbar^2}{mk} \left\{ \frac{1}{\zeta\left(\frac{3}{2}\right)} \frac{N}{V} \right\}^{2/3},$$

and since $\zeta\left(\frac{3}{2}\right) = 2.612$, we may write $T_B$ as

$$
\begin{aligned}
T_B &= \frac{2\pi\hbar^2}{mk} \left\{ \frac{1}{\zeta\left(\frac{3}{2}\right)} \frac{N}{V} \right\}^{2/3} \\
&= \frac{2\pi\hbar^2}{mk} \left\{ \frac{1}{2.612} \frac{N}{V} \right\}^{2/3} \quad\quad (2.6.23) \\
&= 3.313 \frac{\hbar^2}{mk} \left\{ \frac{N}{V} \right\}^{2/3}.
\end{aligned}
$$

A rough, intuitive, understanding of the Bose–Einstein transition temperature may be obtained by considering the thermal de Broglie wavelength

$$\Lambda = \sqrt{\frac{2\pi\hbar^2}{mkT}} \quad\quad (2.6.24)$$

and the mean separation of particles $l$

$$l = \left(\frac{V}{N}\right)^{1/3}. \quad\quad (2.6.25)$$

We saw that $\Lambda$ represents the "quantum size" of the particle. When this becomes comparable with the mean separation of the particles, when

$$\Lambda \approx l, \tag{2.6.26}$$

the particles' wave functions "see" each other and we would expect quantum effects to become significant, such as Bose–Einstein condensation. In other words, this criterion would suggest

$$\sqrt{\frac{2\pi\hbar^2}{mkT_\mathrm{B}}} \approx \left(\frac{V}{N}\right)^{1/3} \tag{2.6.27}$$

or

$$T_\mathrm{B} \approx \frac{2\pi\hbar^2}{mk}\left(\frac{N}{V}\right)^{2/3}. \tag{2.6.28}$$

This is quite good; it is approximately twice the actual calculated temperature of Eq. (2.6.23).

We note the precise relation between $\Lambda$ and $l$, instead of Eq. (2.6.26), would be

$$\Lambda = \zeta\left(\frac{3}{2}\right)^{1/3} l = 1.38\,l. \tag{2.6.29}$$

### 2.6.5 Heat capacity of Bose gas

*Behaviour below $T_B$*

To find the heat capacity of the Bose gas, we must first obtain an expression for the internal energy. To start with, we shall do this all within the approximation of neglecting the chemical potential. In other words, these results will be valid for temperatures *below* the Bose–Einstein condensation temperature.

The internal energy is given by

$$E = \int_0^\infty \varepsilon g(\varepsilon)n(\varepsilon)\mathrm{d}\varepsilon. \tag{2.6.30}$$

Here, we do not need to worry about the ground state occupation since it contributes nothing to the internal energy. The expression for $E$ is then

$$E = \frac{V}{4\pi^2\hbar^3}(2m)^{3/2}\int_0^\infty \frac{\varepsilon^{3/2}}{e^{\varepsilon/kT}-1}\mathrm{d}\varepsilon, \tag{2.6.31}$$

and as in the previous case, the integral may be "tidied" by the substitution $x = \varepsilon/kT$ whereupon

$$E = \frac{V}{4\pi^2\hbar^3}(2m)^{3/2}(kT)^{5/2}\int_0^\infty \frac{x^{3/2}}{e^x - 1}\,dx. \tag{2.6.32}$$

The integral in this expression is a pure number; we recognise it as $\mathscr{B}_{3/2}$ from the definition in Eq. (2.6.14). From Eq. (2.6.15), we have

$$\mathscr{B}_{3/2} = \Gamma\left(\frac{5}{2}\right)\zeta\left(\frac{5}{2}\right), \tag{2.6.33}$$

and these have values

$$\begin{aligned}
\Gamma\left(\frac{5}{2}\right) &= 3\sqrt{\pi}/4, \\
\zeta\left(\frac{5}{2}\right) &= 1.342,\ldots,
\end{aligned} \tag{2.6.34}$$

so that $\mathscr{B}_{3/2} = \frac{3\sqrt{\pi}}{4}\zeta\left(\frac{5}{2}\right)$.

Then the internal energy of the assembly of Bosons below $T_B$ is

$$\begin{aligned}
E &= V\left(\frac{mkT}{2\pi\hbar^2}\right)^{3/2}(kT)\frac{2}{\sqrt{\pi}}\mathscr{B}_{3/2} \\
&= V\left(\frac{mkT}{2\pi\hbar^2}\right)^{3/2}(kT)\frac{3}{2}\zeta\left(\frac{5}{2}\right).
\end{aligned} \tag{2.6.35}$$

But since the number of particles in the system is given by Eq. (2.6.19)

$$N = V\left(\frac{mkT_B}{2\pi\hbar^2}\right)^{3/2}\zeta\left(\frac{3}{2}\right), $$

then we may write the internal energy as

$$E = Nk\frac{3}{2}\frac{\zeta\left(\frac{5}{2}\right)}{\zeta\left(\frac{3}{2}\right)}\frac{T^{5/2}}{T_B^{3/2}}. \tag{2.6.36}$$

Upon differentiation, the heat capacity (at constant volume) is

$$C_V = \frac{15}{4}Nk\frac{\zeta\left(\frac{5}{2}\right)}{\zeta\left(\frac{3}{2}\right)}\left(\frac{T}{T_B}\right)^{3/2}. \tag{2.6.37}$$

And since $\zeta\left(\frac{5}{2}\right) = 1.342$ and $\zeta\left(\frac{3}{2}\right) = 2.612$, we may write the heat capacity below $T_B$ as

$$C_V = \frac{15}{4}Nk\left(\frac{T}{T_B}\right)^{3/2} \times \frac{1.342}{2.612}$$

$$= 1.926Nk\left(\frac{T}{T_B}\right)^{3/2}. \tag{2.6.38}$$

This vanishes as $T \to 0$ in accordance with the Third Law. This should be compared with the classical value of $\frac{3}{2}Nk$.

*Behaviour above $T_B$*

For temperatures above $T_B$, the chemical potential $\mu$ is no longer zero. Then the calculation of internal energy and heat capacity is more complicated. In Appendix C.5, we have evaluated $\mu$ above $T_B$ as a series in powers of $T - T_B$ and in Appendix C.6, we have evaluated $E$ and $C_V$, similarly, as power series. Here, for completeness, we shall simply quote the leading order results.

From Appendix C.5, we find the chemical potential just above $T_B$ as

$$\mu = -kT_B\left\{\frac{9\zeta\left(\frac{3}{2}\right)^2}{16\pi}\left(\frac{T - T_B}{T_B}\right)^2 + \cdots\right\}; \tag{2.6.39}$$

we see that $\mu$ decreases quadratically from zero as $T - T_B$ increases.

The series expression for $\mu$ is used in Appendix C.6 for the power series expansions for $E$ and $C_V$. The internal energy, to leading order, is

$$E = NkT_B\left(\frac{T}{T_B}\right)^{5/2}\frac{3}{2}\frac{\zeta\left(\frac{5}{2}\right)}{\zeta\left(\frac{3}{2}\right)}\left\{1 - \frac{9\zeta\left(\frac{3}{2}\right)^3}{16\pi\zeta\left(\frac{5}{2}\right)}\left(\frac{T - T_B}{T_B}\right)^2 + \cdots\right\}. \tag{2.6.40}$$

Observe the first term, at $T = T_B$ corresponds to Eq. (2.6.36) at $T = T_B$. Then upon differentiation the heat capacity, to leading order, is

$$C_V = \frac{15\zeta\left(\frac{5}{2}\right)}{4\zeta\left(\frac{3}{2}\right)}Nk\left\{1 - \left(\frac{3\zeta\left(\frac{3}{2}\right)^3}{20\pi\zeta\left(\frac{5}{2}\right)} - \frac{1}{2}\right)\left(\frac{T - T_B}{T_B}\right) + \cdots\right\}. \tag{2.6.41}$$

Observe the first term, the $T = T_B$ value, corresponds to Eq. (2.6.37) at $T = T_B$. And as $T$ increases above $T_B$, the next term in $C_V$ indicates that the heat capacity will decrease from its peak value.

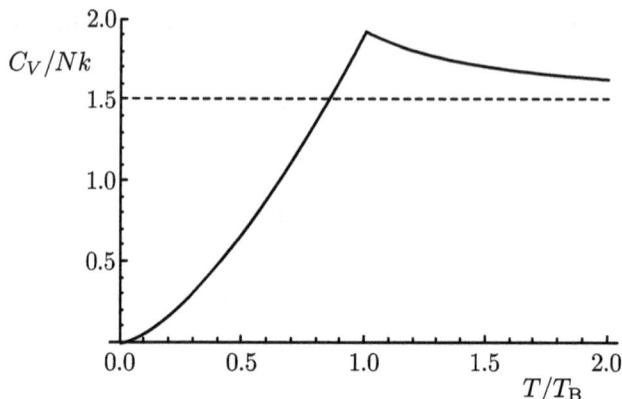

Fig. 2.8.   Heat capacity of a Bose gas. The dashed line shows the classical value.

The full heat capacity is shown in Fig. 2.8; it drops from the peak value, at the transition, approaching the classical value as $T$ increases.

### 2.6.6   *Comparison with superfluid $^4$He*

The common isotope of helium, $^4$He, is found to go superfluid at a temperature of 2.172 K. In 1924, Kamerlingh Onnes observed that *something strange* happened to liquid $^4$He at this temperature, and in 1932, Keesom and Clusius observed a cusp in the heat capacity, indicative of a phase transition. Subsequently, the low-temperature phase was observed to flow with no viscosity — superfluid behaviour. The calculated Bose–Einstein condensation transition temperature for particles whose mass and density corresponds to those of liquid $^4$He is 3.13 K. Since these temperatures are similar, it is possible that the superfluidity is related to the Bose–Einstein condensation. This was first suggested in 1938 by London [7] and by Tisza [8]. The possibility is supported by the fact that at these temperatures $^3$He, a fermion, does not become superfluid; there is no such transition. The small difference in transition temperatures is ascribed to the interatomic interactions in liquid $^4$He, which are neglected in the Bose gas model.

Bose–Einstein condensation in *low-density gases*, where the interactions are very small, will be treated in Section 2.8.2.

The following discussions will consider *interacting* bosons, so strictly this is a topic for Chapter 3. However the system may be regarded as a gas of non-interacting *elementary excitations*. So, we shall place the discussion here as it follows naturally from our previous considerations.

### 2.6.7 *Two-fluid model of superfluid $^4$He*

The two-fluid model is a very successful thermodynamic/macroscopic model of superfluid helium. It was proposed by Tisza in 1938. The Bose–Einstein condensation gives a microscopic *motivation* for the model, but it cannot be regarded as a microscopic *justification* — as we shall see.

According to the two-fluid model, the superfluid comprises mixture of two interpenetrating fluids. One is the *normal* component and the other is the *superfluid* component. The density of the liquid is then expressed as the sum:

$$\rho = \rho_s + \rho_n. \tag{2.6.42}$$

The normal component is assumed to have properties similar to the fluid above the transition. The superfluid component is assumed to have *zero entropy*. As the temperature is lowered below the transition temperature, the superfluid fraction grows at the expense of the normal fraction. And at $T = 0$, the normal component has reduced to zero.

Originally, as proposed by Tisza, the connection with Bose–Einstein condensation was made by identifying the superfluid component with the macroscopically occupied ground state of the Bose gas. Since these particles are all in the same single-particle state there is only one microstate corresponding to this and its entropy is thus zero. However, subsequently, Landau proposed that the normal component should be associated with the elementary excitations of the system.

An experiment by Andronikashvili [9] measured the normal component of superfluid helium using a stack of discs oscillating in the liquid. This is shown in Fig. 2.9.

The normal component will move with the discs while the superfluid component will be decoupled. The resonant frequency of the disc assembly will depend on its effective mass. Thus, the normal component can be measured through the resonant frequency of the discs.

The normal fraction, as measured by Andronikashvili, is compared with the fractional ground state occupation for the Bose–Einstein condensation in Fig. 2.10. There is a *qualitative* similarity, but the discrepancy is serious.

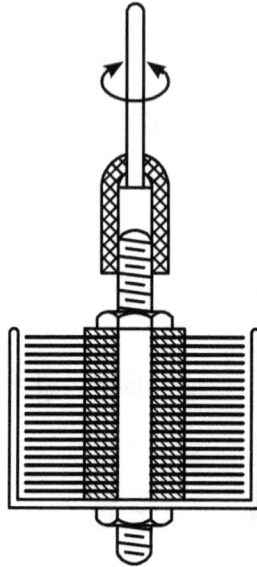

Fig. 2.9.   Andronikashvili's experiment.

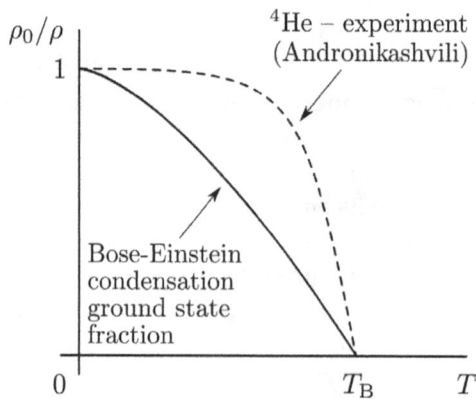

Fig. 2.10.   Ground state fraction.

### 2.6.8   *Elementary excitations*

The difference between superfluid helium and the Bose–Einstein condensed fluid is made more apparent by measurements of heat capacity.

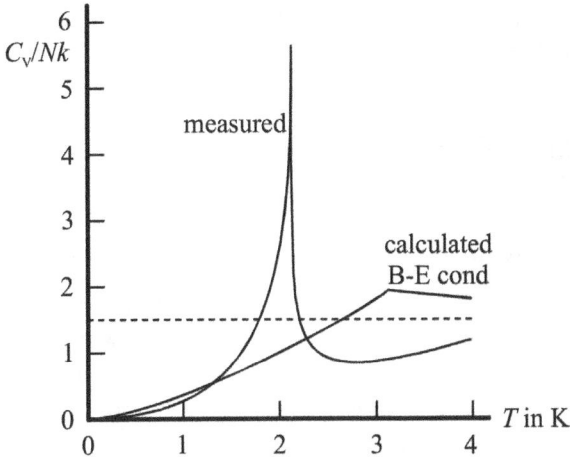

Fig. 2.11.   Heat capacity of liquid helium.

In Fig. 2.11, we show $C_V$ as measured in liquid helium compared with that calculated for an ideal Bose gas. (Most books compare the calculated $C_V$ with the observed heat capacity under the liquid's saturated vapour pressure. This has a qualitatively different behaviour so the comparison is invalid.) Because of characteristic shape of the heat capacity, this is called a *lambda* transition and the transition point the lambda point.

The low temperature behaviour of these two are different. The calculated $C_V$ increases with temperature as $T^{3/2}$, but the observed heat capacity varies as $T^3$. Now, a $T^3$ heat capacity indicates *phonon excitations*. Thus, the interactions between the particles seem to quench the single-particle behaviour and leads to collective motion of the particles. This is supported by calculations of Bogoliubov [10], who showed that even very weak repulsive interactions change the elementary excitations of the system from free particle motion to phonon-like collective motion.

The conclusion is that the interactions cannot be regarded as causing a slight change to the properties of the Bose–Einstein condensate; the two-fluid model cannot be built on the Bose–Einstein condensate idea. In reality, the normal component comprises the elementary excitations in the system. The spectrum of elementary excitations is more complex than just phonons, but that is another story. A good account of the properties of superfluid helium-4 and their understanding is given in the book by

Khalatnikov [11]. This book also contains translated reprints of the original papers by Landau on the subject.

## 2.7 Quantum Gas at High Temperatures — The Classical Limit

### 2.7.1 *General treatment for Fermi, Bose and Maxwell cases*

At high temperatures, we shall see that the difference between Fermi and Bose gases disappears; the quantum distinction becomes negligible. It will therefore prove sensible to consider the Fermi and Bose cases together.

When the temperature is high, the probability of occupation of the individual single-particle states becomes significantly less than unity. Then in the Fermi and Bose distribution functions,

$$n(\varepsilon) = \frac{1}{e^{(\varepsilon-\mu)/kT} \pm 1}, \tag{2.7.1}$$

$n(\varepsilon)$ is much smaller than unity; the denominator is much greater than unity; the exponential is then large compared with the ±1. So, in that limit, the ±1 in the denominator becomes negligible. Thus, the distinction between fermions and bosons disappears. Indeed, the classical case corresponds to dropping the ±1: setting it to zero. The (hypothetical) classical particles, obeying such a distribution function are called *maxwellons*. You should recall our discussion in Section 2.2.2, where we argued that quantum effects become insignificant when the probability of occupation of the states is small.

*Mathematica* Notebooks for the following calculations are given in Appendix C.

In this limit, we treat the Fermi, Bose and Maxwell/classical cases together, by writing the distribution function as

$$n(\varepsilon) = \frac{1}{e^{(\varepsilon-\mu)/kT} + a}, \tag{2.7.2}$$

where

$$a = \begin{cases} +1 & \text{for fermions,} \\ 0 & \text{for maxwellons,} \\ -1 & \text{for bosons,} \end{cases} \tag{2.7.3}$$

and the exponential is much greater than unity. We then re-express $n(\varepsilon)$ as

$$n(\varepsilon) = \frac{e^{(\mu-\varepsilon)/kT}}{1 + ae^{(\mu-\varepsilon)/kT}}, \tag{2.7.4}$$

so that the exponential in the denominator is now small and we can then perform a binomial expansion:

$$\begin{aligned} n(\varepsilon) &= e^{(\mu-\varepsilon)/kT} \left\{ 1 + ae^{(\mu-\varepsilon)/kT} \right\}^{-1} \\ &= e^{(\mu-\varepsilon)/kT} \left\{ 1 - ae^{(\mu-\varepsilon)/kT} + a^2 e^{2(\mu-\varepsilon)/kT} - \cdots \right\}. \end{aligned} \tag{2.7.5}$$

### 2.7.2 Quantum energy parameter

Recall that for fermions we introduced the idea of Fermi energy. In order to treat the Bose, Fermi and indeed Maxwell cases in a parallel fashion, we shall introduce a "quantum energy" $\varepsilon_q$; this is identical to the Fermi energy of the Fermi case, Eq. (2.5.6), but it is now extended to the Bose and Maxwell cases as well. Of course, for bosons and maxwellons, it doesn't have the interpretation as a maximum filled energy state at $T = 0$, but it does provide a convenient energy scale. However, the analogue of the Fermi wave vector Eq. (2.5.10) is more intuitive; the "quantum wave vector" $k_q$ is still essentially the inverse particle spacing:

$$k_q = \left( \frac{6\pi^2}{\alpha} \frac{N}{V} \right)^{1/3}. \tag{2.7.6}$$

So, in *this* spirit, we shall use the quantum energy parameter

$$\varepsilon_q = \frac{\hbar^2 k_q^2}{2m} = \frac{\hbar^2}{2m} \left( \frac{6\pi^2}{\alpha} \frac{N}{V} \right)^{2/3}. \tag{2.7.7}$$

In effect, we are using $\varepsilon_q$ simply as a characteristic energy that conveniently parameterises the number density $N/V$.

Note that the dimensionless ratio $\varepsilon_q/kT$ is related simply to the thermal de Broglie wavelength $\Lambda$ (Eq. (2.3.5)) and the number density $N/V$:

$$\frac{\varepsilon_q}{kT} = \Lambda^2 \left( \frac{3\sqrt{\pi}}{4\alpha} \frac{N}{V} \right)^{2/3}. \tag{2.7.8}$$

### 2.7.3 *Chemical potential*

In order to find the chemical potential, we must invert (the analogue of) Eq. (2.5.43),

$$\varepsilon_q^{3/2} = \frac{3}{2} \int_0^\infty n(\varepsilon)\varepsilon^{1/2}d\varepsilon, \tag{2.7.9}$$

which we shall do in terms of the above series expansion, Eq. (2.7.5), for $n(\varepsilon)$

$$\varepsilon_q^{3/2} = \frac{3}{2} \int_0^\infty \left\{ e^{(\mu-\varepsilon)/kT} - ae^{2(\mu-\varepsilon)/kT} + a^2 e^{3(\mu-\varepsilon)/kT} - \cdots \right\} \varepsilon^{1/2}d\varepsilon. \tag{2.7.10}$$

This can be integrated term by term. It will be convenient, in the first instance, to work in terms of the *fugacity* $z = e^{\mu/kT}$ rather than the chemical potential. Then upon integrating term by term, we obtain $\varepsilon_q^{3/2}$ as a power series in $z$:

$$\varepsilon_q^{3/2} = (kT)^{3/2} \left\{ \frac{3}{4}\sqrt{\pi}z - a\frac{3}{8}\sqrt{\frac{\pi}{2}}z^2 + a^2\frac{1}{4}\sqrt{\frac{\pi}{3}}z^3 + \cdots \right\}. \tag{2.7.11}$$

And then $\varepsilon_q$ is obtained by taking the 2/3 power, which is sensibly written as

$$\frac{\varepsilon_q}{kT} = \frac{1}{2}3^{2/3}\left(\frac{\pi}{2}\right)^{1/3}z^{2/3} - a\frac{1}{2\times2^{5/6}}\left(\frac{\pi}{3}\right)^{1/3}z^{5/3}$$

$$+ a^2\frac{16\sqrt{3}-3}{144}\left(\frac{\pi}{6}\right)^{1/3}z^{8/3} + \cdots. \tag{2.7.12}$$

This is an expression for $\varepsilon_q/kT$ as a power series in $z$. We invert this to give the fugacity $z$ as a power series in $\varepsilon_q/kT$:

$$z = e^{\mu/kT} = \frac{4}{3\sqrt{\pi}}\left(\frac{\varepsilon_q}{kT}\right)^{3/2} + a\frac{4\sqrt{2}}{9\pi}\left(\frac{\varepsilon_q}{kT}\right)^3 + a^2\frac{16(9-4\sqrt{3})}{243\pi^{3/2}}\left(\frac{\varepsilon_q}{kT}\right)^{9/2} + \cdots. \tag{2.7.13}$$

In the classical case, we take $a = 0$, whereupon

$$z_c = e^{\mu_c/kT} = \frac{4}{3\sqrt{\pi}}\left(\frac{\varepsilon_q}{kT}\right)^{3/2} \tag{2.7.14}$$

or

$$\mu_c = kT\ln\left[\frac{4}{3\sqrt{\pi}}\left(\frac{\varepsilon_q}{kT}\right)^{3/2}\right]. \tag{2.7.15}$$

Upon substituting for $\varepsilon_q$, we obtain the classical ideal gas chemical potential as

$$\mu_c = -kT \ln\left[\alpha\left(\frac{mkT}{2\pi\hbar^2}\right)^{3/2}\frac{V}{N}\right] \qquad (2.7.16)$$

in agreement with Eq. (2.3.16) in Section 2.3.3.

We can express the series for the fugacity, Eq. (2.7.13), as

$$z = z_c\left\{1 + a\frac{1}{3}\sqrt{\frac{2}{\pi}}\left(\frac{\varepsilon_q}{kT}\right)^{3/2} + a^2\frac{4(9-4\sqrt{3})}{81\pi}\left(\frac{\varepsilon_q}{kT}\right)^3 + \cdots\right\}. \qquad (2.7.17)$$

Then, upon taking the logarithm, we can write the high-temperature/low-density chemical potential in terms of its classical value plus a series of correction terms

$$\mu = \mu_c + kT\left\{a\frac{1}{3}\sqrt{\frac{2}{\pi}}\left(\frac{\varepsilon_q}{kT}\right)^{3/2} - a^2\frac{16\sqrt{3}-27}{81\pi}\left(\frac{\varepsilon_q}{kT}\right)^3 + \cdots\right\}. \qquad (2.7.18)$$

In the Fermi case ($a = +1$), the chemical potential is above the classical value while in the Bose case ($a = -1$), the chemical potential falls below the classical value.

Observe that $\mu/\varepsilon_q$ may be written as a function of the single variable $kT/\varepsilon_q$. In particular, the classical/maxwellon chemical potential is

$$\frac{\mu_c}{\varepsilon_q} = \frac{kT}{\varepsilon_q}\ln\left[\frac{4}{3\sqrt{\pi}}\left(\frac{\varepsilon_q}{kT}\right)^{3/2}\right]. \qquad (2.7.19)$$

### 2.7.4 *Internal energy*

The internal energy of a quantum gas is

$$E = \alpha\int_0^\infty \varepsilon n(\varepsilon)g(\varepsilon)d\varepsilon. \qquad (2.7.20)$$

We follow the procedure of the previous section. We write $n(\varepsilon)$ for the general Fermi–Maxwell–Bose case as in Eq. (2.7.2)/Eq. (2.7.4), so that $E$ is given by

$$E = \frac{3N}{2\varepsilon_q^{3/2}} \int_0^\infty \frac{\varepsilon^{3/2}}{e^{(\varepsilon-\mu)/kT} + a} d\varepsilon$$

$$= \frac{3N}{2\varepsilon_q^{3/2}} \int_0^\infty \frac{\varepsilon^{3/2} e^{(\mu-\varepsilon)/kT}}{1 + ae^{(\mu-\varepsilon)/kT}} d\varepsilon.$$

(2.7.21)

We perform the binomial expansion and integrate term by term, as in the previous section, giving $E$ as a power series in $z = e^{\mu/kt}$:

$$E = NkT \left(\frac{kT}{\varepsilon_q}\right)^{3/2} \left\{ \frac{9\sqrt{\pi}}{8} z - a \frac{9}{32} \sqrt{\frac{\pi}{2}} z^2 + a^2 \frac{1}{8} \sqrt{\frac{\pi}{3}} z^3 + \cdots \right\}. \qquad (2.7.22)$$

We need the expansion for $z$ from Eq. (2.7.13), which we substitute into the above series. This gives

$$E = \frac{3}{2} NkT \left\{ 1 + a \frac{1}{3\sqrt{2\pi}} \left(\frac{\varepsilon_q}{kT}\right)^{3/2} - a^2 \frac{2(16\sqrt{3} - 27)}{243\pi} \left(\frac{\varepsilon_q}{kT}\right)^3 + \cdots \right\}.$$

(2.7.23)

### 2.7.5  *Equation of state*

From the general relation, Eq. (2.1.35), $pV = \frac{2}{3}E$ we obtain the equation of state for the gases as

$$pV = NkT \left\{ 1 + a \frac{1}{3\sqrt{2\pi}} \left(\frac{\varepsilon_q}{kT}\right)^{3/2} - a^2 \frac{2(16\sqrt{3} - 27)}{243\pi} \left(\frac{\varepsilon_q}{kT}\right)^3 + \cdots \right\}.$$

(2.7.24)

Or, substituting for $\varepsilon_q$,

$$pV = NkT \left\{ 1 + a \frac{\pi^{3/2}}{2\alpha} \frac{N}{V} \frac{\hbar^3}{(mkT)^{3/2}} - a^2 \frac{\pi^3}{\alpha^2} \frac{16\sqrt{3} - 27}{27} \left(\frac{N}{V}\right)^2 \frac{\hbar^6}{(mkT)^3} + \cdots \right\}.$$

(2.7.25)

When $a = 0$, we have the conventional ideal gas equation of state. The pressure of the fermion gas is increased over that of the classical gas, while the pressure of the boson gas is decreased.

Some tables and series expressions for the thermodynamic properties of Fermi gases are given by Ebner and Fu [12] (beware of errors) and references therein. See also the author's more recent paper [13].

## 2.8 Gas in a Harmonic Trap

Bose–Einstein condensation was discussed in Section 2.6. And in Section 2.6.6, we examined the idea that the superfluid state of liquid $^4$He, discovered in 1924, was a Bose–Einstein condensed phase. The problem is that the relatively strong interatomic interactions in the liquid call the ideal Bose gas model into question. So, can one observe Bose–Einstein condensation in a gas?

The condensation temperature is proportional to $(N/V)^{2/3}$. So, going to lower densities — to study a gas — requires going to lower temperatures. Observation of Bose–Einstein condensation in a gas remained a "holy grail" until the advent of laser cooling of trapped gases. Finally, in 1995, Bose–Einstein condensation was observed in rubidium vapour [14] and in lithium vapour [15]. Laser cooling allowed temperatures below $1\mu$K; see Fig. 1.15 outlining the progress in achieving lower and lower temperatures.

We note important differences from the discussion of Section 2.6:

1. Here, we consider gases trapped in a harmonic potential rather than free particles confined to a box.
2. While present experiments on gases involve, perhaps, millions of particles, this is still small compared with macroscopic systems, so the thermodynamic limit is not truly reached.

In this section, we will consider particles trapped in a harmonic potential. For simplicity, we will not be overly concerned with the thermodynamic limit.

### 2.8.1 *Enumeration and counting of states*

The energy states of a one-dimensional harmonic oscillator are given by

$$\varepsilon = \left(n + \frac{1}{2}\right)\hbar\omega, \qquad (2.8.1)$$

where $\omega$ is the (angular) frequency of the oscillator and the quantum number $n$ takes integer values

$$n = 0, 1, 2 \ldots. \qquad (2.8.2)$$

When $n = 0$, the oscillator is in its ground state with ground state energy $\hbar\omega/2$. It will be convenient in what follows, however, to measure energy from the ground state.

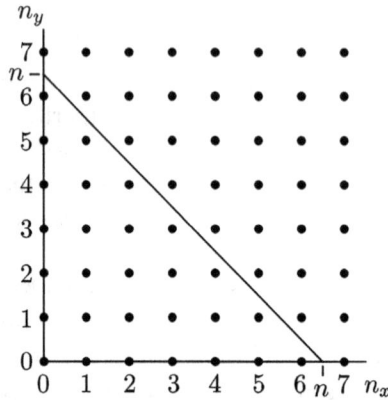

Fig. 2.12.   Counting of quantum states.

In three dimensions, we will need three quantum numbers and we write the energy states of the 3d isotropic oscillator as

$$\varepsilon = (n_x + n_y + n_z)\hbar\omega. \tag{2.8.3}$$

In this expression, it is the triple of quantum numbers $\{n_x, n_y, n_z\}$ which specifies the quantum state. Now, each triple defines a point on a cubic grid as in Fig. 2.12. If we put

$$n = n_x + n_y + n_z, \tag{2.8.4}$$

then the energy is given by

$$\varepsilon = n\hbar\omega. \tag{2.8.5}$$

The number of states of energy up to $\varepsilon = n\hbar\omega$ denoted by $\mathcal{N}(\varepsilon)$ is given by the number of points $\{n_x, n_y, n_z\}$, satisfying

$$n_x + n_y + n_z < n. \tag{2.8.6}$$

In the 2d case, this is the number of points below the sloping line. In the 3d case, it will be the number of points within the oblique pyramid bounded by the $x$ axis, the $y$ axis, the $z$ axis and the plane, Fig. 2.13. This is (approximately) the volume of the pyramid.

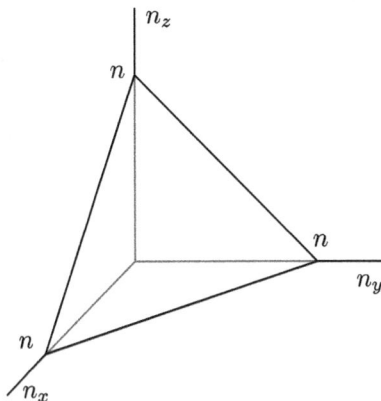

Fig. 2.13.  Counting of quantum states.

The volume of a pyramid is given by $\frac{1}{3}$ base × height $= \frac{1}{6}n^3$, thus

$$\mathcal{N}(\varepsilon) = \frac{1}{6}n^3$$

$$= \frac{1}{6}\left(\frac{\varepsilon}{\hbar\omega}\right)^3.$$

(2.8.7)

The density of states $g(\varepsilon)$ is the derivative of $d\mathcal{N}/d\varepsilon$, so

$$g(\varepsilon) = \frac{1}{2}\frac{\varepsilon^2}{(\hbar\omega)^3}.$$

(2.8.8)

Similar calculations for 2d and 1d give

$$g(\varepsilon) = \frac{\varepsilon}{(\hbar\omega)^2} \quad \text{(two dimensions)}$$

$$= \frac{1}{\hbar\omega} \quad \text{(one dimension)}.$$

(2.8.9)

### 2.8.2  Trapped bosons at low temperatures — Bose–Einstein condensation

*Chemical potential*

The chemical potential is determined from fixing the number of particles in the system

$$N = \int_0^\infty \frac{g(\varepsilon)d\varepsilon}{e^{(\varepsilon-\mu)/kT}+1},$$

(2.8.10)

and we have $g(\varepsilon)$ from Eq. (2.8.8). Note that expression (proportional to $\varepsilon^2$) discounts the ground state, but at low temperatures, the ground state occupation will be significant. For this reason, we must, as with the free boson case, add the ground state occupation separately, so we write $N$ as

$$N = N_0 + \frac{1}{2}\frac{1}{(\hbar\omega)^3}\int_0^\infty \frac{\varepsilon^2 d\varepsilon}{e^{(\varepsilon-\mu)/kT}+1}, \qquad (2.8.11)$$

where $N_0$ is given by

$$N_0 = \frac{1}{e^{-\mu/kT}-1}. \qquad (2.8.12)$$

When the ground state occupation is large, the denominator will be small, so that the exponential can be expanded, and

$$N_0 \sim -kT/\mu. \qquad (2.8.13)$$

Indeed, when the ground state occupation is macroscopic, this corresponds to $\mu = 0$.

So, when there is macroscopic occupation of the ground state, put $\mu = 0$ in Eq. (2.8.11):

$$N = N_0 + \frac{1}{2}\frac{1}{(\hbar\omega)^3}\int_0^\infty \frac{\varepsilon^2 d\varepsilon}{e^{\varepsilon/kT}+1}. \qquad (2.8.14)$$

The integral is evaluated by change of variables to $x = \varepsilon/kT$

$$N = N_0 + \frac{1}{2}\left(\frac{kT}{\hbar\omega}\right)^3\int_0^\infty \frac{x^2 dx}{e^x+1}$$

$$= N_0 + \frac{1}{2}\left(\frac{kT}{\hbar\omega}\right)^3 \mathscr{B}_2, \qquad (2.8.15)$$

where $\mathscr{B}_2$ is a number, a particular case of the integral specified in Eq. (2.6.14), with value given by Eq. (2.6.15)

$$\mathscr{B}_2 = \Gamma(3)\zeta(3) = 2\zeta(3) = 2.404, \qquad (2.8.16)$$

so that

$$N = N_0 + \left(\frac{kT}{\hbar\omega}\right)^3\zeta(3). \qquad (2.8.17)$$

At $T = 0$, all the particles will be in the ground state. As the temperature increases, the ground state becomes depleted as the particles enter the excited states. The number of excited particles is

$$N_{ex} = \left(\frac{kT}{\hbar\omega}\right)^3 \zeta(3). \tag{2.8.18}$$

The point at which the ground state is fully depleted is the transition point; this occurs at a temperature $T_B$ such that

$$N = \left(\frac{kT_B}{\hbar\omega}\right)^3 \zeta(3) \tag{2.8.19}$$

or

$$T_B = \frac{\hbar\omega}{k} \left(\frac{N}{\zeta(3)}\right)^{1/3}$$
$$= 0.940 \frac{\hbar\omega}{k} N^{1/3}. \tag{2.8.20}$$

We note that, in terms of $T_B$, the number of excited particles, $N_{ex}$, may be written as

$$N_{ex} = N \left(\frac{T}{T_B}\right)^3, \tag{2.8.21}$$

and the ground state occupation is

$$N_0 = N \left\{1 - \left(\frac{T}{T_B}\right)^3\right\}. \tag{2.8.22}$$

This is shown in Fig. 2.14.

*Particle density*

With the free gas, the particle density is fixed; for $N$ particles, it is determined by the volume $V$ of the containing box. With the harmonically trapped gas, this is not so. At low temperatures, the particles are confined close to the minimum of the potential; the density is high. But at higher temperatures, the particles make greater excursions in the potential well and the effective density decreases.

At temperature $T$, a particle has a mean thermal energy $\frac{1}{2}kT$ (for motion in one direction). Then in a confining potential $V = \frac{1}{2}\kappa x^2$, it will

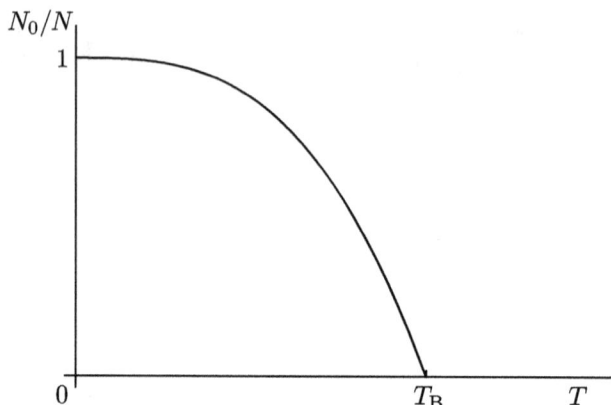

Fig. 2.14.   Ground state occupation for a trapped Bose gas.

have a mean-square displacement $\langle x^2 \rangle_T = kT/\kappa$. It is better to express this in terms of the oscillator frequency $\omega = \sqrt{\kappa/m}$, so that

$$\langle x^2 \rangle_T = \frac{kT}{m\omega^2}. \tag{2.8.23}$$

This corresponds to an effective volume $V = \left(\frac{kT}{m}\right)^{3/2} \omega^{-3}$, an effective density of

$$\frac{N}{V} = \left(\frac{m}{kT}\right)^{3/2} N\omega^3, \tag{2.8.24}$$

or an effective mean particle separation, $l = (V/N)^{1/3}$, of

$$l = \left(\frac{kT}{m}\right)^{1/2} \frac{1}{\omega N^{1/3}}. \tag{2.8.25}$$

Recall, at the end of Section 2.6.4, we argued that at the transition point the thermal de Broglie wavelength should be comparable with the mean particle spacing, Eq. 2.6.26. For the trapped gas, this would give

$$\sqrt{\frac{2\pi\hbar^2}{mkT_c}} \approx \left(\frac{kT_B}{m}\right)^{1/2} \frac{1}{\omega N^{1/3}} \tag{2.8.26}$$

or

$$T_B \approx \frac{\hbar\omega}{k}\sqrt{2\pi}N^{1/3}. \tag{2.8.27}$$

This is about 2.4 times the exact result of Eq. (2.8.20).

We note the precise relation between $\Lambda$ and $l$, instead of Eq. (2.6.26), would be

$$\Lambda = \sqrt{2\pi}\zeta(3)^{1/3}\, l = 2.67\, l. \tag{2.8.28}$$

*Spatial extent*

We saw the mean-square displacement, from the minimum, of a particle at temperature $T$ was

$$\langle x^2 \rangle_T = \frac{kT}{m\omega^2}. \tag{2.8.29}$$

This is the mean-square spatial extent of (a normal) confined gas.

For the particles in the SHO ground state, with wave function

$$\Psi(x) = \left(\frac{m\omega}{\pi\hbar}\right)^{1/4} e^{-m\omega x^2/2\hbar}, \tag{2.8.30}$$

the probability of being found a distance $x$ from the minimum is

$$|\Psi(x)|^2 = \left(\frac{m\omega}{\pi\hbar}\right)^{1/2} e^{-m\omega x^2/\hbar}, \tag{2.8.31}$$

a gaussian distribution with a mean-square displacement

$$\langle x^2 \rangle_0 = \frac{\hbar}{2m\omega}. \tag{2.8.32}$$

This is the mean-square spatial extent of confined gas when the particles are in the ground state.

The ratio of these is

$$\frac{\langle x^2 \rangle_T}{\langle x^2 \rangle_0} = 2\frac{kT}{\hbar\omega}, \tag{2.8.33}$$

the ratio of the thermal energy to the quantum energy. This may be re-expressed using $T_B$ rather than $\omega$:

$$\frac{\langle x^2 \rangle_T}{\langle x^2 \rangle_0} = 2\zeta(3)^{-1/3}N^{1/3}\frac{T}{T_B}$$

$$= 1.88N^{1/3}\frac{T}{T_B}. \tag{2.8.34}$$

This indicates that as one cools through the transition, a much narrower (ground state) peak will appear in the particle density. But note, when you look at Fig. 2.15, that is an image of *velocity*, not position.

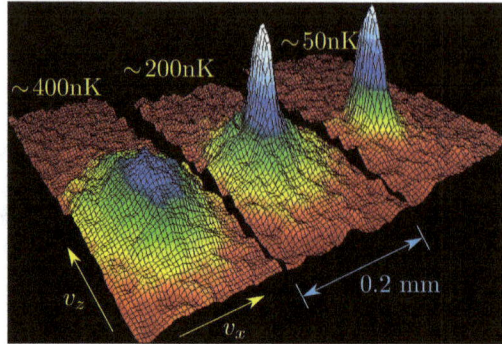

Fig. 2.15.   Bose–Einstein condensation in rubidium vapour. Velocity distributions of the atoms at three different temperatures.

*Source*: After Cornell [17] (image credit: Mike Matthews, JILA, University of Colorado Boulder).

*Observation*

In 1995, Bose–Einstein condensation was observed in rubidium vapour [14] and in lithium vapour [15]. An accessible account of these discoveries is given in the article by Meacher and Ruprecht [16]. In order to have very small inter-particle interactions, the density of these gases was kept very low. The condensation temperature was, correspondingly very low and temperatures in the nanokelvin range were required, achieved by laser confinement and laser cooling of the trapped vapour cloud.

The Bose–Einstein condensation was observed in rubidium vapour by imaging the density and momentum distribution of the atoms. Figure 2.15 shows the velocity distribution of rubidium atoms at three different temperatures. The left image shows the velocity distribution at 400 nK, just before the appearance of the Bose–Einstein condensate. The centre image is at 200 nK, just after the appearance of the condensate. And the right image is at 50 nK, after further evaporation leaves a sample of nearly pure condensate. The size of each image is 200 nm by 270 nm. The appearance of the peak indicates the existence of a significant number of atoms in the ground state.

The interest in the two different materials was because rubidium has a remnant repulsive interaction while the remnant interaction in lithium is attractive; in both these cases, the Bose–Einstein condensation was observed. However, in both these cases, the small amount of material involved precluded the measurement of any thermodynamic properties.

## 2.9   Black Body Radiation — The Photon Gas

### 2.9.1   *Photons as quantised electromagnetic waves*

The harmonic oscillator has the very important property that its energy eigenvalues are equally spaced:

$$\varepsilon_n = n\hbar\omega + \text{zero point energy} \qquad n = 0, 1, 2, \ldots \qquad (2.9.1)$$

You will have calculated the internal energy of the harmonic oscillator in Problem 1.19. We shall write this result here as

$$E = \frac{\hbar\omega}{e^{\hbar\omega/kT} - 1} + \text{zero point contribution.} \qquad (2.9.2)$$

And we shall, in the following sections, ignore the (constant) zero point energy contribution. This is acceptable since the zero of energy is, to a certain extent, arbitrary.

We now see that the expression for $E$ can be reinterpreted in terms of the Bose distribution. The internal energy has the form

$$E = \bar{n}\hbar\omega + \text{zero point contribution,} \qquad (2.9.3)$$

where $\bar{n}$ is the Bose distribution, the mean number of bosons of energy $\hbar\omega$. But here we observe in the Bose distribution that the chemical potential is zero.

The conclusion is that we can regard a harmonic oscillator of (angular) frequency $\omega$ as a collection of bosons of energy $\hbar\omega$, having zero chemical potential.

The fact that we can regard the harmonic oscillator as a collection of bosons is a consequence of the equal spacing of the oscillator energy levels. The vanishing of the chemical potential is due to the fact that the number of these bosons is not conserved.

We shall explore this by considering an isolated system of particles. If $N$ is conserved, then the number is determined — it is given and it will remain constant for the system. On the other hand, if the number of particles is not conserved, then one must determine the equilibrium number by maximising the entropy:

$$\left.\frac{\partial S}{\partial N}\right|_{E,V} = 0, \qquad (2.9.4)$$

where we note that $E$ and $V$ are constant since the system is isolated. From the differential expression for the first law, we see that

$$dS = \frac{1}{T}(dE + pdV - \mu dN). \tag{2.9.5}$$

So, the entropy derivative is

$$\left.\frac{\partial S}{\partial N}\right|_{E,V} = -\frac{\mu}{T}, \tag{2.9.6}$$

and we conclude that the equilibrium condition for this system, at finite temperatures, is simply

$$\mu = 0. \tag{2.9.7}$$

In other words, $\mu = 0$ for non-conserved particles.

### 2.9.2 *Photons in thermal equilibrium — Black body radiation*

It is known that bodies glow and emit light when heated sufficiently. In 1859, Kirchhoff laid down a challenge: to derive the mathematical formula for the radiation spectrum. This challenge was one of the key factors in the development of quantum theory.

The spectrum of radiation emitted by a body depends on its temperature. The spectrum also depends on the nature of its surface — to a more or lesser extent. However, Kirchhoff appreciated that there was an idealisation that could be considered, when the surface absorbs all the radiation that falls upon it. Such a surface, when cold, appears to be black and Kirchhoff speculated on the spectrum of the radiation emitted by such a black body. He appreciated that the radiation was in thermal equilibrium with the body and that therefore the spectrum depended only on the temperature. The intensity of the spectrum was thus some function $U(\omega, T)$ and Kirchhoff's challenge was to find the form of this function.

The spectrum is then a property of the radiation and not of the body under consideration. Thus, we can make a model system, an idealisation of the situation which retains the important features of the problem, but which is possible to solve.

Our model is simply a cavity, which is connected to the outside world by a small hole. We shall look through the hole at the spectrum of the radiation in the cavity, Fig. 2.16.

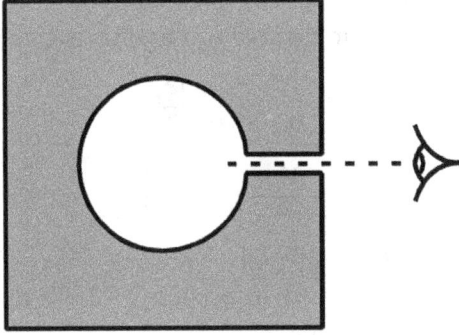

Fig. 2.16.   Radiation from a cavity.

We consider the electromagnetic waves to be in thermal equilibrium with the walls. The photons will have a distribution given by the Bose–Einstein formula, but with zero chemical potential. And we then calculate the properties of this photon gas by the methods of statistical mechanics.

### 2.9.3   Planck's formula

In Section 2.1.5, we saw that the general expression for the energy density of states in three dimensions is given by Eq. (2.1.23):

$$g(\varepsilon) = \frac{V}{2\pi^2}k^2 \left| \frac{d\varepsilon}{dk} \right|, \qquad (2.9.8)$$

where $k$ is to be eliminated in terms of $\varepsilon$. For photons, there is a linear relation between $\varepsilon$ and $k$:

$$\varepsilon = \hbar ck, \qquad (2.9.9)$$

where $c$ is the speed of light. Thus, the density of states is

$$g(\varepsilon) = \frac{1}{2}\frac{V\varepsilon^2}{c^3\pi^2\hbar^3}. \qquad (2.9.10)$$

Now, although photons have a spin $S = 1$, they have zero mass. Quantum-mechanically, this means that only the $S_z = -1$ and $+1$ orientations occur; the $S_z = 0$ orientation is suppressed. Classically, this is understood because electromagnetic waves propagate as transverse disturbances; there is no longitudinal mode. The photons' two quantum states correspond to the two polarisation states of the electromagnetic waves. This means that we must take a degeneracy factor $\alpha$ for photons to be 2. And so the internal energy for the photon gas in thermal equilibrium is given by

$$E = 2 \int_0^\infty \frac{g(\varepsilon)\varepsilon d\varepsilon}{e^{\varepsilon/kT} - 1}$$

$$= \frac{V}{c^3 \pi^2 \hbar^3} \int_0^\infty \frac{\varepsilon^3 d\varepsilon}{e^{\varepsilon/kT} - 1}. \tag{2.9.11}$$

Before we evaluate this integral, we shall make connection with Kirchhoff's $U(\omega, T)$ introduced in Section 2.9.2. To this end, we shall change variables to $\omega = \varepsilon/\hbar$ so that

$$E = \frac{V\hbar}{c^3 \pi^2} \int_0^\infty \frac{\omega^3 d\omega}{e^{\hbar\omega/kT} - 1}. \tag{2.9.12}$$

Now, the energy can be expressed in terms of the spectrum $U(\omega, T)$ by integrating over all frequencies

$$E = \int_0^\infty U(\omega, T) d\omega. \tag{2.9.13}$$

And from these two expressions, we can then identify

$$U(\omega, T) = \frac{V\hbar}{c^3 \pi^2} \frac{\omega^3}{e^{\hbar\omega/kT} - 1}. \tag{2.9.14}$$

This is Planck's formula for black body radiation. Some examples are shown in Fig. 2.17, but plotted as a function of wavelength, which is more popular with spectroscopists.

The spectrum from the sun indicates that its temperature is about 5800 K. It is also of interest to note that the peak of the sun's spectrum corresponds to the visible spectrum, that is, the region of greatest sensitivity of the human eye.

A remarkable example of black body radiation is the spectrum of electromagnetic radiation arriving from outer space, the cosmic microwave background. It is found that when looking into space with radio telescopes, a uniform background electromagnetic "noise" is seen. The spectrum of this is found to fit the black body curve — for a temperature of approximately 2.7 K. The conclusion is that the equilibrium temperature of the universe is 2.7 K, which is understood as the remnant "glow" from the Big Bang. The data shown in Fig. 2.18 come from the COBE satellite and it fits the Planck black body curve for a temperature of 2.74 K. The quality of the fit of the data to the theoretical curve is quite remarkable.

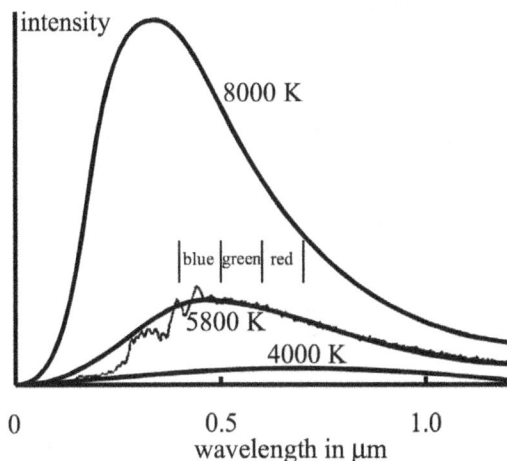

Fig. 2.17.   Black body radiation at three different temperatures; the wiggly line is the spectrum from the sun.

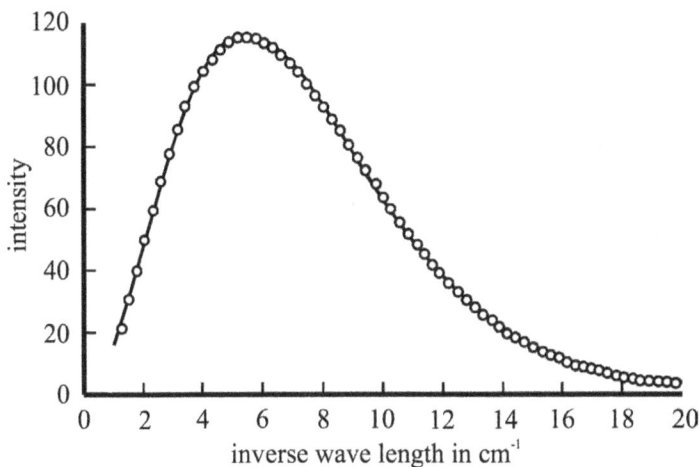

Fig. 2.18.   Cosmic background radiation plotted on 2.74 K black body curve.

### 2.9.4  *Internal energy and heat capacity*

We now integrate the expression for the photon internal energy in the previous section, Eq. (2.9.11):

$$E = \frac{V}{c^3 \pi^2 \hbar^3} \int_0^\infty \frac{\varepsilon^3 \mathrm{d}\varepsilon}{e^{\varepsilon/kT} - 1}. \tag{2.9.15}$$

By changing the variable of integration to $x = \hbar\omega/kT$, the integral becomes a dimensionless number

$$E = \frac{V\hbar}{\pi^2 c^3}\left(\frac{kT}{\hbar}\right)^4 \int_0^\infty \frac{x^3}{e^x - 1}\,dx. \tag{2.9.16}$$

The integral is $\mathscr{B}_3$ from the definition Eq. (2.6.14), so that

$$E = \frac{V\hbar}{\pi^2 c^3}\left(\frac{kT}{\hbar}\right)^4 \mathscr{B}_3 \tag{2.9.17}$$

and

$$\mathscr{B}_3 = \Gamma(4)\zeta(4). \tag{2.9.18}$$

These have values

$$\Gamma(4) = 6,$$
$$\zeta(4) = \pi^4/90, \tag{2.9.19}$$

so that

$$\mathscr{B}_3 = \pi^4/15. \tag{2.9.20}$$

Then the internal energy is

$$E = \frac{\pi^2 V\hbar}{15c^3}\left(\frac{kT}{\hbar}\right)^4. \tag{2.9.21}$$

The heat capacity is found by differentiating the internal energy

$$C_V = \left.\frac{\partial E}{\partial T}\right|_V, \tag{2.9.22}$$

giving

$$C_V = \frac{4\pi^2 V k^4}{15\hbar^3 c^3}T^3. \tag{2.9.23}$$

We see that the photon heat capacity is proportional to $T^3$. Also note that the heat capacity goes to zero as $T \to 0$ in conformity with the Third Law.

### 2.9.5 *Black body radiation in one dimension — Johnson noise*

We shall see that by considering black body radiation in one dimension, we can make an important connection with electrical noise, a topic that will be treated again in Chapter 5. We may regard a waveguide as a one-dimensional black body cavity, and to start with we shall imagine this waveguide to connect black body cavities at either end as in Fig. 2.19. We imagine the cavities and the waveguide to be at a common temperature $T$ and we consider the equilibrium thermal radiation in the waveguide.

In a one-dimensional cavity of length $l$, the energy density of states is given by Eq. (2.1.25):

$$g(\varepsilon) = \frac{l}{\pi} \left| \frac{d\varepsilon}{dk} \right|, \tag{2.9.24}$$

where $\varepsilon = c\hbar k$ for photons. Then the density of states is

$$g(\varepsilon) = \frac{l}{\pi c\hbar}, \tag{2.9.25}$$

independent of energy. The internal energy is then found by integrating

$$E = 2 \int_0^\infty \frac{\varepsilon g(\varepsilon)}{e^{\varepsilon/kT} - 1} d\varepsilon$$

$$= \frac{2l}{\pi c\hbar} \int_0^\infty \frac{\varepsilon}{e^{\varepsilon/kT} - 1} d\varepsilon. \tag{2.9.26}$$

We substitute $x = \varepsilon/kT$ to obtain the integral in terms of the $\mathscr{B}_n$ integrals, Eq. (2.6.14):

$$E = \frac{2lk^2 T^2}{\pi c\hbar} \int_0^\infty \frac{x}{e^x - 1} dx$$

$$= \frac{2lk^2 T^2}{\pi c\hbar} \mathscr{B}_1. \tag{2.9.27}$$

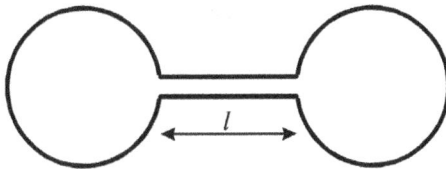

Fig. 2.19.   Waveguide joining two black body cavities.

*Topics in Statistical Mechanics*

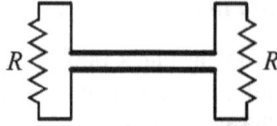

Fig. 2.20.   Transmission line terminated with its characteristic resistance.

and since, from Eq. (2.6.15), $\mathscr{B}_1 = \Gamma(2)\zeta(2) = 1 \times \pi^2/6$, it follows that

$$E = \frac{\pi}{3} \frac{l}{c\hbar} k^2 T^2. \tag{2.9.28}$$

In the three-dimensional case, we made connection with the radiation spectrum by changing variables to *angular* frequency $\omega = \varepsilon/\hbar$. However, in the electrical case, it is conventional to work in terms of *cyclic* frequency $f$; we change variables in Eq. (2.9.26) to $f = \varepsilon/2\pi\hbar = \varepsilon/h$ so that

$$E = \frac{4lh}{c} \int_0^\infty \frac{f \, df}{e^{hf/kT} - 1}. \tag{2.9.29}$$

This enables us to identify the (one-dimensional) radiation spectrum $U(f, T)$ as

$$U(f, T) = \frac{4l}{c} \frac{hf}{e^{hf/kT} - 1}. \tag{2.9.30}$$

From the electrical point of view, an infinitely long transmission line is equivalent to a finite transmission line terminated with its characteristic resistance, Fig. 2.20. In this case, the spectrum of radiation in the transmission line will be the same. Thus, the resistors may be regarded as emitting and absorbing the radiation. The time for the radiation to propagate along the waveguide is $l/c$. Thus, the rate of energy flow, the power, in a frequency range $\Delta f$ is given by

$$P_{\Delta f} = 4 \frac{hf}{e^{hf/kT} - 1} \Delta f. \tag{2.9.31}$$

This may be interpreted as a random current or voltage in the resistors, in a frequency range $\Delta f$, of mean-square value

$$\langle V^2 \rangle_{\Delta f} = 4R \frac{hf}{e^{hf/kT} - 1} \Delta f,$$

$$\langle I^2 \rangle_{\Delta f} = \frac{4}{R} \frac{hf}{e^{hf/kT} - 1} \Delta f. \tag{2.9.32}$$

In the high-temperature/equipartition limit, these expressions become

$$\langle V^2 \rangle_{\Delta f} = 4kTR\Delta f,$$

$$\langle I^2 \rangle_{\Delta f} = \frac{4}{R}kT\Delta f,$$

$$(2.9.33)$$

independent of $h$ and frequency. These random voltage/current fluctuations are a consequence of the thermal motion of the current carriers in the resistors. As we shall see in Chapter 5, this phenomenon was predicted by Einstein in 1906 and observed by Johnson in 1927. The spectrum of the fluctuations was explained theoretically by Nyquist in the same year.

## 2.10   Ideal Paramagnet

### 2.10.1   *Partition function and free energy*

Our treatment of the ferromagnet and the paramagnet–ferromagnet transition in Chapter 4 will be based upon an extension of the conventional model for the paramagnet, that is, an assembly of non-interacting, stationary magnetic moments. For this reason, we shall present, in the following few sections, a summary of the properties of the ideal paramagnet. For convenience, we shall consider magnetic moments of spin ½. This will correspond to electrons, and the treatment is simplified in that case. In a magnetic field $\mathbf{B}$, a magnetic moment $\mu$ will have an energy $-\mu \cdot \mathbf{B}$. Now, a spin ½ may point either parallel or antiparallel to the applied magnetic field. Then the energy of these states will be

$$\varepsilon_\uparrow = -\mu B = -\gamma \hbar B/2,$$

$$\varepsilon_\downarrow = +\mu B = +\gamma \hbar B/2,$$

$$(2.10.1)$$

where $\gamma$ is the magnetogyric (gyromagnetic) ratio of the moments.
   The partition function $Z$ for the entire system is

$$Z = z^N,$$

$$(2.10.2)$$

where $z$ is the partition function for a single one of the $N$ spins. This is the expression for a collection of *distinguishable* identical subsystems. Although the spins themselves are certainly indistinguishable, we are considering a solid assembly where the subsystems can be distinguished by their positions. Now, $z$ is the sum over the two states

$$z = \sum_{i=\uparrow,\downarrow} e^{-\varepsilon_i/kT}$$

$$= e^{\varepsilon/kT} + e^{-\varepsilon/kT} \tag{2.10.3}$$

$$= 2\cosh(\varepsilon/kT),$$

where we denote $\varepsilon = \mu B$. Then

$$Z = \left[2\cosh(\varepsilon/kT)\right]^N \tag{2.10.4}$$

and the connection with thermodynamics is made through the expression for the (magnetic) free energy

$$F = -kT \ln Z. \tag{2.10.5}$$

Thus, we have

$$F = -NkT \ln \left[2\cosh(\mu B/kT)\right]. \tag{2.10.6}$$

### 2.10.2   *Thermodynamic properties*

The various thermodynamic properties are found through differentiation. The differential of $F$, in the magnetic case, is

$$dF = -SdT - MdB. \tag{2.10.7}$$

*Magnetisation and Curie's law*

The magnetic moment $M$ is given by

$$M = -\left.\frac{\partial F}{\partial B}\right|_T, \tag{2.10.8}$$

giving

$$M = N\mu \tanh\left(\frac{\mu B}{kT}\right); \tag{2.10.9}$$

this is the equation of state for the ideal paramagnet.

Figure 2.21 shows the magnetisation as a function of $\varepsilon/kT = \mu B/kT$. At small fields $B$, we have linear behaviour, but saturation sets in at higher fields/lower temperatures when the dipoles are all aligned along the field so that no more magnetisation can be created.

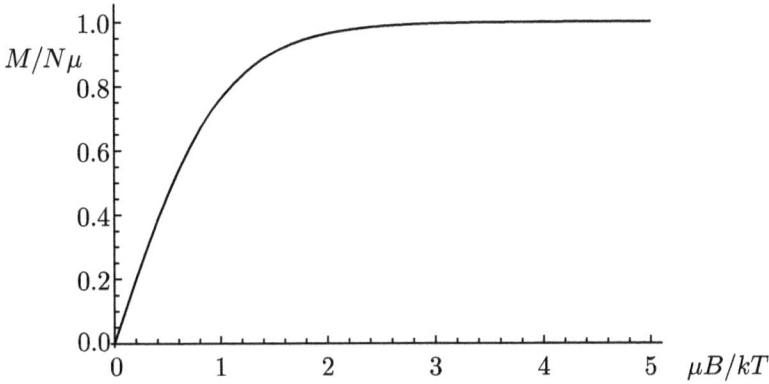

Fig. 2.21. Magnetisation of a paramagnet.

In the low-field/high-temperature limit, the tanh may be expanded to first order: $\tanh x \sim x$. Then

$$M = N\mu^2 \frac{B}{kT}.$$  (2.10.10)

In this limit, we have $M$ proportional to $B$ and inversely proportional to $T$; this result is known as Curie's law. The magnetic susceptibility $\chi$ is

$$\chi = \frac{\mu_0}{V} \frac{M}{B}$$  (2.10.11)

(this ensures that $\chi$ is dimensionless and intensive) and then

$$\chi = \mu_0 \frac{N}{V} \frac{\mu^2}{kT}.$$  (2.10.12)

One conventionally writes

$$M = C\frac{B}{T}, \quad \chi = \frac{\mu_0 C}{V} \frac{1}{T},$$  (2.10.13)

where $C$ is known as the Curie constant.

Observe that Curie's law is like the ideal gas equation of state:

$$M = \text{const} \times \frac{B}{T} \quad \text{is like} \quad \frac{1}{V} = \text{const} \times \frac{p}{T}.$$  (2.10.14)

In this respect, we may regard Curie's law as the equation of state of the *ideal* paramagnet.

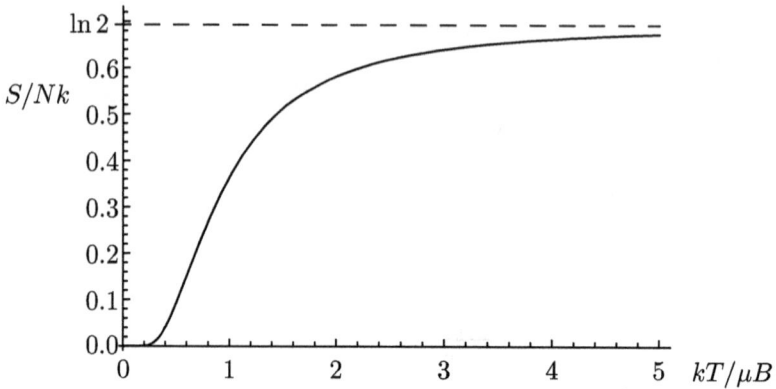

Fig. 2.22.   Entropy of a paramagnet.

*Entropy*

The other variable we find from d$F$ is the entropy:

$$
\begin{aligned}
S &= -\left.\frac{\partial F}{\partial T}\right|_B \\
&= Nk\left[\ln\left\{2\cosh\left(\frac{\varepsilon}{kT}\right)\right\} - \frac{\varepsilon}{kT}\tanh\left(\frac{\varepsilon}{kT}\right)\right].
\end{aligned}
\tag{2.10.15}
$$

This is shown in Fig. 2.22 as a function of temperature ($kT/B$). At high temperatures, the entropy approaches $k\ln 2$ per particle as each moment has a choice of two states (two orientations).

We observe that the entropy tends to zero as $T/B$ tends to zero. When the field $B$ is finite, this indicates behaviour in accordance with the Third Law. The case of zero external fields must be treated carefully, as discussed in Section 1.7.6. The reality is that there will always be internal magnetic fields present: those produced by the dipoles themselves. Thus, the Third Law is safe.

*Internal energy and heat capacity*

The internal energy is found as

$$
E = F + TS = -N\mu B\tanh\left(\frac{\mu B}{kT}\right),
\tag{2.10.16}
$$

which, we see, corresponds to $-MB$.

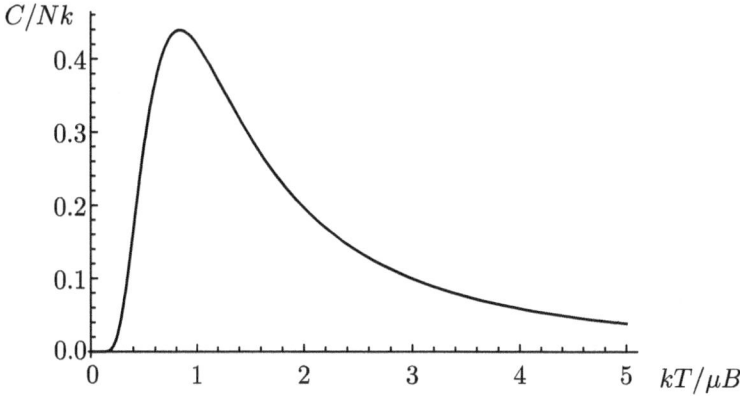

Fig. 2.23.   Heat capacity of a paramagnet.

The heat capacity is given by

$$C_B = \left. \frac{\partial E}{\partial T} \right|_B$$

$$= Nk \left( \frac{\mu B}{kT} \right)^2 \operatorname{sech}^2 \left( \frac{\mu B}{kT} \right),$$

(2.10.17)

which we plot in Fig. 2.23.

The heat capacity of the paramagnet has a characteristic peak at $kT = \mu B$. The peak is a consequence of the magnetic energy being bounded from above (unlike mechanical energy). This is known as the Schottky peak.

We conclude this section by considering the equilibrium fluctuations in the magnetisation of a paramagnet. The magnetic moment of each spin ½ particle is plus or minus $\mu$. The square of the magnetic moment of each particle is then $\mu^2$, so that the mean-square magnetisation for the assembly of $N$ particles is then

$$\langle M^2 \rangle = N\mu^2.$$

(2.10.18)

This may be regarded as the mean square of the fluctuations in the magnetisation from its mean value of zero. This may be related to the Curie law susceptibility in Eq. (2.10.12), thus

$$\chi = \mu_0 \frac{1}{VkT} \langle M^2 \rangle.$$

(2.10.19)

Here, we see how the magnetisation response function, the magnetic susceptibility, is related to the equilibrium fluctuations in the magnetisation.

This is analogous to our result of Section 1.4.6 where we saw that the heat capacity (energy response function) was related to the equilibrium fluctuations in the energy. Indeed, these are both special cases of the *fluctuation–dissipation theorem*, which we will encounter in Chapter 5.

### 2.10.3   *Negative temperatures*

The occupation of the up state and the down state are given by the appropriate Boltzmann factors. As we have seen the energies of the two states are

$$\varepsilon_\uparrow = -\mu B,$$

$$\varepsilon_\downarrow = +\mu B,$$

(2.10.20)

it follows that since

$$p_\uparrow = e^{-\varepsilon_\uparrow/kT}\Big/ z \quad p_\downarrow = e^{-\varepsilon_\downarrow/kT}\Big/ z,$$

(2.10.21)

then the numbers of spins in the up and down states are

$$N_\uparrow = N\, e^{-\varepsilon_\uparrow/kT}\Big/ z \quad N_\downarrow = N\, e^{-\varepsilon_\downarrow/kT}\Big/ z.$$

(2.10.22)

And then we see that the ratio of occupation of the up and down states is

$$\frac{N_\uparrow}{N_\downarrow} = e^{2\mu B/kT}.$$

(2.10.23)

Thus, ordinarily there will be more spins in the up state than in the down state; this follows since the up state has a lower energy than the down state.

It is clear that by pumping energy into this system, we will flip spins from the lower energy "up" state to the higher energy "down" state. And we can conceive of pumping energy in, in this way until the populations of the two states becomes equalised. Moreover, we could continue to pump even more energy into the system so that there are more spins in the (higher energy) down state than the (lower energy) up state. This would be a strange configuration for our system, but in the interests of simplicity, we could still apply Eq. (2.10.23) to describe it. In particular, we could use this equation to specify the temperature of our system. Then we would reach the following conclusions:

1. The "normal" situation, where the occupation of the lower energy state is higher than the occupation of the upper energy state corresponds to a situation where the temperature is a positive quantity.

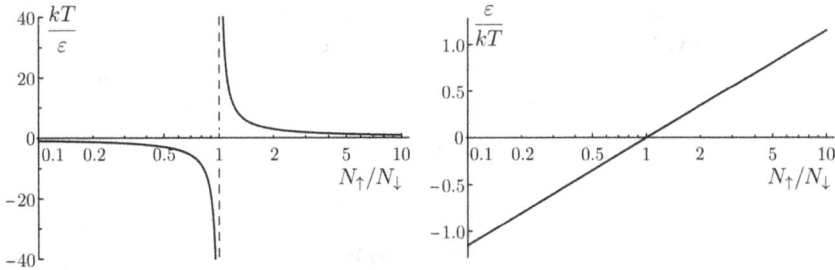

Fig. 2.24.   Temperature and inverse temperature as a function of population ratio (logarithmic scale).

2. When the populations of the two states are equalised, this corresponds to an *infinite* temperature.
3. When the occupation of the higher energy state is higher than the occupation of the lower energy state, the system would have a *negative* temperature.

Things become clearer when we plot the resultant temperature against the population ratio. We observe that there is a linear relation between *inverse* temperature and the logarithm of the population ratios.

$$\frac{\varepsilon}{kT} = \frac{1}{2} \ln \frac{N_\uparrow}{N_\downarrow}, \qquad (2.10.24)$$

So, in this context, the inverse temperature is a more natural variable to use. In Fig. 2.24, we see that starting at large occupation of the up state corresponds to a low (positive) temperature — far right-hand side. Then feeding energy in, so that the up state occupation decreases, the temperature increases — moving to the left. The temperature diverges to plus infinity as the populations equalise. And then with greater occupation of the down state, the temperature returns from minus infinity and then it increases towards zero from the negative side.

It may be argued that the discussion justifying the concept of negative temperatures is somewhat artificial. It relies upon inversion of the Boltzmann factor expression for the population probabilities. Now, the Boltzmann factor $\exp -\varepsilon/kT$ gives the population probabilities for a system in thermal equilibrium at a temperature $T$. But to what extent can we regard the states discussed in the section above as being in thermal equilibrium?

We were able to sidestep this question in the introductory discussion by restricting the consideration to spin ½ particles where each has the choice of two orientations. Then any values for $N_\uparrow$ and $N_\downarrow$ will give a temperature value through Eq. (2.10.12). The situation is different, however, for spins $S > 1/2$. If there are more than two possible spin states, then only specific numbers of particles occupying the different states will correspond to the Boltzmann distribution and thus to states of thermal equilibrium.

As an example, consider the case of $S = 1$, where there are three spin orientations: $S_z = -1, 0$ and $+1$. The energies of these states, in a magnetic field $B$, are

$$\varepsilon_1 = -\mu B,$$

$$\varepsilon_0 = 0, \qquad\qquad (2.10.25)$$

$$\varepsilon_{-1} = +\mu B,$$

so that the populations of the three states will be

$$N_1 = N e^{\mu B/kT}/z,$$
$$N_0 = N/z, \qquad\qquad (2.10.26)$$
$$N_{-1} = N e^{-\mu B/kT}/z.$$

And only when the populations obey a relation such as this can one ascribe a temperature to the system.

If we use the ratio $N_1/N_{-1}$ as the independent variable (by analogy with the spin ½ case), then in this case we have the supplementary condition $N_0 = \sqrt{N_1 N_{-1}}$. Only when this is satisfied will the system be in thermal equilibrium and we can ascribe a temperature to it. In real systems, the (small) interactions between the spins will be such to establish this thermal equilibrium. We will return to this point at the end of the following section.

### 2.10.4   *Thermodynamics of negative temperatures*

The thermodynamic definition of temperature, from the First Law, is given by

$$\frac{1}{T} = \frac{\partial S}{\partial E}. \qquad\qquad (2.10.27)$$

Now, in "normal" systems, the entropy is a monotonically increasing function of the energy. In this case, the derivative will be positive and so

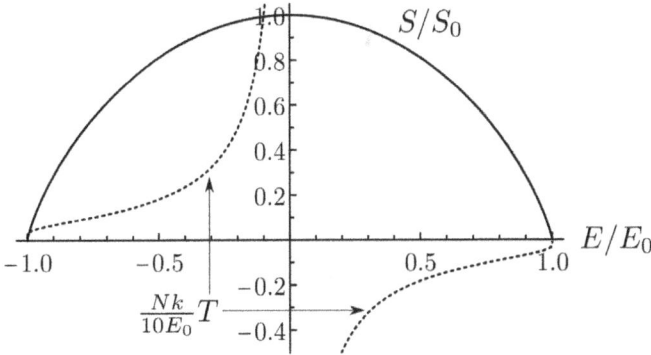

Fig. 2.25. Entropy and temperature of a spin ½ paramagnet as a function of energy.

negative temperatures will not arise. We see that to have a negative temperature the entropy must be a *decreasing* function of the energy. This is possible with a spin system.

At the lowest energy, all spins will be in their lowest energy state; all spins in the same state then means low (zero) entropy. Then as the energy increases, an increasing number of spins will be promoted to the higher state. Here, the disorder is increasing and so the entropy is increasing. With sufficient energy, the populations will become equalised; this corresponds to the maximum entropy possible $S_0 = Nk \ln 2$. And then as the energy continues to increase, an increasing number of spins will find themselves in the higher energy state; the entropy is decreasing. Finally, at the maximal energy, all spins will be in the higher energy state and the entropy once again will become zero. This behaviour is shown in Fig. 2.25; we now calculate this.

The entropy is most conveniently expressed in terms of the Gibbs expression

$$S = -Nk \sum_j p_j \ln p_j, \qquad (2.10.28)$$

where the $p_j$ are the probabilities of the single-particle states. Restricting our discussion to the spin ½ case, there are two states to consider:

$$S = -Nk \left\{ p_\uparrow \ln p_\uparrow + p_\downarrow \ln p_\downarrow \right\}. \qquad (2.10.29)$$

These probabilities may be expressed in terms of the magnetisation as

$$p_\uparrow = \frac{1 + M/N\mu}{2}, \quad p_\downarrow = \frac{1 - M/N\mu}{2}, \quad (2.10.30)$$

or, since the energy is given by $E = -MB$,

$$p_\uparrow = \frac{1 - E/E_0}{2}, \quad p_\downarrow = \frac{1 + E/E_0}{2}, \quad (2.10.31)$$

where $E_0$ is the energy corresponding to the saturation magnetisation: $E_0 = N\mu B$. Thus, the entropy is given by

$$S = \frac{Nk}{2}\left\{2\ln 2 - \left(1 + \frac{E}{E_0}\right)\ln\left(1 + \frac{E}{E_0}\right) - \left(1 - \frac{E}{E_0}\right)\ln\left(1 - \frac{E}{E_0}\right)\right\}, \quad (2.10.32)$$

and correspondingly the temperature $T = 1/(\partial S/\partial E)$ is

$$T = -\frac{E_0}{Nk}\frac{1}{\text{arctanh}(E/E_0)}. \quad (2.10.33)$$

These are shown in Fig. 2.25.

For $-E_0 < E < 0$, we have $\partial S/\partial E > 0$, so the temperature is positive. When $E = E_0$, we have $\partial S/\partial E = 0$, so here the temperature is infinite. And when $0 < E < E_0$, we have $\partial S/\partial E < 0$, so now the temperature is negative.

It should be appreciated that a fundamental requirement for the existence of negative temperatures is that the system's energy must be bounded from above.

In the previous section, we showed the heat capacity of the paramagnet for positive temperatures. The same expression holds for negative temperatures. Thus, we obtain the heat capacity shown in Fig. 2.26. Note that the heat capacity remains positive at negative temperatures. This ensures stability of the system.

We finish this section with some general comments about negative temperature states. Since a fundamental requirement for states of negative temperature is that the energy must be bounded from above, this means that in any real system the degree of freedom with such bounded energy states must be thermally isolated from the normal (unbounded) degrees of freedom. Then the negative temperature applies to the subsystem and it is, in reality, quasi-equilibrium state.

Such quasi-equilibrium states are of interest only to the extent that they are reasonably long-lived. And they are only meaningful to the extent that the quasi-equilibrium state is established on a timescale much shorter than this. The first observations of negative temperatures were in the nuclear

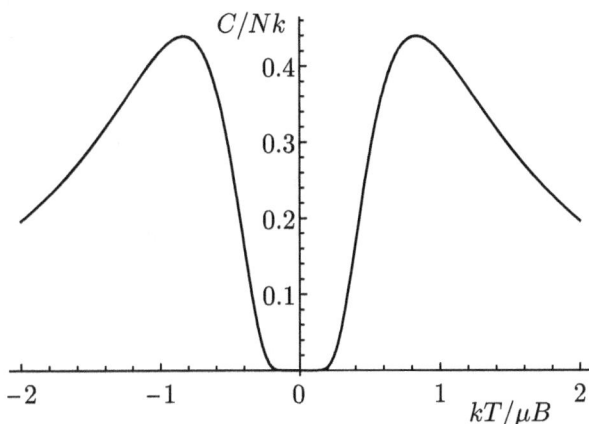

Fig. 2.26. Heat capacity of a paramagnet at positive and negative temperatures.

spins of lithium in LiF crystals by Purcell and Pound [18]. There the relaxation time for the interaction between the spin degrees of freedom and the normal degrees of freedom was of the order of five minutes while the internal equilibrium of the Li nuclear spins was established in less than $10^{-5}$ second.

## Problems

2.1   In Section 2.1, we saw that the density of free-particle states for a three-dimensional volume $V$ was shown to be

$$g(\varepsilon) = \frac{1}{4}\frac{V}{\pi^2\hbar^3}(2m)^{3/2}\varepsilon^{1/2};$$

this followed from counting the number of states in the octant of radius

$$R = \sqrt{n_x^2 + n_y^2 + n_z^2}.$$

(a)   By similar arguments, show that in two dimensions, by counting the number of states in the quadrant of radius

$$R = \sqrt{n_x^2 + n_y^2},$$

the density of states is given by

$$g(\varepsilon) = \frac{mA}{2\pi\hbar^2},$$

where $A$ is the area. Note in two dimensions the density of states is independent of energy.

(b)   And, similarly, show that in one dimension the density of states is

$$g(\varepsilon) = \frac{L}{\pi\hbar} \left(\frac{m}{2}\right)^{1/2} \varepsilon^{-1/2}.$$

2.2   Show, using arguments similar to those in Section 2.1.3, that the energy levels of an ultra-relativistic or a massless particle with energy–momentum relation $E = cp$ are given by

$$\varepsilon = \frac{c\pi\hbar}{V^{1/3}} \left(n_x^2 + n_y^2 + n_z^2\right)^{1/2}.$$

Hence, show that the pressure of a gas of such particles is one-third of the (internal) energy density.

2.3   In Sections 2.3.1 and 2.3.2, the ideal gas partition function was calculated quantum-mechanically and classically. Although the calculations were quite different, they both resulted in (different) gaussian integrals. By writing the gaussian integral of the classical case as

$$\int_{-\infty}^{\infty} dx \int_{-\infty}^{\infty} dy \int_{-\infty}^{\infty} dz \, e^{-(x^2+y^2+z^2)}$$

and transforming to spherical polar coordinates, you can perform the integration over $\theta$ and $\phi$ trivially. Show that the remaining integral can be reduced to that of the quantum case.

2.4   The Sackur–Tetrode equation, discussed in Section 2.3.3, and written as

$$S = Nk \ln V - Nk \ln N + \frac{3}{2} Nk \ln T + Nks_0$$

is often interpreted as indicating different contributions to the entropy: the volume contribution is in the first term, the number contribution in the second term and the temperature contribution in the third term. Show that such an identification is fallacious by demonstrating that the various contributions depend on the choice of units adopted, even though the total sum is independent. Discuss the origin of the fallacy.

2.5 The Gibbs–Duhem relation tells us that the intensive variables $T, p, \mu$ are not independent so that, for example, $\mu$ depends on $T$ and $p$. Demonstrate this by showing that

$$\mu = -kT \ln \left[ \left( \frac{m}{2\pi\hbar^2} \right)^{3/2} \frac{(kT)^{5/2}}{p} \right]$$

for an ideal gas.

2.6 Show that the Fermi energy for a two-dimensional gas of Fermions is

$$\varepsilon_F = \frac{2\pi\hbar^2}{am} \frac{N}{A},$$

where $A$ is the area of the system.

2.7 Show that the chemical potential of a two-dimensional gas of fermions may be expressed analytically as

$$\mu = kT \ln \{ e^{\varepsilon_F/kT} - 1 \}.$$

2.8 Calculate the low-temperature chemical potential of a two-dimensional gas of fermions by the Sommerfeld expansion method of Section 2.5.3. Observe that the temperature series expansion terminates. Compare this result with the exact result of the previous question. Discuss the difference between the two results.

2.9 Obtain expressions for the chemical potential $\mu$, the internal energy $E$ and the heat capacity $C_V$ for a system of fermions with general density of states $g(\varepsilon)$. That is, show that these are given in terms of the behaviour of the density of states at the Fermi surface.

2.10 Evaluate the Fermi temperature for liquid $^3\mathrm{He}$, assuming it to be a Fermi "gas". Its molar volume is $36\,\mathrm{cm}^3$. Calculate the thermal de Broglie wavelength at $T = T_F$ and show that it is comparable with the inter-particle spacing as expected.

2.11    In Problem 2.1, we found the expression for the energy density of states $g(\varepsilon)$ for a gas of fermions confined to two dimensions and we saw that it was independent of energy. What surface density of electrons is necessary in order that $T_F = 100\,\text{mK}$? Show that, for a given area, the heat capacity is independent of the number of electrons.

2.12    Use the Sommerfeld expansion method of Section 2.5.3 to show that the Fermi–Dirac distribution function may be approximated, at low temperatures, by

$$\frac{1}{e^{(\varepsilon-\mu)kT}+1} \sim \Theta(\mu-\varepsilon) - \frac{\pi^2}{6}(kT)^2\,\delta'(\varepsilon-\mu)+\cdots,$$

where $\Theta$ is the unit step function and $\delta'$ is the first derivative of the Dirac delta function.

Can you write down the general term of the series?

2.13    Liquid $^4$He has a molar volume of $27\,\text{cm}^3$ at saturated vapour pressure. Treating the liquid as an ideal gas of bosons, find the temperature at which Bose–Einstein condensation will occur. How will this temperature change as the pressure on the fluid is increased?

The superfluid transition temperature of liquid helium decreases with increasing pressure; very approximately $\partial T_c/\partial p \sim -0.015$ $\text{K}\,\text{bar}^{-1}$. How does this compare with the behaviour predicted for ideal gas Bose–Einstein condensation?

2.14    Consider the Bose gas at low temperatures. You saw in Sections 2.6.3 and 2.6.4 that when the occupation of the ground state is appreciable, then the chemical potential $\mu$ is very small and it may be ignored, compared with $\varepsilon$ in the integral for the number of excited states.

(a)    Show that when the ground state occupation $N_0$ is appreciable, then $\mu$ may be approximated by

$$\mu \sim -kT/N_0.$$

(b)    Now, consider the requirement that $\mu$ may be neglected in comparison with $\varepsilon$ in the integral for the number of excited states.

This will be satisfied if $|\mu|$ is much less than the energy $\varepsilon_1$ of the first excited state. The expression for $\varepsilon_1$ is

$$\varepsilon_1 = \frac{\pi^2 \hbar^2}{2mV^{2/3}}.$$

Where does this expression come from?

(c) Show that the condition $|\mu| \ll \varepsilon_1$ is satisfied right up to $T_B$ in the thermodynamic limit. **Hint:** Demonstrate that $|\mu|/\varepsilon_1 \sim N^{-1/3}$.

(d) For a finite system comprising $1 \, \text{cm}^3$ of $^4\text{He}$ (molar volume $27 \, \text{cm}^3$) show that the expression

$$N_0 = N \left\{ 1 - \left( \frac{T}{T_B} \right)^{3/2} \right\}$$

is valid to temperatures below $T_B$ right up to within $\sim 10^{-8} T_B$ of the Bose temperature.

2.15 (a) Show that the pressure of a Bose gas below $T_B$ depends *only* on temperature, that is, it is independent of other thermodynamic variables, such as $N$ and $V$.

(b) Comment on this result in the context of the Gibbs–Duhem relation (asserting that the intensive variables $T, p, \mu$ are not independent).

2.16 (a) Show that below the transition temperature the entropy of a Bose gas is given by

$$S = \frac{5}{2} Nk \frac{\zeta\left(\frac{5}{2}\right)}{\zeta\left(\frac{3}{2}\right)} \left( \frac{T}{T_B} \right)^{3/2}.$$

(b) Since the number of excited particles is given by

$$N_{\text{ex}} = N \left( \frac{T}{T_B} \right)^{3/2},$$

show that the entropy per excited particle is given by

$$\frac{S}{N_{\text{ex}}} = \frac{5}{2} \frac{\zeta\left(\frac{5}{2}\right)}{\zeta\left(\frac{3}{2}\right)} k \approx 1.28 \, k.$$

(c) Discuss the connection between this result and the two-fluid model of superfluid $^4\text{He}$.

2.17  Show that the chemical potential of a two-dimensional gas of bosons may be expressed analytically as

$$\mu = kT \ln \{1 - e^{-\varepsilon_q/kT}\},$$

where $\varepsilon_q$ is the two-dimensional quantum energy, the analogue of the 2d Fermi energy in Problem 2.6.

This problem is the Bose analogue of the Fermi case treated in Problem 2.7.

2.18  Show that the Bose–Einstein transition temperature of a gas of bosons and the Fermi temperature for a gas of "similar" fermions are of comparable magnitude. Discuss why this should be.

2.19  The formula for the Bose integrals $\mathscr{B}_n$ of Section 2.6.4 was quoted as

$$\mathscr{B}_n = \int_0^\infty \frac{x^n}{e^x - 1}\,dx$$

$$= \Gamma(n+1)\zeta(n+1).$$

Derive this result.

**Hint:** Use the sequence of transformations below

$$\frac{1}{e^x - 1} = e^{-x}\frac{1}{1 - e^{-x}} = e^{-x}\sum_{m=0}^{\infty} e^{-mx} = \sum_{m=0}^{\infty} e^{-(m+1)x} = \sum_{m=1}^{\infty} e^{-mx}$$

and then change the variable of integration to $y = mx$.

2.20  The general formula for the Fermi integrals $\mathscr{F}_n$ of Section 2.5.3 was quoted as

$$\mathscr{F}_n = \int_{-\infty}^{\infty} \frac{e^x}{(e^x + 1)^2}\,x^n\,dx$$

$$= 2(1 - 2^{1-n})\zeta(n)\,n!,$$

where $\zeta(n)$ is the Riemann zeta function. Derive this result. This is difficult; *Mathematica* v.12 can't do it!

**Hint:** The first step is to integrate by parts, essentially the reverse of Eq. (2.5.28), to obtain

$$\mathscr{F}_n = 2n \int_0^\infty \frac{x^{n-1}}{e^x + 1} dx,$$

noting that this holds for even $n$, while $\mathscr{F}_n = 0$ for odd $n$. Next, follow a sequence of transformations *similar* to those for the Fermi calculation of Problem 2.19.

2.21 The Riemann zeta function and the Dirichlet eta function are defined as infinite sums:

$$\zeta(n) = \sum_{m=1}^\infty \frac{1}{m^n}, \quad \eta(n) = \sum_{m=1}^\infty \frac{(-1)^{m-1}}{m^n}.$$

Show that $\zeta(n) - \eta(n) = 2^{1-n}\zeta(n)$ and hence that $\eta(n) = (1 - 2^{1-n})\zeta(n)$.

2.22 A dilute gas of $10^6$ rubidium atoms (relative atomic mass 86.9) is confined to an isotropic harmonic trap of frequency 150 Hz.

(a) Calculate the Bose–Einstein condensation (BEC) temperature $T_B$.

(b) Estimate the width of the thermal cloud at $T_B$.

(c) Estimate the width of the condensate cloud.

2.23 This is the trapped-gas analogue of Problem 2.14. Consider a Bose gas confined to an isotropic harmonic trap at low temperatures, as in the previous problem. You know that when the occupation of the ground state is appreciable, then the chemical potential $\mu$ is very small and it may be ignored, compared with $\varepsilon$ in the integral for the number of excited states. Moreover, you know that when the ground state occupation $N_0$ is appreciable, then $\mu \sim -kT/N_0$.

(a) The requirement that $\mu$ may be neglected in comparison with $\varepsilon$ in the integral for the number of excited states will be satisfied if $|\mu|$ is much less than the energy $\varepsilon_1$ of the first excited state. The expression for $\varepsilon_1$ is

$$\varepsilon_1 = \hbar\omega,$$

where $\omega$ is the (angular) frequency of the trap. Where does this expression come from?

(b)   Show that the condition $|\mu| \ll \varepsilon_1$ is satisfied right up to $T_B$ in the thermodynamic limit. **Hint:** Demonstrate that $|\mu|/\varepsilon_1 \sim N^{-2/3}$.

(c)   For a finite system, as in the previous problem, comprising $10^6$ atoms, show that the expression

$$N_0 = N \left\{ 1 - \left( \frac{T}{T_B} \right)^3 \right\}$$

is valid to temperatures below $T_B$ right up to within $\sim 10^{-5} T_B$ of the Bose temperature.

2.24   In Section 2.10, we studied a paramagnetic *solid*: a collection of essentially *distinguishable* magnetic moments. If we were to consider a (classical) gas of indistinguishable magnetic moments, how would the partition function be modified? What would be the observable consequences of this modification?

2.25   According to Curie's law, the magnetic susceptibility of a paramagnet is inversely proportional to temperature.

(a)   Show that a different behaviour is implied by the Third Law.

(b)   What is wrong with the Curie law model?

# References

[1]   B. I. Bleaney and B. Bleaney, *Electricity and Magnetism* (OUP, Oxford, 1989).
[2]   A. M. Guénault, *Statistical Physics* (Springer, 2007).
[3]   A. Sommerfeld, Zur Elektronentheorie der Metalle auf Grund der Fermischen Statistik, *Z. Physik*, **47**, 1–2 (1928) 1–32.
[4]   L. D. Landau and E. M. Lifshitz, *Statistical Physics, Part 2* (Pergamon Press, 1980).
[5]   F. Reif, *Fundamentals of Statistical and Thermal Physics* (McGraw-Hill, 1965).
[6]   F. London, *Superfluids, Vol II: Macroscopic Theory of Superfluid Helium* (John Wiley, 1954). Reprinted (Dover, 1964).
[7]   F. London, The $\lambda$-phenomenon of liquid helium and the Bose–Einstein degeneracy, *Nature*, **141** (1938) 643–644.

[8]  L. Tisza, Transport phenomena in helium II, *Nature*, **141** (1938) 913.

[9]  E. L. Andronikashvili, A direct observation of two kinds of motion in helium II, *J. Phys.* (USSR), **10** (1946) 201–206.

[10]  N. Bogoliubov, On the theory of superfluidity, *J. Phys.* (USSR), **11** (1947) 23.

[11]  I. M. Khalatnikov, *An Introduction to the Theory of Superfluidity* (Benjamin, 1965).

[12]  C. Ebner and H.-H. Fu, Thermodynamic functions of ideal two- and three-dimensional Fermi gases, *J. Low Temp. Phys.*, **16** (1974) 1–8.

[13]  B. Cowan, On the chemical potential of ideal Fermi and Bose gases, *J. Low Temp. Phys.*, **197**, 5–6 (2019) 412–444.

[14]  M. H. Anderson, J. R. Ensher, M. R. Matthews, C. E. Weiman, and E. A. Cornell, Observation of Bose–Einstein condensation in a dilute atomic vapor, *Science*, **269** (1995) 198–201.

[15]  C. C. Bradley, C. A. Sackett, J. J. Tollett, and R. G. Hulet, Evidence of Bose–Einstein condensation in an atomic gas with attractive interactions, *Phys. Rev. Lett.*, **75**, 9 (1995) 1687–1690.

[16]  D. Meacher and P. Ruprecht, Atoms lose their identity, *Phys. World*, **8**, 8 (1995) 21–21.

[17]  E. Cornell, Very cold indeed: The nanokelvin physics of Bose–Einstein condensation, *J. Res. Natl. Inst. Stand. Technol.*, **101** (1996) 419–434.

[18]  E. M. Purcell and R. V. Pound, A nuclear spin system at negative temperature, *Phys. Rev.*, **81** (1951) 279–280.

# CHAPTER 3

# NON-IDEAL GASES

This chapter is devoted to considering systems where the interactions between particles can no longer be ignored. We note that in the previous chapter we did indeed consider, albeit briefly, the effects of interactions in fermion and in boson gases. This chapter is concerned more with a systematic treatment of interatomic interactions. Here the quantum aspect is but a complication and most of the discussions will thus take place within the context of a classical description.

## 3.1 Statistical Mechanics of Interacting Particles

### 3.1.1 *The partition function*

We are now considering gases where the interactions between particles cannot be ignored. Our starting point is that everything can be found from the partition function. We will work, initially, in the classical framework where the hamiltonian of the system is

$$H(p_i, q_i) = \sum_i \frac{p_i^2}{2m} + \sum_{i<j} U(q_i, q_j). \tag{3.1.1}$$

Because of the interaction term $U(q_i, q_j)$ the partition function can no longer be factorised into the product of single-particle partition functions. The many-body partition function is

$$Z = \frac{1}{N! h^{3N}} \int e^{-\left(\sum_i p_i^2/2m + \sum_{i<j} U(q_i, q_j)\right)/kT} d^{3N}p\, d^{3N}q, \tag{3.1.2}$$

where the factor $1/N!$ is used to account for the particles being indistinguishable.

While the partition function cannot be factorised into the product of single-particle partition functions, we can factor out the partition function for the non-interacting case since the energy is a sum of a momentum-dependent term (kinetic energy) and a coordinate-dependent term (potential energy). The non-interacting partition function is

$$Z_{id} = \frac{V^N}{N!h^{3N}} \int e^{-(\sum_i p_i^2/2m)/kT} d^{3N}p, \qquad (3.1.3)$$

where the $V$ factor comes from the integration over the $q_i$. Thus, the interacting partition function may be written as

$$Z = Z_{id} \frac{1}{V^N} \int e^{-(\sum_{i<j} U(q_i,q_j))/kT} d^{3N}q. \qquad (3.1.4)$$

The "correction term" is referred to as the configuration integral. We denote this by $Q$,

$$Q = \frac{1}{V^N} \int e^{-(\sum_{i<j} U(q_i,q_j))/kT} d^{3N}q. \qquad (3.1.5)$$

(Different authors have different prefactors such as $V$ or $N!$, but that is not important.) The partition function for the interacting system is then

$$Z = \frac{1}{N!} \left( \frac{V}{\Lambda^3} \right)^N Q \qquad (3.1.6)$$

and the corresponding free energy is

$$F = F_{id} - kT \ln Q. \qquad (3.1.7)$$

Attention thus focuses on evaluation/approximation of the configuration integral $Q$.

### 3.1.2  *Cluster expansion*

We need a quantity in terms of which to perform an expansion. To this end we define the dimensionless

$$f_{ij} = e^{-U(q_i,q_j)/kT} - 1, \qquad (3.1.8)$$

called Mayer's $f$-function, shown in Fig. 3.1. This has the property that $f_{ij}$ is zero when $U(q_i, q_j)$ is zero and it is appreciable only when the particles are close together.

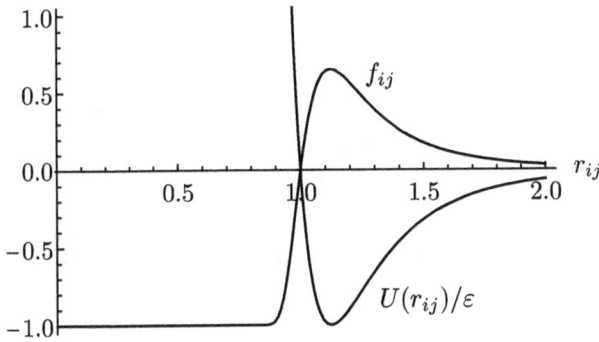

Fig. 3.1.   Lennard-Jones interaction potential $U(r_{ij})$ and the corresponding Mayer $f$-function (for $kT = 2\varepsilon$).

In terms of the Mayer $f$-function the configuration integral is

$$Q = \frac{1}{V^N} \int \prod_{i<j} (1 + f_{ij}) d^{3N} q_i, \tag{3.1.9}$$

where the exponential of the sum has been factored into the product of exponentials.

Next, we expand the product as follows:

$$\prod_{i<j} (1 + f_{ij}) = 1 + \sum_{i<j} f_{ij} + \sum_{i<j} \sum_{k<l} f_{ij} f_{kl} + \sum_{i<j} \sum_{k<l} \sum_{m<n} f_{ij} f_{kl} f_{mn} + \cdots. \tag{3.1.10}$$

The contributions to the second term are significant whenever pairs of particles are close together. Diagrammatically we may represent the contributions to the second term as

Contributions to the third term are significant, (a) if $i$, $j$, $k$, $l$ are distinct, when pairs $i - j$ and $k - l$ are simultaneously close together, or (b) if $i = k$ in the sums, when triples $i$, $j$, $l$ are close together, or (c) if $i = k$ and $j = l$, when $i$, $j$, $k$ and $l$ are close together. The contributions to the third term may be represented as follows:

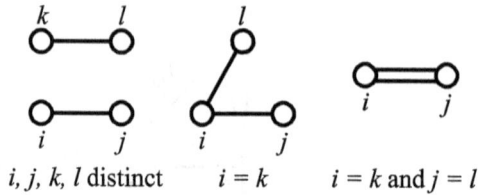

$i, j, k, l$ distinct          $i = k$          $i = k$ and $j = l$

The contributions to the higher-order terms may be represented in a similar way. The general expansion in this way is called a "cluster expansion" for obvious reasons.

### 3.1.3  *Low-density approximation*

In the case of a *dilute* gas, we need only consider the effect of pairwise interactions — the first two terms of Eq. (3.1.10). This is because while the probability of two given particles being simultaneously close is small, the probability of three atoms being close is considerably smaller. Then we have

$$\prod_{i<j}(1 + f_{ij}) \approx 1 + \sum_{i<j} f_{ij} \tag{3.1.11}$$

so that, within this approximation,

$$Q = \frac{1}{V^N} \int \left\{ 1 + \sum_{i<j} f_{ij} \right\} \mathrm{d}^{3N} q_i$$

$$= 1 + \int \sum_{i<j} f_{ij}\, \mathrm{d}^{3N} q_i. \tag{3.1.12}$$

There are $N(N-1)/2$ terms in the sum since we take all pairs without regard to order. And for large $N$ this may be approximated by $N^2/2$. Since the particles are identical, each integral in the sum will be the same, so that

$$Q = 1 + \frac{N^2}{2V} \int f_{12}\, \mathrm{d}^3 r_{12}. \tag{3.1.13}$$

The $V^N$ in the denominator has now become $V$ since the integration over $i, j \neq 1, 2$ gives a factor $V^{N-1}$ in the numerator.

Finally, then, we have the partition function for the interacting gas:

$$Z = Z_{id}\left\{1 + \frac{N^2}{2V}\int\left[e^{-U(\mathbf{r})/kT} - 1\right]d^3r\right\} \qquad (3.1.14)$$

and on taking the logarithm, the free energy is the sum of the non-interacting gas free energy and the new term

$$F = F_{id} - kT\ln\left\{1 + \frac{N^2}{2V}\int\left[e^{-U(\mathbf{r})/kT} - 1\right]d^3r\right\}. \qquad (3.1.15)$$

Since $U(r)$ may be assumed to be spherically symmetric, in spherical polars we can integrate over the angular coordinates as follows:

$$\int\ldots d^3r \quad\rightarrow\quad 4\pi\int_0^\infty r^2\ldots dr \qquad (3.1.16)$$

to give

$$F = F_{id} - kT\ln\left\{1 + \frac{N^2}{2V}4\pi\int_0^\infty r^2\left[e^{-U(r)/kT} - 1\right]dr\right\}. \qquad (3.1.17)$$

In this low-density approximation, the second term in the logarithm, which accounts for pairwise interactions, is much less than the first term. Otherwise, the third- and higher-order terms would also be important. And if the second term is small, then the logarithm may be expanded. Thus, we obtain

$$F = F_{id} - 2\pi kT\frac{N^2}{V}\int_0^\infty r^2\left[e^{-U(r)/kT} - 1\right]dr. \qquad (3.1.18)$$

A more rigorous treatment of the cluster expansion technique, including the systematic incorporation of the higher-order terms, is given in the article by Mullin [1].

## 3.1.4 *Equation of state*

The pressure is found by differentiating the free energy:

$$p = -\left.\frac{\partial F}{\partial V}\right|_{T,N}$$

$$= kT\frac{N}{V} - kT\frac{N^2}{V^2}2\pi\int_0^\infty r^2\left[e^{-U(r)/kT} - 1\right]dr. \qquad (3.1.19)$$

We see that the effect of the interaction $U(r)$ can be regarded as modifying the pressure from the ideal gas value. The net effect can be either attractive or repulsive, decreasing or increasing the pressure. This will be examined, for various model interaction potentials $U(r)$. However, before that we consider a systematic way of generalising the gas equation of state.

## 3.2  The Virial Expansion

### 3.2.1  *Virial coefficients*

At low densities we know that the equation of state reduces to the ideal gas equation. A systematic procedure for generalising the equation of state would therefore be as a power series in the number density $N/V$. Thus, we write

$$\frac{p}{kT} = \frac{N}{V} + B_2(T)\left(\frac{N}{V}\right)^2 + B_3(T)\left(\frac{N}{V}\right)^3 + \cdots . \tag{3.2.1}$$

The $B$ factors are called *virial coefficients*; $B_n$ is the $n$th virial coefficient. By inspecting the equation of state derived above, Eq. (3.1.19), we see that it is equivalent to an expansion up to the second virial coefficient. And by comparison the second virial coefficient is

$$B_2(T) = -2\pi \int_0^\infty r^2 \left[e^{-U(r)/kT} - 1\right] dr, \tag{3.2.2}$$

which should be "relatively" easy to evaluate once the form of the interparticle interaction $U(r)$ is known. It is also possible to calculate higher order virial coefficients, but it becomes more difficult.

From the practical perspective, measurements of the second virial coefficient $B_2(T)$ can give information about the form of the interaction potential $U(r)$. To this end we shall find $B_2(T)$ for various interaction potential models. The reader is referred to Reichl [2] for further details of some of the models in the following sections.

### 3.2.2  *Hard-core potential*

The hard-core potential is specified by

$$U(r) = \infty \quad r < \sigma$$
$$= 0 \quad r > \sigma. \tag{3.2.3}$$

This is shown in Fig. 3.2.

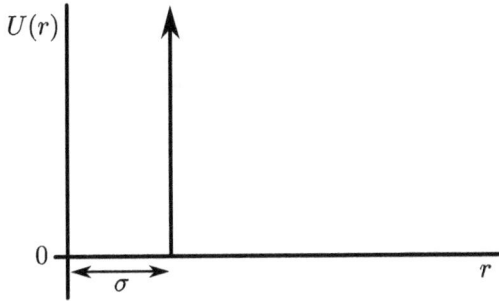

Fig. 3.2.   Hard-core potential.

Here, the single parameter $\sigma$ is the hard-core diameter, the closest distance between the centres of two particles. This is modelling the particles as impenetrable spheres. There is no interaction when the particles are separated greater than $\sigma$ and they are prevented, by the interaction, from getting any closer than $\sigma$. It should, however, be noted that this model interaction is un-physical since it only considers the repulsive part; there is no attraction at any separation.

The gas of hard sphere particles is considered in further detail in Section 3.6. For the present we are concerned solely with the second virial coefficient.

For this potential, we have

$$e^{-U(r)/kT} = 0 \quad r < \sigma$$
$$= 1 \quad r > \sigma,$$

(3.2.4)

so that the expression for $B_2(T)$ is

$$B_2(T) = 2\pi \int_0^\sigma r^2 dr$$
$$= \frac{2}{3}\pi\sigma^3.$$

(3.2.5)

In this case, we see that the second virial coefficient is independent of temperature; it is a positive constant, essentially the hard-core volume of a particle. The (low-density) equation of state is then

$$pV = NkT \left\{ 1 + \frac{2}{3}\pi\sigma^3 \frac{N}{V} \right\}$$

(3.2.6)

which indicates that the effect of the hard core is to increase the $pV$ product over the ideal gas value.

It is instructive, however, to rearrange this equation of state. We write it as

$$pV \left\{ 1 + \frac{2}{3}\pi\sigma^3 \frac{N}{V} \right\}^{-1} = NkT \qquad (3.2.7)$$

and we note that the "correction" term $\frac{2}{3}\pi\sigma^3 N/V$ is small within the validity of the derivation; it is essentially the hard-core volume of a particle divided by the total volume per particle. So, performing a binomial expansion, we find to the same leading power of density

$$pV \left\{ 1 - \frac{2}{3}\pi\sigma^3 \frac{N}{V} \right\} = NkT \qquad (3.2.8)$$

or

$$p \left\{ V - \frac{2}{3}N\pi\sigma^3 \right\} = NkT. \qquad (3.2.9)$$

In this form, we see that the effect of the hard core can be interpreted as simply reducing the available volume of the system.

### Excluded volume

By how much is the volume reduced? Two spheres of *diameter* $\sigma$ cannot approach each other closer than this distance. As indicated in Fig. 3.3, this means that the effect of one particle is to exclude a sphere of *radius* $\sigma$ from the other particle.

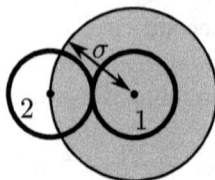

Fig. 3.3.   Volume of space excluded to particle 2 by particle 1.

Two particles exclude a volume $\frac{4}{3}\pi\sigma^3$. Thus, the "excluded volume" per particle is *one half* of this. And the excluded volume of $N$ particles is then

$$V_{\text{ex}} = \frac{2}{3}N\pi\sigma^3, \qquad (3.2.10)$$

that is, *four* times the volume of the particles. This is precisely the volume reduction of Eq. (3.2.9).

We conclude that the second virial coefficient for the hard sphere gas is simply the excluded volume per particle.

### 3.2.3 Square-well potential

The square-well potential is somewhat more realistic than the simple hard sphere interaction; it includes a region of attraction as well as the repulsive hard core. The potential is specified by

$$
\begin{aligned}
U(r) &= \infty & 0 < r < \sigma \\
&= -\varepsilon & \sigma < r < R\sigma \qquad (3.2.11) \\
&= 0 & R\sigma < r < \infty
\end{aligned}
$$

so we see it depends on three parameters: $\sigma$, $\varepsilon$, and the dimensionless $R$. This is shown in Fig. 3.4.

For this potential, we have

$$
\begin{aligned}
e^{-U(r)/kT} &= 0 & 0 < r < \sigma \\
&= e^{\varepsilon/kT} & \sigma < r < R\sigma \qquad (3.2.12) \\
&= 1 & R\sigma < r < \infty
\end{aligned}
$$

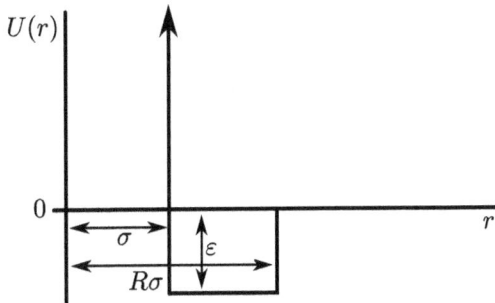

Fig. 3.4. Square-well potential.

so that the expression for $B_2(T)$ is

$$B_2(T) = -2\pi \left\{ (-1) \int_0^\sigma r^2 dr + (e^{\varepsilon/kT} - 1) \int_\sigma^{R\sigma} r^2 dr \right\}$$

$$= \frac{2}{3}\pi\sigma^3 \left\{ 1 - (R^3 - 1)(e^{\varepsilon/kT} - 1) \right\}$$

(3.2.13)

or

$$B_2(T) = \frac{2}{3}\pi\sigma^3 \left\{ R^3 - (R^3 - 1)e^{\varepsilon/kT} \right\}$$

$$= \frac{2}{3}\pi\sigma^3 R^3 - \frac{2}{3}\pi\sigma^3 (R^3 - 1)e^{\varepsilon/kT}.$$

(3.2.14)

In this case, using the more realistic potential, we see that the second virial coefficient depends on temperature, varying as

$$B_2(T) = A - Be^{\varepsilon/kT}.$$

(3.2.15)

The second virial coefficient for nitrogen is shown in Fig. 3.5. The square-well $B_2(T)$ curve of Eq. (3.2.14) has been fitted through the data with $\varepsilon/k = 88.3\,\mathrm{K}$, $\sigma = 3.27\,\text{Å}$ $(1\,\text{Å} = 10^{-10}\,\mathrm{m})$, and $R = 1.62$. Observe that this crude approximation to the inter-particle interaction gives a remarkably good fit through the experimental data. The figure also shows the hard sphere asymptote $B_2^{hs} = \frac{2}{3}\pi\sigma^3 = 44.28\,\mathrm{cm}^3/\mathrm{mol}$.

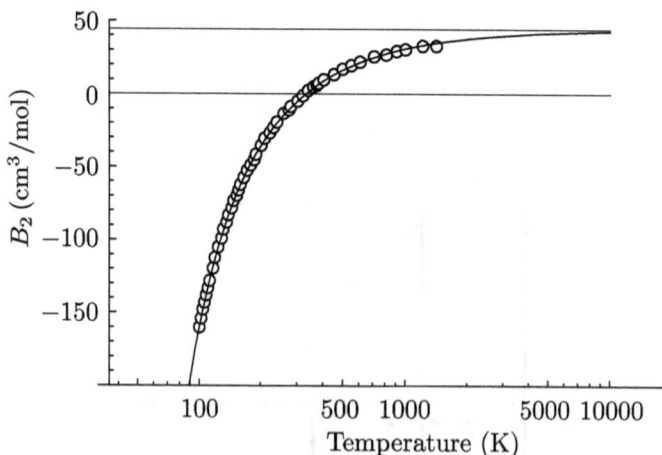

Fig. 3.5. Second virial coefficient of nitrogen as a function of temperature with the square-well functional form Eq. (3.2.14). Square-well parameters $\varepsilon/k = 88.3\,\mathrm{K}$, $\sigma = 3.27\,\text{Å}$ and $R = 1.62$.

At low temperatures, where $B_2(T)$ is negative, this indicates that the attractive part of the potential is dominant and the pressure is reduced compared with the ideal gas case. And at higher temperatures, where it is intuitive that the small attractive part of the potential will have negligible effect, $B_2(T)$ will be positive and the pressure will be increased, as in the hard sphere case. The temperature at which $B_2(T)$ goes through zero is called the *Boyle temperature*, denoted by $T_B$. For the square-well potential

$$T_B = \frac{-\varepsilon/k}{\ln\left(1 - \frac{1}{R^3}\right)}. \qquad (3.2.16)$$

At very high temperatures, we see from the expression for $B_2(T)$ that it saturates at the hard-core excluded volume.

### 3.2.4 *Lennard-Jones potential*

The Lennard-Jones potential is a very realistic representation of the inter-atomic interaction. It comprises an attractive $1/r^6$ term and a repulsive $1/r^{12}$ term. The form of the attractive part is well-justified as a description of the interaction arising from fluctuating electric dipole moments. The repulsive term is simply a power law approximation to the effect of the overlap of the atoms' external electron clouds. We write the Lennard-Jones potential as

$$U(r) = 4\varepsilon\left\{\left(\frac{\sigma}{r}\right)^{12} - \left(\frac{\sigma}{r}\right)^6\right\}; \qquad (3.2.17)$$

this depends on the two parameters: $\varepsilon$ and $\sigma$ as shown in Fig. 3.6.

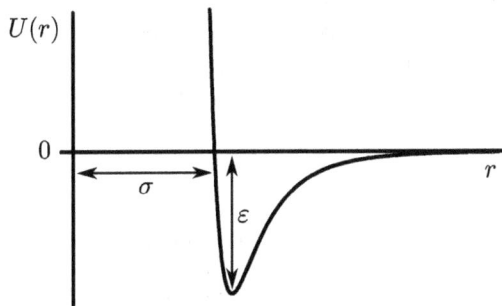

Fig. 3.6.   Lennard-Jones 6–12 potential.

The expression for $B_2(T)$ in terms of $U(r)$ is given by Eq. (3.2.2). However, when the interaction potential is a smooth function, it is more convenient to integrate this equation by parts, giving

$$B_2(T) = -\frac{2\pi}{3kT} \int_0^\infty e^{-U(r)/kT} \frac{dU(r)}{dr} r^3 dr. \qquad (3.2.18)$$

Then for the Lennard-Jones $U(r)$, upon substituting $x = r/\sigma$, we obtain

$$B_2(T) = \frac{2}{3}\pi\sigma^3 \times 24\frac{\varepsilon}{kT} \int_0^\infty e^{-4(x^{-12}-x^{-6})\varepsilon/kT}(2x^{-10} - x^{-4}) dx. \qquad (3.2.19)$$

This is instructive. The $\sigma$-dependence is all in the "hard-core" prefactor and the integral depends on temperature solely through the combination $kT/\varepsilon$.

It is possible to express the integral of Eq. (3.2.19) in terms of a Hermite function $H_\nu(x)$[1]. In this way, we obtain:

$$B_2(T) = \frac{2}{3}\pi\sigma^3 \sqrt{2\pi} \left(\frac{\varepsilon}{kT}\right)^{1/4} H_{\frac{1}{2}}\left(-\sqrt{\frac{\varepsilon}{kT}}\right). \qquad (3.2.20)$$

This is an elegant closed-form expression for the second virial coefficient of the Lennard-Jones gas.

This is plotted in Fig. 3.7 together with the data from nitrogen. The curve has been fitted with parameters $\varepsilon/k = 95.5\,\text{K}$, $\sigma = 3.76\,\text{Å}$. The figure also shows the "hard sphere" asymptote $B_2^{hs} = 67.00\,\text{cm}^3/\text{mol}$. The fit is good. We see that the Lennard-Jones form exhibits a maximum, with a reduction in $B_2$ at higher temperatures. Here, the energetic collisions can cause the atoms to come even closer together; the "hard core" is not so hard. This effect is observed in helium, shown in Fig. 3.13. Some insight into this high-temperature reduction of $B_2(T)$ is provided by Problem 3.8.

---

[1] The Hermite *polynomials* $H_n(x)$ should be familiar, from the quantum mechanics of the harmonic oscillator. For integer order $n$, $H_n(x)$ is a polynomial of power $n$ in $x$. Hermite *functions* generalise to the case of non-integer order $\nu$; such functions are no longer finite polynomials. We follow the *Mathematica* definition and terminology whereby the same symbol is used for both: HermiteH[ν, x].

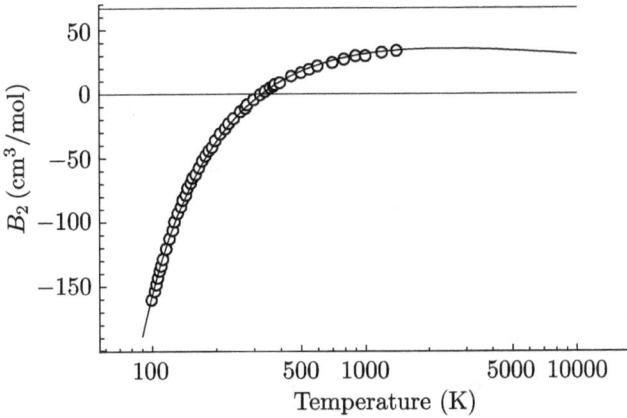

Fig. 3.7. Second virial coefficient of nitrogen plotted with the Lennard-Jones functional form, Eq. (3.2.20). Lennard-Jones parameters $\varepsilon/k = 95.5\,\text{K}$, $\sigma = 3.76\,\text{Å}$.

The maximum in $B_2(T)$ occurs at what is called the *Joule inversion temperature*, or Joule temperature, $T_J$ (different from the Joule–Kelvin inversion temperature $T_i$ of the throttling process, Section 3.3.4). The free expansion of a gas is accompanied by cooling when $T < T_J$ and by warming when $T > T_J$ [3]. We find, from Eq. (3.2.20), that the maximum value, $B_2^{max} = 0.529 \times \frac{2}{3}\pi\sigma^3$ occurs at $T_J = 25.13\,\varepsilon/k$.

From the zero of the Hermite function, we find the Boyle temperature to be $T_B = 3.418\varepsilon/k$.

At high temperatures we have, by expanding the Hermite function,

$$B_2(T) = \frac{2}{3}\pi\sigma^3 \left\{ \frac{2\pi}{\Gamma(1/4)} \left(\frac{\varepsilon}{kT}\right)^{1/4} - \frac{\pi}{\Gamma(3/4)} \left(\frac{\varepsilon}{kT}\right)^{3/4} \right.$$

$$\left. - \frac{\pi}{4\Gamma(5/4)} \left(\frac{\varepsilon}{kT}\right)^{5/4} - \frac{\pi}{8\Gamma(7/4)} \left(\frac{\varepsilon}{kT}\right)^{7/4} - \frac{5\pi}{64\Gamma(9/4)} \left(\frac{\varepsilon}{kT}\right)^{9/4} + \cdots \right. \tag{3.2.21}$$

And at low temperatures we have the expansion[2]

$$B_2(T) = -\frac{2}{3}\pi\sigma^3 \times e^{\varepsilon/kT}$$

$$\times \sqrt{\frac{\pi}{2}} \left\{ \left(\frac{kT}{\varepsilon}\right)^{1/2} + \frac{15}{16} \left(\frac{kT}{\varepsilon}\right)^{3/2} + \frac{945}{512} \left(\frac{kT}{\varepsilon}\right)^{5/2} + \cdots \right. \tag{3.2.22}$$

---

[2]This is quoted from a calculation by Gutiérrez [4].

*A comment on scaling*

The Lennard-Jones potential has two parameters: an energy $\varepsilon$ and a length $\sigma$. We note that these happen to correspond to the vertical and the horizontal axes of the plot of $U(r)$ against $r$. This means that the Lennard-Jones potential has the form of a *universal* function that just needs the appropriate scaling in the $U$ and $r$ directions. And by extension, this tells us that for any system of particles which interact with a Lennard-Jones potential, those properties that depend on the inter-particle potential, similarly, will have a universal form when the energies are scaled by $\varepsilon$, the distances by $\sigma$, and other variables in the corresponding way.

It follows that any inter-particle potential which has only two adjustable (system-specific) parameters with different dimensions, will possess this scaling property. A special case of this is the hard-sphere interaction which has only one parameter; we may regard this as having an energy parameter of zero. But we recognise immediately that the square-well potential, with *three* parameters $\varepsilon$, $\sigma$, and the dimensionless distance ratio $R$ does *not* possess the scaling property. But if $R$ were to be regarded as *fixed*, then the two parameters $\varepsilon$ and $\sigma$ would lead to universal behaviour. Indeed, for the inert gases neon, argon, krypton, and xenon, the fitted values of $R$ are close: approximately 1.65.

It is clear that two-parameter potentials, and their resultant universal system properties, are particularly convenient in statistical mechanics. This is one of the reasons for the popularity of the Lennard-Jones function where the dipolar attraction and the electron shell repulsion — two very different phenomena — are parameterised in similar ways: both having energies scaling with the same $\varepsilon$ and distances scaling with the same $\sigma$.

In this vein, the Sutherland potential of the following section and the "soft-sphere" interaction treated in Problem 3.5 both possess the scaling property.

### 3.2.5    *The Sutherland potential*

The interaction between atoms or molecules comprises a repulsive part at short distances and an attractive part at large distances. The Lennard-Jones potential of the previous section is often used as an analytical representation of the interaction. As we explained, the attractive tail is well-described by the $r^{-6}$ law, while the $r^{-12}$ description of the repulsive core is but a simple approximation to the actual short-range interaction. The popularity of the 6–12 potential lies principally in its mathematical elegance.

The Sutherland potential treats the short-distance interaction in a different way; it approximates the repulsion as a hard core. The attractive tail is described by the conventional dipolar $r^{-6}$ law.

The form of the Sutherland potential is shown in Fig. 3.8; it is specified by

$$U(r) = \infty \qquad r < \sigma$$
$$= -\varepsilon \left(\frac{\sigma}{r}\right)^6 \qquad r > \sigma. \tag{3.2.23}$$

As with the Lennard-Jones potential, the Sutherland potential has a universal form, scaled vertically with an energy parameter $\varepsilon$ and horizontally with a distance parameter $\sigma$.

The second virial coefficient is given by

$$B_2(T) = -2\pi \int_0^\infty r^2 (e^{-U(r)/kT} - 1) dr,$$

so using the mathematical form for $U(r)$, the integral splits into two parts

$$B_2(T) = 2\pi \int_0^\sigma r^2 dr - 2\pi \int_\sigma^\infty r^2 \left(e^{\frac{\varepsilon}{kT}\left(\frac{\sigma}{r}\right)^6} - 1\right) dr$$
$$= \frac{2}{3}\pi\sigma^3 - 2\pi \int_\sigma^\infty r^2 \left(e^{\frac{\varepsilon}{kT}\left(\frac{\sigma}{r}\right)^6} - 1\right) dr. \tag{3.2.24}$$

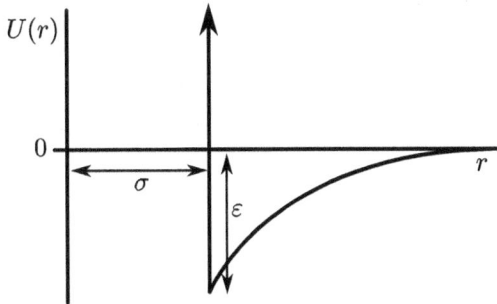

Fig. 3.8. Sutherland potential.

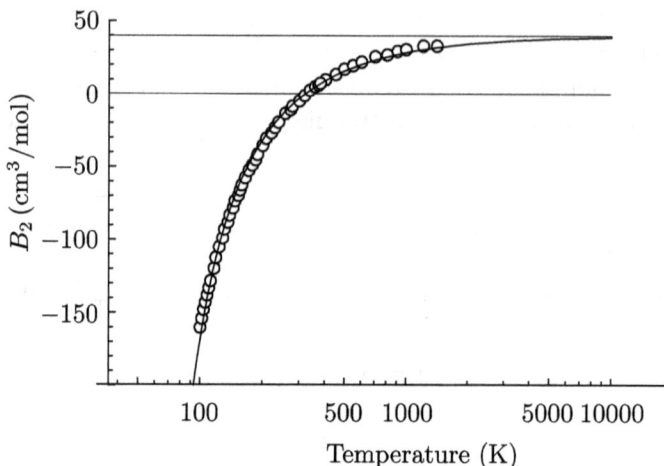

Fig. 3.9. Second virial coefficient of nitrogen plotted with the Sutherland functional form, Eq. (3.2.20). Sutherland parameters $\varepsilon/k = 274.2\,\text{K}$, $\sigma = 3.16\,\text{Å}$.

We substitute $x = r/\sigma$, so that

$$B_2(T) = \frac{2}{3}\pi\sigma^3\left\{1 - 3\int_1^{\infty} x^2(e^{\frac{\varepsilon}{kT}x^{-6}} - 1)dx\right\}. \qquad (3.2.25)$$

This may be expressed analytically, in terms of the imaginary error function erfi[3]:

$$B_2(T) = \frac{2}{3}\pi\sigma^3\left(e^{\varepsilon/kT} - \sqrt{\pi}\sqrt{\frac{\varepsilon}{kT}}\,\text{erfi}\,\sqrt{\frac{\varepsilon}{kT}}\right). \qquad (3.2.26)$$

This is plotted in Fig. 3.9 together with data from nitrogen. The curve has been fitted with parameters $\varepsilon/k = 274.2\,\text{K}$, $\sigma = 3.16\,\text{Å}$. The figure also shows the hard-sphere asymptote $B_2^{\text{hs}} = 39.81\,\text{cm}^3/\text{mol}$. Observe that this is not such a good fit to the data.

---

[3]The *standard* error function erf(z) is an area under the Gaussian distribution function: $\text{erf}(z) = \frac{2}{\sqrt{\pi}}\int_0^z e^{-t^2}\,dt$. The imaginary error function is defined as $\text{erfi}(z) = \text{erf}(iz)/i$. The *Mathematica* symbols for these are `Erf[z]` and `Erfi[z]`. We note that erfi(z) is real for real z.

The Boyle temperature for the Sutherland gas is

$$T_B = 1.171\varepsilon/k. \tag{3.2.27}$$

At high temperatures we have the series expansion

$$B_2(T) = -\frac{2}{3}\pi\sigma^3 \sum_{n=0}^{\infty} \frac{(kT/\varepsilon)^{-n}}{n!(2n-1)}$$

$$= \frac{2}{3}\pi\sigma^3 \left\{ 1 - \frac{\varepsilon}{kT} - \frac{1}{6}\left(\frac{\varepsilon}{kT}\right)^2 - \frac{1}{30}\left(\frac{\varepsilon}{kT}\right)^3 - \cdots \right\}, \tag{3.2.28}$$

while at low temperatures, we have

$$B_2(T) = -\frac{2}{3}\pi\sigma^3 e^{\varepsilon/kT} \left\{ \frac{1}{2}\frac{kT}{\varepsilon} + \frac{3}{4}\left(\frac{kT}{\varepsilon}\right)^2 + \frac{15}{8}\left(\frac{kT}{\varepsilon}\right)^3 + \cdots \right\}. \tag{3.2.29}$$

The interesting point about the Sutherland potential is that it gives the high-temperature behaviour of the $B_2(T)$ as

$$B_2(T) \sim \frac{2}{3}\pi\sigma^3 \left( 1 - \frac{\varepsilon}{kT} - \cdots \right); \tag{3.2.30}$$

the limiting value at high temperatures is the hard-core $2\pi\sigma^3/3$, while the leading deviation goes as $T^{-1}$.

(Compare with square-well potential:

$$B_2(T) \sim \frac{2}{3}\pi\sigma^3 \left( 1 - \frac{(R^3 - 1)\,\varepsilon}{kT} - \cdots \right). \tag{3.2.31}$$

Here also the limiting high-temperature value is the hard-core expression and the leading deviation goes as $T^{-1}$. Note $R$ is dimensionless, greater than unity. And $\varepsilon$ is different in the two cases, i.e.

$$\varepsilon_S = (R^3 - 1)\varepsilon_{sw}. \tag{3.2.32}$$

### 3.2.6  *Comparison of models*

For high temperatures, the square-well and the Sutherland $B_2$ both saturate at the hard-core value. By contrast, the second virial coefficient for the Lennard-Jones gas does not have such a simple high-temperature

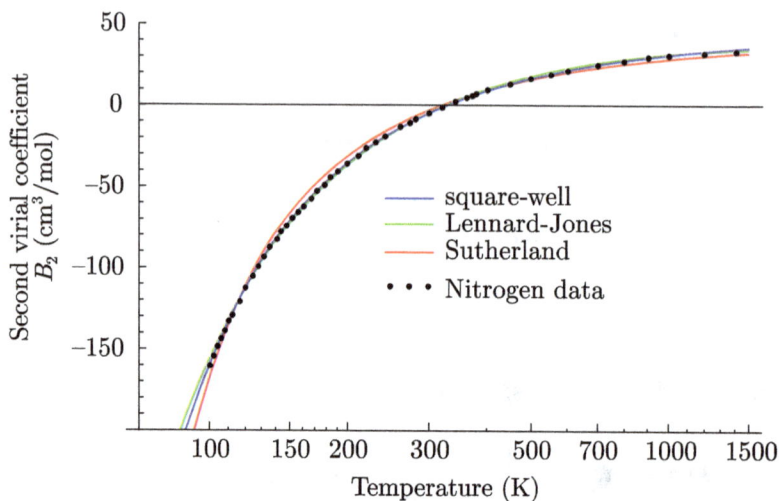

Fig. 3.10. Second virial coefficient of nitrogen compared with fits corresponding to the square-well, Lennard-Jones and Sutherland potentials.

behaviour — a consequence of the "softness" of the core. In the high temperature limit

$$B_2(T) \sim \frac{2}{3}\pi\sigma^3 \times \frac{2\pi}{\Gamma(1/4)}\left(\frac{\varepsilon}{kT}\right)^{1/4}$$

$$\sim \frac{2}{3}\pi\sigma^3 \times 1.73\left(\frac{\varepsilon}{kT}\right)^{1/4},$$

(3.2.33)

so that in this case $B_2(t) \to 0$ as $T \to \infty$; the Lennard-Jones second virial coefficient tends to zero (but very slowly) rather than the hard-core limiting value. The high-temperature $T^{-1/4}$ behaviour is considered in Problem 3.8.

We plot the nitrogen second virial coefficient again, in Fig. 3.10, now showing the best fit curves corresponding to the square-well, the Lennard-Jones, and the Sutherland potentials.

It will be observed that there is not much to choose between them. The Lennard-Jones is better than the Sutherland potential — clearly the latter is too crude. However, the square-well potential provides the best fit. This is made clearer in Fig. 3.11 where the deviations of the square-well, Lennard-Jones, and Sutherland potential fits from the experimental measurements are shown.

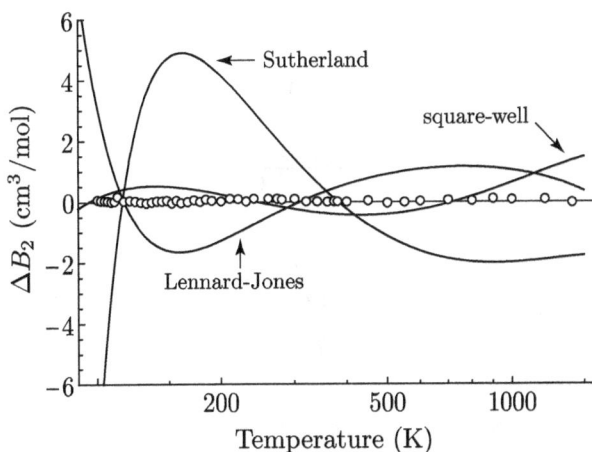

Fig. 3.11. Second virial coefficient of nitrogen — Deviations of the square-well, Lennard-Jones and Sutherland potential fits.

One should not, however, conclude that the square-well potential is the best physical model. Its better mathematical fit is simply a consequence of the fact that that model has *three* adjustable parameters, as compared with the two adjustable parameters of the Lennard-Jones and the Sutherland models.

From the practical perspective, the quality of the square-well expression for $B_2$ of Eq. (3.2.15) is such that databases of second virial coefficients often simply give the equation's three parameters $A$, $B$, and $\varepsilon$, rather than extensive tables.[4]

The inter-particle potentials corresponding to the fits of the various models through the nitrogen $B_2$ data are shown in Fig 3.12.

### 3.2.7 *Universal behaviour*

The second virial coefficient $B_2(T)$ was obtained as an integral over the interaction energy $U(r)$ in Eq. (3.2.2). Different gases will have different values for $B_2(T)$ because they will have different interaction energies.

We shall make the assumption that the interaction energy may be expressed in the following way:

$$U(r) = \varepsilon u(r/\sigma). \tag{3.2.34}$$

---

[4]See, for instance, Kaye and Laby [5], also archived on the NPL website.

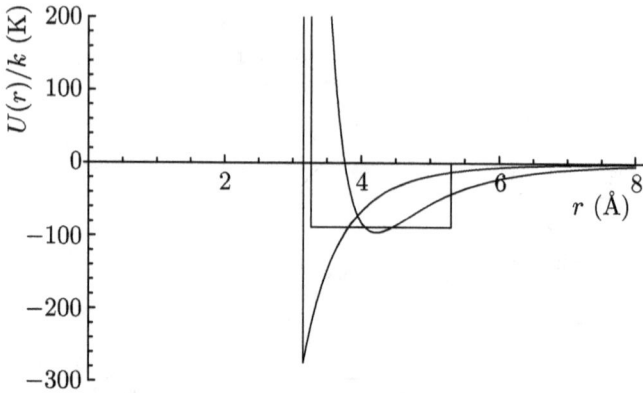

Fig. 3.12. Square-well, Lennard-Jones and Sutherland potentials corresponding to nitrogen second virial coefficient fits.

What we are saying here is that $U(r)$ has a universal shape and, for a given gas, it is scaled vertically by a specific energy $\varepsilon$ and horizontally by a specific length $\sigma$. We note that the Lennard-Jones interaction is of this form and that values of $\varepsilon$ and $\sigma$ are tabulated, see for example, Kittel [6] and our Table 4.2 that follows.

Using this expression for $U(r)$ in Eq. (3.2.2), by a change of variable we obtain

$$B_2/\sigma^3 = -2\pi \int_0^\infty x^2 \left[ e^{-u(x)/\tau} - 1 \right] \mathrm{d}x, \qquad (3.2.35)$$

where $\tau = kT/\varepsilon$ is a reduced temperature. Then to the extent that our scaling assumption for $U(r)$ is valid, plots of $B_2/\sigma^3$ against $kT/\varepsilon$ would fall on a single curve. In Fig. 3.13, we show such data for the noble gases, scaled with their Lennard-Jones $\varepsilon$ and $\sigma$. Apart from helium, the points do indeed collapse onto a single curve. This provides support for our scaling assumption; this will be formalised into the Law of Corresponding States in Chapter 4. The points in Fig. 3.13 fall very close to the curve calculated from the Lennard-Jones potential, Eq. (3.2.20).

Helium is special: because of its light mass, quantum effects are important. This is discussed in the following section.

### 3.2.8 *Quantum gases — The special case(s) of helium*

The main aim of this section is to explain why helium's second virial coefficient points, in Fig. 3.13, do not fall on the universal curve, which

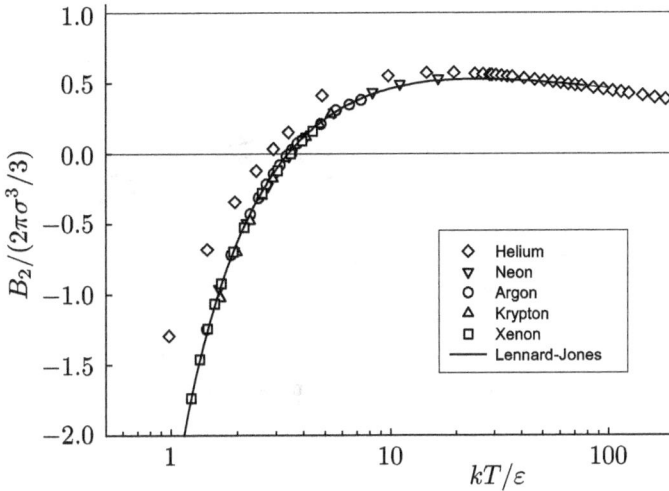

Fig. 3.13. Reduced second virial coefficient of the noble gases, together with Lennard-Jones form. The helium data are for the abundant isotope $^4$He.

follows from the scaling property of the interaction potential. But first we shall see that a quantum gas has a non-zero $B_2$ even in the absence of any interactions.

### Non-interacting quantum gas

The second virial coefficient is a measure of the deviation from the ideal gas equation of state. But at low temperatures, the equation of state for a *non-interacting* gas of quantum particles deviates from the ideal gas equation of state. We saw this in Section 2.7.5. This deviation, when expressed as a series in powers of density may be regarded as a virial expansion. This is precisely what we did in Eq. (2.7.25), which we shall write here as

$$pV = NkT \left\{ 1 \pm \frac{\pi^{3/2}}{2\alpha} \frac{\hbar^3}{(mkT)^{3/2}} \frac{N}{V} - \frac{\pi^3}{\alpha^2} \frac{16\sqrt{3} - 27}{27} \frac{\hbar^6}{(mkT)^3} \left(\frac{N}{V}\right)^2 + \cdots \right\},$$

(3.2.36)

where the + is for fermions and the − is for bosons, and $\alpha$ is the spin degeneracy factor. From this expression, we can read off directly the second and

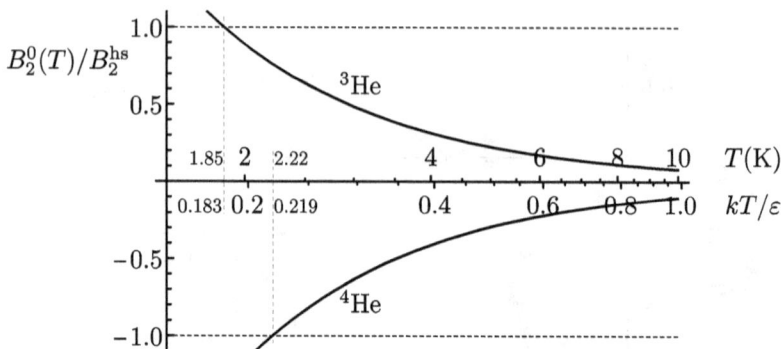

Fig. 3.14.    Second virial coefficient of a quantum ideal gas.

third virial coefficients. In particular,

$$B_2^0(T) = \pm \frac{\pi^{3/2}}{2\alpha} \frac{\hbar^3}{(mkT)^{3/2}}$$

$$= \pm \frac{1}{2^{5/2}\alpha} \Lambda^3(T),$$ 

(3.2.37)

where $\Lambda(T)$ is the thermal de Broglie wavelength, Eq. (2.3.5), and the "0" superscript indicates this is in the absence of any interaction potential. $B_2^0$ is monotonic in $T$; it is positive for fermions and negative for bosons. This is shown in Fig. 3.14, for the fermion $^3$He and the boson $^4$He. I have plotted $B_2^0$ in units of the hard-sphere $B_2^{hs} = \frac{2}{3}\pi\sigma^3$, showing both the absolute temperature scale and the reduced temperature scale $kT/\varepsilon$.

This result has an interesting interpretation. Recall the hard-sphere expression for $B_2$ was essentially the volume of a particle. Here we see that this quantum contribution to $B_2$ is essentially the thermal de Broglie volume of the particle — positive for fermions and negative for bosons.

It is often stated crudely that requirements of quantum statistics lead to "exchange forces" on classical "Maxwell" particles; the exclusion principle for fermions gives a repulsive force while bosons experience an attraction. Such a view can, however, be seriously misleading [7].

*Interacting quantum gas*

Quantum mechanics may be regarded as influencing the second virial coefficient of a gas in two different ways. Particles are delocalised over a length scale $\Lambda$, and particle "statistics" will determine the symmetry of the states that are included in the partition function sum. The Lee-Yang approach to quantum statistical mechanics [8] allows one to treat these contributions as separate and additive. We shall therefore write

$$B_2(T) = B_2^{ex}(T) + B_2^d(T), \tag{3.2.38}$$

where $B_2^{ex}(T)$ is the statistics or "exchange" term and $B_2^d(T)$ is the delocalisation or "direct" term.

Of course a full quantum treatment will incorporate both contributions, but such calculations are rather complex and tedious [2]. Instead, we shall examine the two contributions separately. This will provide an intuitive understanding of the way quantum effects influence the second virial coefficient of the helium gases. We shall denote these two different contributions as the *exchange* and the *direct* contributions.

## Exchange contribution: $B_2^{ex}$

For non-interacting fermions or bosons, there is only the exchange contribution to $B_2$, given by Eq. (3.2.37). However, in the presence of interactions the repulsive core causes a dramatic attenuation of this [9],

$$
\begin{aligned}
B_2^{ex}(T) &= B_2^0(T) e^{-\frac{\pi^3}{2}\left(\frac{\sigma}{\lambda}\right)^2} \\
&= \pm \frac{\pi^{3/2}}{2\alpha} \frac{\hbar^3}{(mkT)^{3/2}} e^{-\frac{\pi^2\sigma^2 mkT}{4\hbar^2}} \\
&= \pm \frac{\Lambda^{*3}}{2^4 \pi^{3/2}\alpha} \sigma^3 \left(\frac{\varepsilon}{kT}\right)^{3/2} e^{-\frac{\pi^4}{\Lambda^{*2}}\frac{kT}{\varepsilon}}
\end{aligned}
\tag{3.2.39}
$$

so that the overwhelming effect is from the delocalisation.

## Direct contribution: $B_2^d$

Quantum mechanics may be regarded as causing a delocalisation of the atomic positions and this may be accommodated by a renormalisation of the inter-atomic interaction. This idea was suggested by Feynman [10] and subsequently the procedure was developed by Young [11] for the case of the Lennard-Jones interaction. Essentially one averages the interaction potential over a probability density whose width is the thermal de Broglie wavelength $\Lambda(T)$. The result is a Lennard-Jones interaction with renormalised $\varepsilon$ and $\sigma$, which depend on temperature. We shall simply quote the result, and refer the interested reader to the original references for the details.

The renormalisation of the interaction is best expressed in terms of the dimensionless variable $\tau$, defined as

$$\tau = \frac{3}{2}\frac{\sigma^2}{\Lambda(T)^2};$$ (3.2.40)

it depends on the ratio of the particle diameter to the thermal de Broglie wavelength. Now, $\tau$ may be written as

$$\tau = \frac{3\sigma^2 mk}{\hbar^2}T$$ (3.2.41)

so we see that, in this form, $\tau$ may be regarded as a reduced temperature.

Then the renormalisation of the Lennard-Jones $\varepsilon$ and $\sigma$ is given, in terms of this reduced temperature, as follows:

$$\varepsilon \to \bar{\varepsilon}(\tau) = \mathcal{E}(\tau)\varepsilon$$
$$\sigma \to \bar{\sigma}(\tau) = \mathcal{S}(\tau)\sigma$$ (3.2.42)

where

$$\mathcal{E}(\tau) = \left[1 + 19.1\tau^{-1} + f(\tau)\tau^{-2}\right]^{-3/4}$$
$$\mathcal{S}(\tau) = \left[1 + g(\tau)\tau^{-1}\right]^{1/2},$$ (3.2.43)

and

$$f(\tau) = 5 + (177.7 - 5)\left[1 - e^{-\tau/250}\right]$$
$$g(\tau) = 4 + (7.54 - 4)\left[1 - e^{-\tau/250}\right].$$ (3.2.44)

These quantum renormalisation factors are shown as a function of reduced temperature $\tau$ in Fig. 3.15.

### Consequences for helium

There are two isotopes of helium: the abundant and heavier $^4$He and the rarer and lighter $^3$He. We note that $^4$He is a boson and $^3$He is a fermion. I have plotted $B_2$ data for the helium isotopes in Fig. 3.16 together with the Lennard-Jones curve, as in Fig. 3.13.

The Lennard-Jones parameters for the two isotopes are the same. But since they have different masses, the renormalisation factors will be different. By applying these renormalisations to the $\varepsilon$ and $\sigma$ in Eq. (3.2.20), we

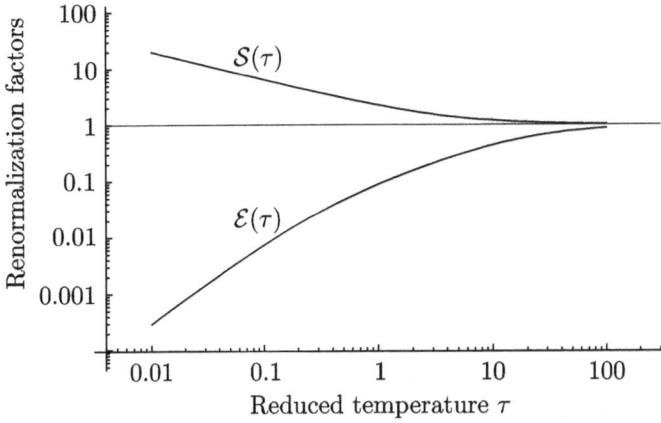

Fig. 3.15. Quantum renormalisation factors for the Lennard-Jones $\sigma$ and $\varepsilon$ parameters.

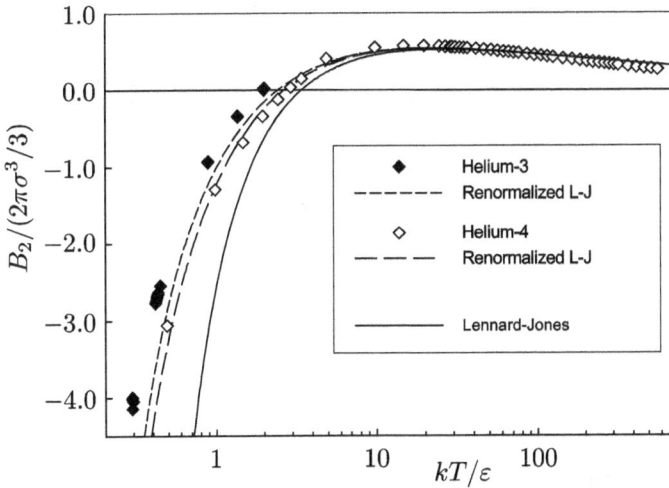

Fig. 3.16. Second virial coefficient of $^3$He and $^4$He.

obtain the two dashed curves in Fig. 3.16. The bare Lennard-Jones curve has been shifted *almost* through the data points.

From this we conclude that delocalisation effect almost accounts for the second virial coefficients of $^3$He and $^4$He. The difference between $B_2$ for the two isotopes is predominantly because of their different masses and not because of their different statistics.

## 3.3 Thermodynamics

### 3.3.1 *Throttling*

In a throttling process, a gas is forced through a flow impedance such as a porous plug. For a continuous process, in the steady state, the pressure will be constant (but different) either side of the impedance. When this happens to a thermally isolated system so that heat neither enters nor leaves the system, then it is referred to as a Joule–Kelvin or Joule–Thomson process. This is fundamentally an irreversible process, but the arguments of thermodynamics are applied to such a system simply by considering the equilibrium initial state and the equilibrium final state which are applied way before and way after the actual process. This throttling process may be modelled by the diagram in Fig. 3.17.

Work must be done to force the gas through the plug. The work done is

$$\Delta W = -\int_{V_1}^{0} p_1 dV - \int_{0}^{V_2} p_2 dV = p_1 V_1 - p_2 V_2. \tag{3.3.1}$$

Since the system is thermally isolated, the change in the internal energy is due entirely to the work done:

$$E_2 - E_1 = p_1 V_1 - p_2 V_2 \tag{3.3.2}$$

or

$$E_1 + p_1 V_1 = E_2 + p_2 V_2. \tag{3.3.3}$$

The *enthalpy H* is defined by

$$H = E + pV \tag{3.3.4}$$

thus, we conclude that in a Joule–Kelvin process the enthalpy is conserved.

The interest in the throttling process is that whereas for an ideal gas the temperature remains constant, it is possible to have either cooling or warming when the process happens to a non-ideal gas. The operation of most domestic refrigerators is based on this.

$$p_1 \, V_1 \, T_1 \qquad\qquad p_2 \, V_2 \, T_2$$

before                           after

Fig. 3.17.   Joule–Kelvin throttling process.

### 3.3.2 *Joule–Kelvin coefficient*

The fundamental differential relation for the enthalpy is

$$dH = TdS + Vdp. \tag{3.3.5}$$

It is, however, rather more convenient to use $T$ and $p$ as the independent variables rather than the natural $S$ and $p$. This is effected by expressing the entropy as a function of $T$ and $p$, whereupon its differential may be expressed as

$$dS = \frac{\partial S}{\partial T}\bigg|_p dT + \frac{\partial S}{\partial p}\bigg|_T dp. \tag{3.3.6}$$

But

$$\frac{\partial S}{\partial T}\bigg|_p = \frac{c_p}{T} \tag{3.3.7}$$

and using a Maxwell relation we have

$$\frac{\partial S}{\partial p}\bigg|_T = -\frac{\partial V}{\partial T}\bigg|_p \tag{3.3.8}$$

so that

$$dH = c_p dT + \left\{ V - T \frac{\partial V}{\partial T}\bigg|_p \right\} dp. \tag{3.3.9}$$

Now, since $H$ is conserved in the throttling process $dH = 0$ so that

$$dT = \frac{1}{c_p} \left\{ T \frac{\partial V}{\partial T}\bigg|_p - V \right\} dp, \tag{3.3.10}$$

this tells us how the temperature change is determined by the pressure change. The *Joule–Thomson* coefficient $\mu_J$ is defined as the derivative

$$\mu_J = \frac{\partial T}{\partial p}\bigg|_H, \tag{3.3.11}$$

giving

$$\mu_J = \frac{1}{c_p} \left\{ T \frac{\partial V}{\partial T}\bigg|_p - V \right\}. \tag{3.3.12}$$

This is zero for the ideal gas (Problem 3.1). When $\mu_J$ is positive, then the temperature decreases in a throttling process when a gas is forced through a porous plug.

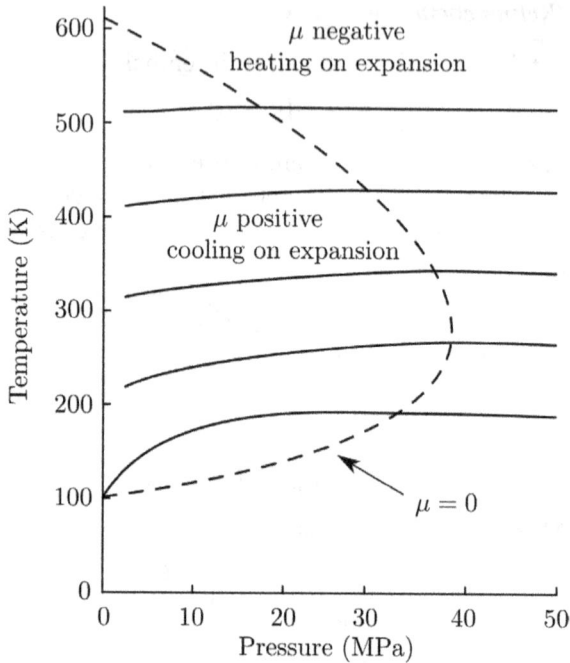

Fig. 3.18. Isenthalps and inversion curve for nitrogen (after Zemansky [12]).

### 3.3.3 *Connection with the second virial coefficient*

We consider the case where the second virial coefficient gives a good approximation to the equation of state. Then we are assuming that the density is low enough so that the third and higher coefficients can be ignored. This means that the second virial coefficient correction to the ideal gas equation is small and then solving for $V$ in the limit of small $B_2(T)$ gives

$$V = \frac{NkT}{p} + NB_2(T), \qquad (3.3.13)$$

so that the Joule–Thomson coefficient is then

$$\mu_J = \frac{NT}{c_p} \left\{ \frac{dB_2(T)}{dT} - \frac{B_2(T)}{T} \right\}. \qquad (3.3.14)$$

Within the low-density approximation it is appropriate to use the ideal gas heat capacity

$$c_p = \frac{5}{2}Nk \qquad (3.3.15)$$

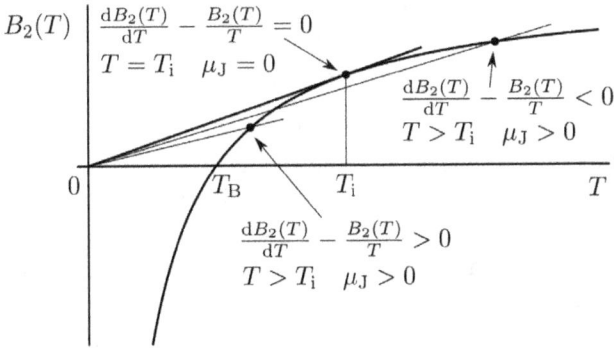

Fig. 3.19.   Behaviour of the Joule–Thomson coefficient.

so that

$$\mu_J = \frac{2T}{5k} \left\{ \frac{dB_2(T)}{dT} - \frac{B_2(T)}{T} \right\}. \tag{3.3.16}$$

### 3.3.4   *Inversion temperature*

The behaviour of the Joule–Thomson coefficient is related to the relative behaviour of $B_2(T)/T$ and $dB_2(T)/dT$. This can be visualised in the following construction, Fig 3.19. We plot the $B_2(T)$ curve together with the tangent line passing through the origin.

The point where the tangent kisses the curve corresponds to

$$\frac{dB_2(T)}{dT} = \frac{B_2(T)}{T}. \tag{3.3.17}$$

Here $\mu_J = 0$; the temperature at which this happens is called the *inversion temperature* $T_i$. (This is the *Joule–Kelvin* inversion temperature, to distinguish from the Joule inversion temperature $T_J$ mentioned in Section 3.2.4.) At lower temperatures, $T < T_i$, the slope of the curve $dB_2/dT$ is greater than $B_2/T$ so that $\mu_J$ is positive, while at high temperatures, $T > T_i$, the slope of the curve $dB_2/dT$ is less than $B_2/T$ so that $\mu_J$ is negative.

The inversion curve for nitrogen is shown as the dashed line in Fig. 3.18. We see that at high temperatures $\mu_J$ is negative, as expected. As the temperature is decreased, the inversion curve is crossed and $\mu_J$ becomes positive. Note, however, that the low-density approximation, implicit in going only to the *second* virial coefficient, keeps us away from the lower temperature region where the gas is close to condensing, where the Joule–Thomson coefficient changes sign again.

## 3.4   Van der Waals Equation of State

### 3.4.1   *Approximating the partition function*

Rather than perform an exact calculation as a series in powers of an expansion parameter — the density or the cluster function $f_{ij}$ — in this section we shall adopt a different approach by making an approximation to the partition function, which should be reasonably valid at *all* densities. Furthermore the approximation we shall develop will be based on the single-particle partition function. We shall, in this way, obtain an equation of state that approximates the behaviour of real gases. This equation was originally proposed by van der Waals in his Ph.D. Thesis in 1873. An English translation is available [13] and it is highly readable; van der Waals's brilliance shines out. There is an encyclopaedic, yet accessible, account of the properties of the van der Waals fluid by Johnston [14].

In the absence of an interaction potential the single-particle partition function is

$$z = \frac{V}{\Lambda^3}. \tag{3.4.1}$$

Recall that the factor $V$ here arises from integration over the position coordinates. The question now is how to account for the inter-particle interactions — in an approximate way. Now, the interaction $U(r)$ comprises a strong repulsive hard core at short separations and a weak attractive long tail at large separations. And the key is to treat these two parts of the interaction in separate ways.

- The repulsive core effectively excludes regions of space from the integration over position coordinates. This may be accounted for by replacing $V$ by $V - V_{\text{ex}}$ where $V_{\text{ex}}$ is the volume excluded by the repulsive core.
- The attractive long tail is accounted for by including a factor in the expression for $z$ of the form

$$e^{-\langle E \rangle / kT}, \tag{3.4.2}$$

where $\langle E \rangle$ is an average of the attractive part of the potential.

Thus, we arrive at the approximation

$$z = \frac{V - V_{\text{ex}}}{\Lambda^3} e^{-\langle E \rangle / kT}. \tag{3.4.3}$$

Note that we have approximated the interaction by a *mean field* assumed to apply to individual particles. This allows us to keep the simplifying feature of the free-particle calculation where the many-particle partition function factorises into a product of single-particle partition functions. This is accordingly referred to as a mean field calculation.

### 3.4.2 Van der Waals equation

The equation of state is found by differentiating the free energy expression:

$$p = kT \left.\frac{\partial \ln Z}{\partial V}\right|_{T,N} = NkT \left.\frac{\partial \ln z}{\partial V}\right|_{T}. \tag{3.4.4}$$

Now, the logarithm of $z$ is

$$\ln z = \ln(V - V_{\text{ex}}) - 3 \ln \Lambda - \langle E \rangle / kT, \tag{3.4.5}$$

so that

$$p = NkT \left.\frac{\partial \ln z}{\partial V}\right|_{T} = \frac{NkT}{V - V_{\text{ex}}} - N \frac{d \langle E \rangle}{dV} \tag{3.4.6}$$

since we allow the average interaction energy to depend on volume (density). This equation may be rearranged as

$$p + N \frac{d \langle E \rangle}{dV} = \frac{NkT}{V - V_{\text{ex}}} \tag{3.4.7}$$

or

$$\left(p + N \frac{d \langle E \rangle}{dV}\right)(V - V_{\text{ex}}) = NkT. \tag{3.4.8}$$

This is similar to the ideal gas equation, except that the pressure is increased and the volume decreased from the ideal gas values. These are constant parameters. They account, respectively, for the attractive long tail and the repulsive hard core in the interaction. Conventionally, we express the parameters as $aN^2/V^2$ and $Nb$, so that the equation of state is

$$\left(p + a \frac{N^2}{V^2}\right)(V - Nb) = NkT, \tag{3.4.9}$$

and this is known as the van der Waals equation.

Some isotherms of the van der Waals equation are plotted in Fig. 3.20 for three temperatures $T_1 > T_2 > T_3$. On the right-hand side of the plot,

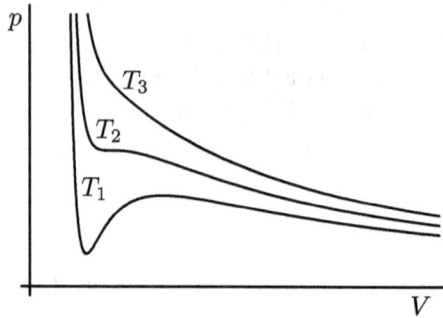

Fig. 3.20.    Van der Waals isotherms.

corresponding to low density, we have gaseous behaviour; here, the van der Waals equation gives small deviations from the ideal gas behaviour. On the left-hand side, particularly at lower temperatures, the steep slope indicates incompressibility. This is indicative of liquid behaviour. The non-monotonic behaviour at low temperature is peculiar and indeed it is non-physical. This will be discussed in detail in Chapter 4.

The van der Waals equation gives a good description of the behaviour of both gases and liquids. In introducing this equation of state we said that the method should treat both low-density and high-density behaviour, and this it has done admirably. For this reason, Landau and Lifshitz [15] refer to the van der Waals equation as an interpolation equation. The great power of the equation, however, is that it also gives a good *qualitative* description of the gas–liquid phase transition, to be discussed in Chapter 4.

### 3.4.3    *Estimation of van der Waals parameters*

In the van der Waals approach, the repulsive and the attractive parts of the inter-particle interaction were treated separately. Within this spirit, let us consider how the two parameters of the van der Waals equation might be related to the two parameters of the Lennard-Jones inter-particle interaction potential. The repulsion is strong; particles are correlated when they are very close together. We accounted for this by saying that there is zero probability of two particles being closer together than $\sigma$. Then, as in the hard-core discussion of Section 3.2.2, the region of co-ordinate space is excluded, and the form of the potential in the excluded region ($U(r)$ very large) does not enter the discussion. Thus, just as in the discussion of the hard-core model, the excluded volume will be

$$V_{ex} = \frac{2}{3} N \pi \sigma^3. \tag{3.4.10}$$

The attractive part of the potential is weak. Here there is very little correlation between the positions of the particles; we therefore treat their distribution as approximately uniform. The mean interaction for a single pair of particles $\langle E_p \rangle$ is then

$$\begin{aligned}
\langle E_p \rangle &= \frac{1}{V} \int_\sigma^\infty 4\pi r^2 U(r) dr \\
&= \frac{1}{V} \int_\sigma^\infty 4\pi r^2 4\varepsilon \left\{ \left(\frac{\sigma}{r}\right)^{12} - \left(\frac{\sigma}{r}\right)^6 \right\} dr \\
&= -\frac{32\pi\sigma^3}{9V} \varepsilon.
\end{aligned} \tag{3.4.11}$$

Now, there are $N(N-1)/2$ pairs, each interacting through $U(r)$, so neglecting the 1, the total energy of interaction is $N^2 \langle E_p \rangle / 2$. This is shared among the $N$ particles, so the mean energy per particle is

$$\begin{aligned}
\langle E \rangle &= \langle E_p \rangle N/2 \\
&= -\frac{16\pi\sigma^3}{9} \frac{N}{V} \varepsilon.
\end{aligned} \tag{3.4.12}$$

In the van der Waals equation, it is the derivative of this quantity we require. Thus, we find

$$N \frac{d\langle E \rangle}{dV} = \frac{16}{9} \pi\sigma^3 \left(\frac{N}{V}\right)^2 \varepsilon. \tag{3.4.13}$$

These results give the correct assumed $N$ and $V$ dependence of the parameters used in the previous section. So, finally, we identify the van der Waals parameters $a$ and $b$ as

$$\begin{aligned}
a &= \frac{16}{9} \pi\sigma^3 \varepsilon, \\
b &= \frac{2}{3} \pi\sigma^3.
\end{aligned} \tag{3.4.14}$$

### 3.4.4   *Virial expansion*

It is a straightforward matter to expand the van der Waals equation as a virial series. We express $p/kT$ as

$$\frac{p}{kT} = \frac{N}{V - Nb} - \frac{aN^2}{kTV^2}$$

$$= \left(\frac{N}{V}\right)\left(1 - b\frac{N}{V}\right)^{-1} - \frac{a}{kT}\left(\frac{N}{V}\right)^2$$

(3.4.15)

and this may be expanded in powers of $N/V$ to give

$$\frac{p}{kT} = \left(\frac{N}{V}\right) + \left(\frac{N}{V}\right)^2\left(b - \frac{a}{kT}\right) + \left(\frac{N}{V}\right)^3 b^2 + \left(\frac{N}{V}\right)^4 b^3 + \cdots.$$

(3.4.16)

Thus, we immediately identify the second virial coefficient as

$$B_2^{VW}(T) = b - \frac{a}{kT}.$$

(3.4.17)

This has the form as sketched for the square-well potential. For this model we can find the Boyle temperature and the inversion temperature:

$$T_B = \frac{a}{bk},$$

$$T_i = \frac{2a}{bk}.$$

(3.4.18)

So we conclude that for the van der Waals gas the inversion temperature is double the Boyle temperature.

Incidentally, we observe that the third and all higher virial coefficients, within the van der Waals model, are constants independent of temperature.

## 3.5   Other Phenomenological Equations of State

### 3.5.1   *The Dieterici equation*

The Dieterici equation of state is one of a number of phenomenological equations crafted to give reasonable agreement with the behaviour of real gases. The interest in the Dieterici is twofold.

1. The equation gives a better description of the behaviour of fluids in the vicinity of the critical point than does the van der Waals equation. This will be discussed in Chapter 4, in Section 4.2.

2. It is claimed the equation is consistent with the Third Law of Thermodynamics [16].

The Dieterici equation may be written as

$$p(V - Nb) = NkTe^{-\frac{Na}{kTV}}. \tag{3.5.1}$$

As with the van der Waals equation, this equation has two parameters, $a$
and $b$, that parametrise the deviation from ideal gas behaviour.

For the present, we briefly examine the virial expansion of the Dieterici
equation. In other words, we will look at the way this equation treats the
initial deviations from the ideal gas.

*Virial expansion*

In order to obtain the virial expansion, we express the Dieterici equation as

$$\frac{p}{kT} = \frac{N}{V - Nb}e^{-\frac{Na}{kTV}}. \tag{3.5.2}$$

And from this we may expand to give the series in $N/V$

$$\frac{p}{kT} = \frac{N}{V} + \left(\frac{N}{V}\right)^2 \left(b - \frac{a}{kT}\right) + \left(\frac{N}{V}\right)^3 \left(b^2 - \frac{a^2}{2k^2T^2} - \frac{ab}{kT}\right) + \cdots . \tag{3.5.3}$$

This gives the second virial coefficient to be

$$B_2^D = b - \frac{a}{kT}. \tag{3.5.4}$$

This is the same as that for the van der Waals gas, and the parameters
$a$ and $b$ may thus be identified with those of the van der Waals model.
As a consequence, we conclude that both the van der Waals gas and the
Dieterici gas have the same values for the Boyle temperature and the inversion temperature.

The third virial coefficient is given by

$$B_3^D(T) = b^2 - \frac{a^2}{2k^2T^2} - \frac{ab}{kT}; \tag{3.5.5}$$

we see that this depends on temperature, unlike that for the van der Waals
equation, which is temperature-independent.

### 3.5.2 *The Berthelot equation*

As with the Dieterici equation, the Berthelot equation is another of phenomenological origin. The equation is given by

$$\left(p + \frac{\alpha N^2}{kTV^2}\right)(V - Nb) = NkT. \tag{3.5.6}$$

The parameters of the Berthelot equation are given by $\alpha$ and $b$. We observe this equation is very similar to the van der Waals equation; there is a slight difference in the pressure-correction term that accounts for the long distance attraction of the intermolecular potential.

Since the Berthelot and van der Waals equations are related by $a = \alpha/kT$, it follows that the Berthelot second virial coefficient is given by

$$B_2^B = b - \frac{\alpha}{(kT)^2}. \tag{3.5.7}$$

### 3.5.3 *The Redlich–Kwong equation*

Most improved phenomenological equations of state involve the introduction of additional parameters. The Redlich–Kwong equation is as good as many multi-parameter equations, but, as with the previous equations we have considered, it has only two parameters. We shall write the Redlich–Kwong equation as

$$p = \frac{NkT}{V - Nb} - \frac{aN^2}{\sqrt{kT}\,V(V + Nb)}. \tag{3.5.8}$$

This should be compared to the similar expression for the van der Waals equation, Eq. (3.4.15).

The virial expansion is

$$\frac{p}{kT} = \frac{N}{V} + \left(\frac{N}{V}\right)^2 \left(b - \frac{a}{(kT)^{3/2}}\right) + \left(\frac{N}{V}\right)^3 \left(b^2 + \frac{ab}{(kT)^{3/2}}\right) + \cdots, \tag{3.5.9}$$

so, in particular, we identify the Redlich–Kwong second virial coefficient to be

$$B_2^{RK} = b - \frac{a}{(kT)^{3/2}}. \tag{3.5.10}$$

## 3.6 Hard-Sphere Gas

The interactions between the atoms or molecules of a real gas comprise a strong repulsion at short distances and a weak attraction at long distances. Both of these are important in determining how the properties of the gas differ from those of an ideal (non-interacting) gas. We have already considered various approximations to the inter-particle interaction when we looked at initial deviations from ideal behaviour in the calculations of the virial expansions in Section 3.2. We considered a sequence of model approximations from the simplest: the hard-core potential, to the most realistic: the Lennard-Jones 6–12 potential.

In this section, we shall return for a deeper study of the hard-core potential. In justification we can do no better than to quote from Chaikin and Lubensky [17] p. 40: "Although this seems like an immense trivialisation of the problem, there is a good deal of unusual and unexpected physics to be found in hard-sphere models".

We recall that the hard-core potential, Eq. (3.2.3), is

$$U(r) = \infty \quad r < \sigma$$
$$= 0 \quad r > \sigma$$

where $\sigma$ is the hard-core diameter. This is indeed a simplification of a real inter-particle interaction — but what behaviour does it predict? What properties of real systems can be understood in terms of the short-distance repulsion? And, indeed, what properties cannot be understood from this simplification?

### 3.6.1 *Possible approaches*

The direct way of solving the problem of the hard-sphere fluid would be to evaluate the partition function; everything follows from that. Even for an interaction as simple as this, it turns out that the partition function cannot be evaluated analytically except in one dimension; this is the so-called Tonks "hard stick" model, which leads to the (one-dimensional) Clausius equation of state

$$p(L - L_{ex}) = NkT. \tag{3.6.1}$$

Certainly, in two and three dimensions no explicit solution is possible. Arguments about why the partition function (really the configuration integral) is so difficult to evaluate are given in Reif [18]. The point is that the excluded volumes appear in nested integrals and these are impossible to untangle,[5] except in one dimension. Accepting that no analytic solution is possible in three dimensions, there is a number of approaches that might be considered.

- Approximating the partition function in some plausible way. This should indicate the general behaviour to be expected.
- Virial expansion — this represents first-principles theoretical calculation.
- Molecular dynamics — this may be viewed as accurate "measurements made by computer".

One might adopt an approximation procedure similar to that used for the van der Waals case. But here there is no attractive mean energy, we have only the excluded volume term. This results in the Clausius equation of state: the ideal gas equation, but with an excluded volume term.

$$p(V - V_{ex}) = NkT. \qquad (3.6.2)$$

This follows by analogy with our treatment of the van der Waals gas, where now there is no attractive term in the interaction. See also Problem 3.18.

There are extensive molecular dynamics simulations, see in particular Bannerman *et al.* [19] and it is even possible to do your own; the applications of Gould and Tobochnik [21] are very instructive for this.

We shall look at virial expansions and see how far they may be "pushed". In other words, our interest is in what *analytical* conclusions may be drawn. We can then compare these conclusions with the molecular dynamics "experimental data".

### 3.6.2 *Hard-sphere equation of state*

The equation of state of a hard-sphere fluid has a very special form. Recall that the Helmholtz free energy $F$ is given in terms of the partition function

---

[5]In this respect, the Carnahan–Starling procedure, to be discussed in the following (Section 3.6.5) may be regarded as a way of attacking the nesting problem.

$Z$ by

$$F = -kT \ln Z. \tag{3.6.3}$$

We saw that the partition function for an interacting gas may be written as

$$Z = Z_{id}Q, \tag{3.6.4}$$

where $Z_{id}$ is the partition function for an ideal (non-interacting) gas

$$Z = \frac{1}{N!} \left( \frac{V}{\Lambda^3} \right)^N \tag{3.6.5}$$

and $Q$ is the configuration integral

$$Q = \frac{1}{V^N} \int e^{-\left( \sum_{i<j} U(q_i, q_j) \right)/kT} d^{3N}q. \tag{3.6.6}$$

To obtain the equation of state we must find the pressure, by differentiating the free energy

$$\begin{aligned}
p &= -\left. \frac{\partial F}{\partial V} \right|_{T,N} \\
&= kT \left. \frac{\partial \ln Z}{\partial V} \right|_{T,N} \\
&= kT \left( \left. \frac{\partial \ln Z_{id}}{\partial V} \right|_{T,N} + \left. \frac{\partial \ln Q}{\partial V} \right|_{T,N} \right).
\end{aligned} \tag{3.6.7}$$

It is important, now, to appreciate that the configuration integral is independent of temperature. This must be so, since there is no energy scale for the problem; the interaction energy is either zero or it is infinite. Thus, the ratio $E/kT$ will be temperature-independent. The pressure of the hard-sphere gas must then take the form

$$p = kT \left( \frac{N}{V} + g(N/V) \right). \tag{3.6.8}$$

The function $g(N/V)$ is found by differentiating $\ln Q$ with respect to $V$. We know it is a function of $N$ and $V$ and in the thermodynamic limit the argument must be intensive. Thus, the functional form, and we have the low-density ideal gas limiting value $g(0) = 0$.

The important conclusion we draw from these arguments, and in particular from Eq. (3.6.8), is that for a hard-sphere gas the combination $p/kT$

is a function of the density $N/V$. This function must depend also on the only parameter of the interaction: the hard-core diameter $\sigma$.

### 3.6.3  *Virial expansion*

The virial expansion, Eq. (3.2.1), is written as

$$\frac{p}{kT} = \frac{N}{V} + B_2\left(\frac{N}{V}\right)^2 + B_3\left(\frac{N}{V}\right)^3 + \cdots, \qquad (3.6.9)$$

where the virial coefficients $B_m$ are, in the general case, functions of temperature. However, as argued above, for the hard-sphere gas the virial coefficients are temperature-independent.

The virial expansion may be regarded as a low-density approximation to the equation of state. Certainly this is the case when only a finite number of coefficients is available. If, however, *all* the coefficients were known, then provided the series were convergent, the sum would give $p/kT$ for all values of the density $N/V$ up to the radius of convergence of the series: the complete equation of state. Now, although we are likely to know the values for but a finite number of the virial coefficients, there may be ways of guessing/inferring/estimating the higher-order coefficients. We shall examine two ways of doing this.

### 3.6.4  *Virial coefficients*

The second virial coefficient for the hard-sphere gas was calculated in Section 3.2.2; we found

$$B_2 = \frac{2}{3}\pi\sigma^3, \qquad (3.6.10)$$

independent of temperature, as expected.

The general term of the virial expansion is $\left(\frac{N}{V}\right)^m B_m$, which must have the dimensions of $N/V$. Thus, $B_m$ will have the dimensions of $V^{m-1}$. Now, the only variable that the hard-sphere virial coefficients depend on is $\sigma$. Thus, it is clear that

$$B_m = \text{const} \times \sigma^{3(m-1)} \qquad (3.6.11)$$

where the constants are dimensionless numbers — which must be determined.

It is increasingly difficult to calculate the higher-order virial coefficients; the second and third were calculated by Boltzmann in 1899; those

Table 3.1. Virial coefficients for the hard-sphere gas. $B_2$ and $B_3$ calculated by Boltzmann, $B_4$ to $B_6$ by Ree and Hoover, $B_7$ to $B_{10}$ by Clisby and McCoy.

| | | |
|---|---|---|
| $B_2/b$ | = | 1 |
| $B_3/b^2$ | = | 0.625 |
| $B_4/b^3$ | = | 0.28694950 |
| $B_5/b^4$ | = | 0.11025210 |
| $B_6/b^5$ | = | 0.03888198 |
| $B_7/b^6$ | = | 0.01302354 |
| $B_8/b^7$ | = | 0.00418320 |
| $B_9/b^8$ | = | 0.00130940 |
| $B_{10}/b^9$ | = | 0.00040350 |
| $B_{11}/b^{10}$ | = | 0.00012300 |
| $B_{12}/b^{11}$ | = | 0.00003700 |

up to sixth order were evaluated by Ree and Hoover [22] in 1964, and terms up to tenth order were found by Clisby and McCoy [23] in 2006. More recently the eleventh and twelfth were obtained by Wheatley in 2013 [24]. These all are listed in the above Table 3.1, in terms of the single parameter $b$:

$$b = B_2 = \frac{2}{3}\pi\sigma^3. \tag{3.6.12}$$

Note/recall that the hard-sphere virial coefficients are independent of temperature (Problem 3.17) and they are all expressed in terms of the hard-core dimension.

We now consider ways of guessing/inferring/estimating the higher-order coefficients, so that the hard-sphere equation of state may be approximated.

### 3.6.5 *Carnahan and Starling procedure*

We start with the remarkable procedure of Carnahan and Starling [25]. They inferred a general (approximate) expression for the $n$th virial coefficient, enabling them to sum the virial expansion and thus obtain an equation of state in *closed form*. The virial expansion is written as

$$\frac{pV}{NkT} = 1 + B_2\left(\frac{N}{V}\right) + B_3\left(\frac{N}{V}\right)^2 + \cdots.$$

(3.6.13)

In 1969, only the first six virial coefficients, from Ree and Hoover, were known. Carnahan and Starling specified the density as the fraction of the volume occupied by the spheres. The volume of a sphere of diameter $\sigma$ is $\frac{1}{6}\pi\sigma^3$ or $b/4$. So, the packing fraction $y$ is given by $y = Nb/4V$ in terms of which Carnahan and Starling wrote the virial expansion as[6]

$$\frac{pV}{NkT} = 1 + 4y + 10y^2 + 18.36y^3 + 28.22y^4 + 39.82y^5 + \cdots.$$

(3.6.14)

It is convenient to introduce "reduced" virial coefficients $\beta_n$ such that Carnahan and Starling series is

$$\frac{pV}{NkT} = 1 + \beta_2 y + \beta_3 y^2 + \cdots + \beta_n y^{n-1} + \cdots.$$

(3.6.15)

Here,

$$\beta_n = 4^{n-1} B_n / b^{n-1}$$

(3.6.16)

and we tabulate the $\beta_n$ in Table 3.2:

Table 3.2.   Reduced virial coefficients for the hard-sphere gas.

| | | |
|---|---|---|
| $\beta_2$ | $=$ | 4 |
| $\beta_3$ | $=$ | 10 |
| $\beta_4$ | $=$ | 18.364768 |
| $\beta_5$ | $=$ | 28.224512 |
| $\beta_6$ | $=$ | 39.81514752 |
| $\beta_7$ | $=$ | 53.34441984 |
| $\beta_8$ | $=$ | 68.5375488 |
| $\beta_9$ | $=$ | 85.8128384 |
| $\beta_{10}$ | $=$ | 105.775104 |
| $\beta_{11}$ | $=$ | 128.974848 |
| $\beta_{12}$ | $=$ | 155.189248 |

---

[6]Actually, Carnahan and Starling had a slight, but insignificant error in their final term's coefficient; Ree and Hoover's $B_6$ was not quite right.

This was the Carnahan and Starling train of argument:

- they observed that the $\beta_n$ coefficients were "close to" integers;
- they noted that if they rounded to whole numbers: 4, 10, 18, 28, 40, then $\beta_n$ was given by $(n-1)(n+2)$;
- they then made the assumption that this expression would work for the higher-order terms as well;
- this enabled them to sum the virial series, to obtain an equation of state in closed form.

So, they had a suggestion for the values of the virial coefficients to *all* orders. We can check their hypothesis, based upon the coefficients known to them, by comparison with the newly known virial coefficients.

| $n$ | 2 | 3 | 4 | 5 | 6 | 7 | 8 | 9 | 10 | 11 | 12 |
|---|---|---|---|---|---|---|---|---|---|---|---|
| rounded true $\beta_n$ | 4 | 10 | 18 | 28 | 40 | 53 | 69 | 86 | 106 | 129 | 155 |
| C+S: $(n-1)(n+2)$ | 4 | 10 | 18 | 28 | 40 | 54 | 70 | 88 | 108 | 130 | 154 |

The agreement is not quite perfect, but it is still rather good.

The Carnahan and Starling virial series is then

$$\frac{pV}{NkT} = 1 + 4y + 10y^2 + 18y^3 + 28y^4 + 40^5 + \cdots$$

$$= 1 + \sum_{n=2}^{\infty} (n-1)(n+2)y^{n-1} \tag{3.6.17}$$

$$= 1 + \sum_{n=1}^{\infty} n(n+3)y^n.$$

The series is summed, to give

$$\frac{pV}{NkT} = 1 + \frac{2y(2-y)}{(1-y)^3}$$

$$= \frac{1 + y + y^2 - y^3}{(1-y)^3}. \tag{3.6.18}$$

This is the Carnahan and Starling equation of state. In Fig. 3.21, we have plotted this equation together with molecular dynamics simulation data from Bannerman *et al.* [19]. The agreement between the Carnahan and Starling equation of state and the molecular dynamics data is highly impressive. The drop in $pV/NkT$ at $y = 0.524$ is "associated" with the transition to a solid phase.

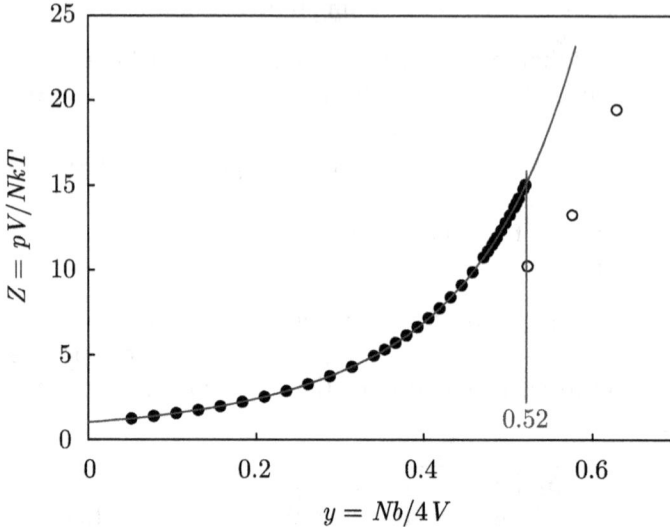

Fig. 3.21. Molecular dynamics simulation data plotted with the Carnahan and Starling equation of state.

From Carnahan and Starling's model the general expression for the $n$th virial coefficient is

$$B_n = \frac{(n-1)(n+2)}{4^{n-1}} b^{n-1}. \tag{3.6.19}$$

These are listed in Table 3.3 together with the true values.

Incidentally, the universal function $g(n)$ of Eq. (3.6.8) is then given by

$$g(n) = \frac{n^2 b}{2} \frac{\left(2 - \frac{1}{4} nb\right)}{\left(1 - \frac{1}{4} nb\right)^3}. \tag{3.6.20}$$

A more systematic way at arriving at an equation of state is the Padé method.

### 3.6.6   *Padé approximants*

The equation of state of the hard-sphere gas takes the form

$$\frac{pV}{NkT} = f(y), \tag{3.6.21}$$

where $y = Nb/4V$ and $f$ is a universal function of its argument. So, if the function is determined, then the hard-sphere equation of state is known.

Table 3.3. Carnahan and Starling's hard-sphere virial coefficients.

|  |  | Calculated | C+S value |
|---|---|---|---|
| $B_2/b$ | = | 1 | 1 |
| $B_3/b^2$ | = | 0.625 | 0.625 |
| $B_4/b^3$ | = | 0.28694950 | 0.28125 |
| $B_5/b^4$ | = | 0.11025200 | 0.10937500 |
| $B_6/b^5$ | = | 0.03888198 | 0.03906250 |
| $B_7/b^6$ | = | 0.01302354 | 0.01318359 |
| $B_8/b^7$ | = | 0.00418320 | 0.00427246 |
| $B_9/b^8$ | = | 0.00130940 | 0.00134277 |
| $B_{10}/b^9$ | = | 0.00040350 | 0.00041199 |
| $B_{11}/b^{10}$ | = | 0.00012300 | 0.00012398 |
| $B_{12}/b^{11}$ | = | 0.00003700 | 0.00003672 |

The virial expansion gives $f$ as a power series in its argument. And in reality, one can only know a finite number of these terms. The Carnahan and Starling procedure took the known terms, it 'guessed' the (infinite number of) higher-order terms and it then summed the series. Figure 3.21 indicates that the result is good, but it relied on guesswork and intuition.

For the Carnahan and Starling equation of state, the function $f$ may be written as

$$f(y) = \frac{1 + y + y^2 - y^3}{1 - 3y + 3y^2 - y^3}. \tag{3.6.22}$$

In this form, we observe that $f(y)$ is the quotient of two polynomials. And this leads us naturally to the Padé method; this is the general framework for making approximations as such quotients [20].

One knows $f(x)$ to a finite number of terms, say $N$. The Padé method provides a systematic procedure for approximating the higher-order terms and summing the series. In the Padé method, the true function $f(x)$ is approximated by the quotient of two polynomials

$$f(x) \approx F_{nm}(x) = \frac{P_n(x)}{Q_m(x)}. \tag{3.6.23}$$

Here, $P_n(x)$ and $Q_m(x)$ are polynomials of degrees $n$ and $m$, respectively:

$$\begin{aligned} P_n(x) &= p_0 + p_1 x + p_2 x^2 + \cdots + p_n x^n, \\ Q_m(x) &= q_0 + q_1 x + q_2 x^2 + \cdots + q_m x^m. \end{aligned} \tag{3.6.24}$$

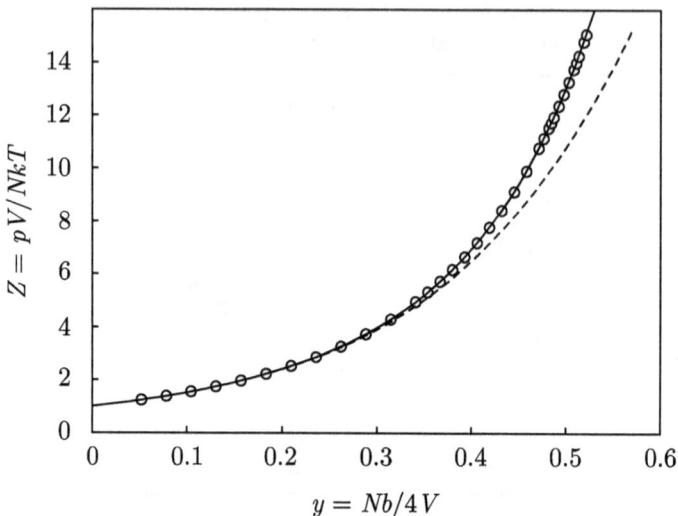

Fig. 3.22.  The 3-2 Padé approximant (solid line) and truncated virial series up to $B_6$ (dashed line) shown with molecular dynamics simulations of the hard-sphere gas.

Without loss of generality we may (indeed it is convenient to) restrict $q_0 = 1$. This will ensure the coefficients $p_i, q_j$ (for a given $n, m$) are unique.

All the coefficients of $P_n(x)$ and $Q_m(x)$ may be determined so long as $f(x)$ is known to at least $n + m$ terms. In other words, if $f(x)$ is known to $N = n + m$ terms, then $F_{nm}(x)$ agrees with the known terms of the series for $f(x)$. However, the quotient generates a sequence of higher-order terms as well. And the hope is that this series will be a good approximation to the true (but unknown) $f(x)$.

The power series of

$$f(x)Q_m(x) - P(x) \qquad (3.6.25)$$

begins with the term in $x^{m+n+1}$. In other words, the coefficient of this and the higher powers are "manufactured" by the Padé procedure.

One can construct approximants with different $m, n$ subject to $m + n = N$. See Reichl [2] for (some) details. Essentially, there will be a sub-set of $m, n$ pairs whose Padé approximants appear similar, usually when $m \sim n \sim N/2$. These "robust" approximants would be expected to provide a good approximation to the true function.

In this way, in 1964 Ree and Hoover [22], using the then known $B_2$ to $B_6$ (i.e. before Clisby and McCoy's extra virial coefficients) constructed the

3-2 Padé approximant:

$$\frac{pV}{NkT} = \frac{1 + 1.81559y + 2.45153y^2 + 1.27735y^3}{1 - 2.18441y + 1.18916y^2}. \tag{3.6.26}$$

This is plotted as the solid line in Fig. 3.22. For comparison, the dotted line shows the truncated virial series up to $B_6$.

Observe the 3-2 Padé gives very good agreement with the molecular dynamics data. By contrast, the corresponding truncated virial series is essentially useless; this indicates the value of the Padé procedure.

## Problems

3.1 Show that the Joule–Kelvin coefficient is zero for an ideal gas.

3.2 Evaluate the constants $A, B$ of Eq. (3.2.15) in terms of the square-well potential parameters $\sigma$ and $R$.

3.3 Show that the Boyle temperature $T_B$ for the square-well gas is given by

$$kT_B = -\varepsilon / \ln(1 - R^{-3}).$$

3.4 For the van der Waals gas, show that $T_B = a/bk$ and $T_i = 2a/bk$.

3.5 The interatomic potential $U(r) = \varepsilon \left(\frac{\sigma}{r}\right)^n$ may be regarded as a "soft sphere" interaction where $1/n$ measures the "softness".

   (a) Show that when $n \to \infty$, this reduces to the hard-sphere potential.

   (b) Plot $U(r)$ for different $n$ showing the hardening of the potential as $n$ increases.

   (c) The second virial coefficient for particles interacting with this potential is

$$B_2(T) = \frac{2}{3}\pi\sigma^3 \left(\frac{\varepsilon}{kT}\right)^{3/n} \Gamma\left(1 - \frac{3}{n}\right).$$

   Show this reduces to the hard-sphere case when $n \to \infty$.

(d)   Sketch the form of this $B_2(T)$ together with the general form for the second virial coefficient. (You might choose the soft-sphere case for $n = 12$ and Lennard-Jones $B_2(T)$, and you might plot these over the temperature range $0 < \frac{kT}{\varepsilon} < 20$). They are very different. Explain why.

3.6   For the *hard-sphere* gas, we saw that the compressibility factor $pV/NkT$ could be expressed as a universal, temperature-independent function of the density $N/V$. By noting that the soft-sphere interaction, for a given value of $n$, where the interaction potential scales with a single parameter $\varepsilon\sigma^n$, argue that the soft-sphere compressibility factor $z$ may be expressed as a universal function of $N/VT^{3/n}$.

3.7   Show (with the aid of a computer) that for the Lennard-Jones gas

$$T_B = 3.418\,\varepsilon/k, \quad T_i = 6.431\,\varepsilon/k.$$

3.8   For a general inter-particle interaction potential $U(r)$, we may define an *effective* temperature-dependent hard-core dimension $\sigma_{\text{eff}}$ by $U(\sigma_{\text{eff}}) = kT$.

(a)   What is the interpretation of this definition?

(b)   At high temperatures, only the *repulsive* part of the interaction is relevant.

Show that for the (repulsive part of the) Lennard-Jones 6–12 potential, $\sigma_{\text{eff}}$ is given by

$$\sigma_{\text{eff}} = \sigma\left(\frac{4\varepsilon}{kT}\right)^{1/12}.$$

(c)   The previous result corresponds to an *effective volume* $v_{\text{eff}} = \frac{2}{3}\pi\sigma_{\text{eff}}^3$ so that $v_{\text{eff}} = \frac{2}{3}\pi\sqrt{2}(\varepsilon/kT)^{1/4}$. Compare this with the high-temperature limit of the Lennard-Jones $B_2$. Discuss the similarities.

(d)   You might care to carry out the calculation of $\sigma_{\text{eff}}$ and $v_{\text{eff}}$ for the full Lennard-Jones potential, comparing the high-temperature expansion of $v_{\text{eff}}$ with the high-temperature expansion of the Lennard-Jones $B_2$.

Table 3.4. Second virial coefficient of xenon.

| $T$(K) | $B_2$(cm$^3$/mol) |
|--------|-------------------|
| 273.16 | $-154.74$ |
| 298.16 | $-130.21$ |
| 323.16 | $-110.62$ |
| 348.16 | $-95.04$ |
| 373.16 | $-82.13$ |
| 423.16 | $-62.10$ |
| 473.16 | $-46.74$ |
| 573.16 | $-25.06$ |
| 673.16 | $-10.77$ |
| 773.16 | $-0.13$ |
| 873.16 | $7.95$ |
| 973.16 | $14.22$ |

3.9 Derive the second virial coefficient expression for the Joule–Kelvin coefficient $\mu_J = \frac{2T}{5k} \left\{ \frac{dB_2(T)}{dT} - \frac{B_2(T)}{T} \right\}$, and find the inversion temperature for the square-well potential gas in the limit $R \to \infty$, $\varepsilon \to 0$. (The full expression for $T_i$ will be found in Problem 3.14.)

3.10 Show (with the aid of a computer) that for the Sutherland gas

$$T_B = 1.171\,\varepsilon/k, \quad T_i = 2.251\,\varepsilon/k.$$

3.11 Compare the square-well and the van der Waals expressions for the second virial coefficient. Show that they become equivalent when the range of the square-well potential tends to infinity while its depth tends to zero.

3.12 Some measurements of the second virial coefficient $B_2(T)$ for xenon at high temperatures are given in Table 3.4.

If you assume that the interaction between two xenon atoms has the form of the square-well potential, what can you deduce about the potential's parameters $\sigma$, $\varepsilon$, and $R$?

3.13 From the expression for the second virial coefficient, Eq. (3.2.2), show that if the interaction potential has a *universal* form $U(r) = \varepsilon u(r/\sigma)$ — in other words $u(x)$ is the same for different species, where just the scaling parameters $\varepsilon$ and $\sigma$ differ — then $B_2(T)$ also has a universal form.

3.14  Show that the inversion temperature $T_i$ for the square-well gas is given by

$$kT_i = \varepsilon / \left\{ W \left[ e/(1 - R^{-3}) \right] - 1 \right\},$$

where $W$ is the Lambert W-function.

(The Lambert W-function $W[y]$ is the solution to the equation $y = We^W$. It is implemented in *Mathematica* as `LambertW[y]`.)

3.15  In Problems 3.3 and 3.14, the Boyle temperature and the inversion temperature for the hard-sphere gas were obtained:

$$kT_B = -\varepsilon / \ln(1 - R^{-3}).$$
$$kT_i = \varepsilon / \left\{ W \left[ e/(1 - R^{-3}) \right] - 1 \right\}.$$

(a)  Plot these (with the aid of a computer) and comment on the behaviours in the vicinity of $R = 1$.

(b)  Plot the ratio $T_i/T_B$ as a function of $R$. Comment on the behaviour in the vicinity of $R = 1$ and in the limit $R \to \infty$.

3.16  (a)  From the van der Waals partition function expression Eq. (3.4.3), show that the internal energy of a van der Waals gas may be written as

$$\frac{E}{N} = \frac{3}{2}kT - a\frac{N}{V}.$$

(b)  Using the relation $pV = \frac{2}{3}E$, Eq. (2.1.35) would give

$$p = \frac{NkT}{V} - \frac{2}{3}a\frac{N^2}{V^2}.$$

This is different from the van der Waals equation of state. Explain.

(c)  Obtain the heat capacity at constant volume $C_V$ for the van der Waals gas. How does this compare with the $C_V$ of an ideal (classical) gas?

3.17 In Section 3.1.1, we saw that the partition function for an interacting gas may be expressed as $Z = Z_{id}Q$ where $Z_{id}$ is the partition function for a non-interacting gas and $Q$ is the configuration integral. Explain why the partition function of a *hard-sphere* gas might be approximated by

$$Z = Z_{id}\left(\frac{V - Nb}{V}\right)^N.$$

3.18 (a) Show that the approximate partition function for the hard-sphere gas in the previous question leads to the equation of state $p(V - Nb) = NkT$. This is sometimes called the Clausius equation of state. Give a physical interpretation of this equation.

(b) Show that the first few virial coefficients are given by $B_2(T) = b$, $B_3(T) = b^2$, $B_4(T) = b^3$, etc. These virial coefficients are independent of temperature. Discuss whether this is a fundamental property of the hard-sphere gas, or whether it is simply a consequence of the *approximated* partition function.

3.19 The one-dimensional analogue of the hard-sphere gas is an assembly of rods constrained to move along a line (the Tonks model). For such a gas of $N$ rods of length $l$ confined to a line of length $L$, evaluate the configuration integral $Q$. Show that in the thermodynamic limit the equation of state is

$$f(L - Nl) = NkT,$$

where $f$ is the force, the one-dimensional analogue of pressure. Comment on the similarities and the differences from the hard-sphere equation of state mentioned in Problem 3.18 (Clausius equation) and the van der Waals equation of state.

3.20 Estimate the van der Waals parameters $a$ and $b$ in terms of the Lennard-Jones parameters $\sigma$ and $\varepsilon$.

3.21 Show that the leading term in the expansion of $g(n)$, of Eq. (3.6.8), is in $n^2$, i.e. show that there is no linear term.

3.22 Evaluate the sum $\sum_{n=1}^{\infty} n(n + 3)y^n$ required in the derivation of the Carnahan and Starling hard-sphere equation of state, Eq. (3.6.18).

**Hint:** You know the sum of the convergent geometric progression $\sum_{n=0}^{\infty} y^n = \frac{1}{1-y}$. The trick of differentiation with respect to $y$ will "bring down" an $n$ into the sum.

3.23    Equation (3.6.26) gives the 3-2 Padé approximation to the hard-sphere equation of state based on knowledge of the first six virial coefficients. Use *Mathematica* to expand this expression in powers of $y$.

     (a)   Show that the expansion is consistent with the first six reduced virial coefficients (Table 3.2).

     (b)   Compare the higher-order terms with those given from the Carnahan and Starling equation of state and compare both with the values corresponding to the Clisby and McCoy virial coefficients.

     Comment on whether (in your opinion) the 3-2 Padé or the Carnahan and Starling is the better equation of state.

## References

[1]   W. J. Mullin, The cluster expansion, *Am. J. Phys.*, **40** (1972) 1473.

[2]   L. E. Reichl, *A Modern Course in Statistical Mechanics* (John Wiley, 1998).

[3]   H. B. Callen, *Thermodynamics and an Introduction to Thermostatistics* (John Wiley, 1985).

[4]   I. Cachadiña Gutiérrez, Comment on 'Exact Analytic Second Virial Coefficient for the Lennard-Jones Fluid' (arXiv:0909.3326v1). arXiv:1310.3586v1, 2013.

[5]   G. W. C. Kaye and T. H. Laby, *Tables of Physical and Chemical Constants* (Longman, 1995).

[6]   C. Kittel, *Introduction to Solid State Physics*, 7th ed. (John Wiley, New York, 1996).

[7]   W. J. Mullin and G. Blaylock, Quantum statistics: Is there an effective fermion repulsion or boson attraction? *Am. J. Phys.*, **71** (2003) 1223.

[8]   T. D. Lee and C. N. Yang, Many-body problem in quantum statistical mechanics. I. General formulation, *Phys. Rev.*, **113** (1959) 1165–1177.

[9]   S. Y. Larsen, J. E. Kilpatrick, E. H. Lieb, and H. F. Jordan, Suppression at high temperature of effects due to statistics in the second virial coefficient of a real gas, *Phys. Rev.*, **140** (1965) A129–A130.

[10]   R. P. Feynman, *Statistical Mechanics* (Benjamin, 1972).

[11]   R. Young, Theory of quantum-mechanical effects on the thermodynamic properties of Lennard-Jones fluids, *Phys. Rev. A*, **23** (1981) 1498–1510.

[12]   M. W. Zemansky and R. H. Dittman, *Heat and Thermodynamics* (McGraw-Hill, 1968).

[13]   J. S. Rowlinson, *J. D. van der Waals: On the Continuity of the Gaseous and the Liquid States, Studies in Statistical Mechanics XIV* (North Holland, 1988).

[14]   D. C. Johnston, *Advances in Thermodynamics of the van der Waals Fluid, IOP Concise Physics* (Morgan and Claypool, 2014).

[15]   L. D. Landau and E. M. Lifshitz, *Statistical Physics, Part 2* (Pergamon Press, 1980).

[16]   H. J. Wintle, More on the equation of state for gases, *Am. J. Phys.*, **42**, 9 (1974) 794.

[17]   P. M. Chaikin and T. C. Lubensky, *Principles of Condensed Matter Physics* (Cambridge University Press, 1995).

[18]   F. Reif, *Fundamentals of Statistical and Thermal Physics* (McGraw-Hill, 1965).

[19]   M. N. Bannerman, L. Lue, and L. V. Woodcock, Thermodynamic pressures for hard spheres and closed-virial equation-of-state, *J. Chem. Phys.*, **132** (2010) 084507.

[20]   G. A. Baker Jr. and P. Graves-Morris, *Padé approximants* (Cambridge university press, Cambridge, 2nd ed., 1996).

[21]   H. Gould and J. Tobochnik, Hard disc simulation. http://stp.clarku.edu/simulations/harddisks/dynamics/index.html.

[22]   F. H. Ree and W. G. Hoover, Fifth and sixth virial coefficients for hard spheres and hard disks, *J. Chem. Phys.*, **40** (1964) 939.

[23]   N. Clisby and B. McCoy, Ninth and tenth order virial coefficients for hard spheres in *D* dimensions, *J. Stat. Phys.*, vol. 122, p. 15, 2006.

[24]   R. J. Wheatley, Calculation of high-order virial coefficients with applications to hard and soft spheres, *Phys. Rev. Lett.*, vol. 110, no. 20, p. 200601, 2013.

[25]   N. F. Carnahan and K. E. Starling, Equation of state for non-attracting rigid spheres, *J. Chem. Phys.*, vol. 51, p. 635, 1969.

# CHAPTER 4

# PHASE TRANSITIONS

---

## 4.1 Phenomenology

### 4.1.1 *Basic ideas*

In Chapter 3, we saw how interactions can affect the behaviour of systems, modifying properties from those of the corresponding ideal (non-interacting) system. Nevertheless, a gas was still recognisable as a gas, albeit with somewhat altered properties; the interactions did not change the fundamental nature of the system. There is, however, another consequence of interactions when, by altering a thermodynamic variable such as temperature, pressure, etc. there can suddenly occur a dramatic change in the system's properties; there is a transition to a qualitatively different state. We refer to this as a phase transition.

Phase transitions present a challenge to statistical mechanics. At the transition point the system exhibits, by definition, singular behaviour. As one passes through the transition, the system moves between analytically distinct parts of the phase diagram. But how can this be? The thermodynamic behaviour is embodied in the partition function. Thus, the partition function must contain the details of any phase transition; it should exhibit singular behaviour. But the partition function is merely a sum of Boltzmann factors — exponentials of the energy, which must therefore be analytic. There is a paradox: *does* the partition function contain the description of a phase transition or does it not? This question was a source of worry to physicists. And it was debated at the van der Waals Centenary Conference in November 1937. When a vote was taken, it transpired that opinion was approximately equally divided between those who believed the partition function did contain details of a sharp transition and those who believed it did not!

The paradox was resolved later by H. Kramers. He suggested that the singular behaviour only appears when the *thermodynamic limit* is taken, that is, when $N$ and $V$ go to infinity while the density $N/V$ remains constant. This view was vindicated through the work of T. D. Lee and C. N. Yang published in 1952 [1, 2], and discussed in Huang's book *Statistical Mechanics* [3]. Subsequently, there was an elegant reformulation by Fisher [4]; see also Jones [5].

Fisher considered the partition function $Z$ as a function of a *complex* inverse temperature $\beta$. For a finite system, there are zeros in $Z$ for certain complex values of $\beta$. Fisher showed that as the size of the system increases, these zeros shift in the complex $\beta$ plane. And as the thermodynamic is approached, one or more of these zeros moves to the real $\beta$ axis. It is this that causes a zero in the partition function, resulting in its singular behaviour, indicating a phase transition.

Interactions are responsible for phase transitions. (The Bose–Einstein condensation is the only exception to this; there a phase transition occurs in the absence of interactions.) As a very rough "rule of thumb", often the temperature of the phase transition is approximately related to the interaction energy by $kT \approx E_{\text{int}}$. Thus, the exchange interaction between electronic spins $-\hbar J S_1 \cdot S_2$ leads to a ferromagnetic transition at the Curie temperature $T_c \approx \hbar J / k$. However, in some cases it might not be immediately clear how to quantify the "interaction". In the BCS theory of superconductivity, it is interactions between lattice vibrations and the electrons that result in the superconducting transition. The transition temperature is given by $kT = 1.14 \hbar \omega_D e^{-1/\rho(0)V}$ where $\omega_D$ is the Debye frequency characterising the phonons, $\rho(0)$ is the density of electron states at the Fermi surface, and $V$ is the volume. Here, $\hbar \omega_D$ *might* be regarded as the characteristic energy, but the effect of the exponential is to shift the naïvely expected transition temperature by orders of magnitude; the rule of thumb is thus useless in this case.

From the reductionist perspective the first question one might ask about phase transitions is: "given the hamiltonian for a system, can we predict whether there will be a phase transition?" The answer to this is: "in general, probably not", for reasons that should become apparent. The second question might then be: "knowing that there is a transition and knowing the nature of that transition, can we predict the temperature of the transition from the hamiltonian?" The answer to this question is that we can usually apply the rule above; often we can find a better approximation to the temperature of the transition; very occasionally we can find it exactly (analytically). The key question is the *nature* of the transition; *that* is difficult to predict.

Working against this reductionist view of phase transitions, where the hamiltonian is regarded as the all-important descriptor of the system, we have the observed phenomena of *universality*. It is found that many properties of systems in the vicinity of phase transitions do *not* depend on microscopic details, but are shared by dissimilar systems. The ultimate explanation for this is found through use of the renormalisation group, sadly outside the scope of this book. However, the scaling arguments of Section 4.1.9 support this idea. The Landau picture of phase transitions in Section 4.5 provides a common language and mathematical description for the phenomena.

### 4.1.2 *Phase diagrams*

Let us start with a very familiar example. The general behaviour of a $p - V$ system is shown in Fig. 4.1. The variables of the thermodynamic

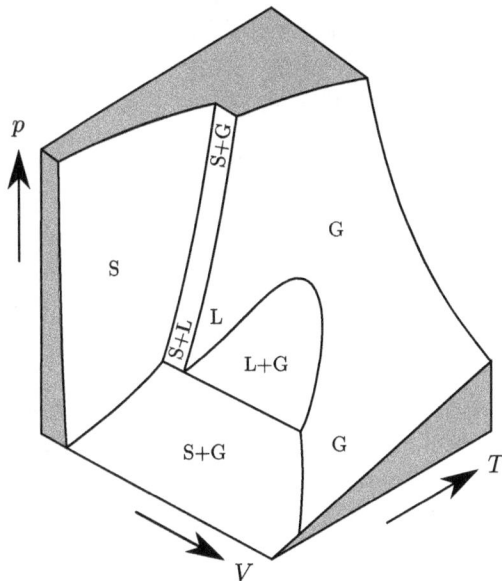

Fig. 4.1. Thermodynamic configuration diagram of a $p - V$ system, S — solid, L — liquid, G — gas.

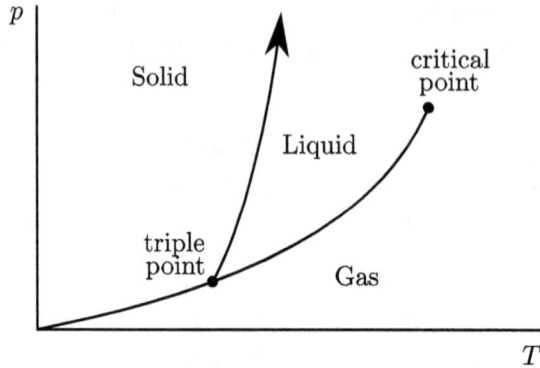

Fig. 4.2.    Phase diagram of a $p$–$V$ system.

configuration space are pressure $p$ and volume $V$ together with temperature $T$. A number of general observations can be made. There are regions where three distinct phases exist: solid, liquid and gas. And there are regions of two-phase coexistence.

If we project the configuration diagram onto the $p$–$T$ plane, then the surfaces of coexistence become lines (why?). Such a phase diagram is shown in Fig. 4.2. Solid, liquid and gas regions are seen, separated by the coexistence lines. All three phases coexist at the *triple point* and the liquid–gas line terminates at the *critical point*, where the distinction between the gas phase and the liquid phase disappears.

Many phase transitions involve a symmetry change. Often it is quite clear what symmetry is involved. Thus, the liquid–solid and the gas–solid transitions both involve the breaking of translational symmetry. In the case of superfluids and superconductors, it took many years until the appropriate symmetry was identified — gauge symmetry. However, in the case of the liquid–gas transition there is no symmetry broken. We will see what symmetries are broken in some different types of phase transitions in the following section.

The corresponding configuration diagram for a ferromagnetic magnetic system is shown in Fig. 4.3. The phase diagram for the system is the projection of this diagram onto the $B$–$T$ plane. This is shown in Fig. 4.4.

When the magnetic field is positive, the magnetisation is pointing up. When the magnetic field is negative, the magnetisation is pointing down. Above the critical temperature there is no magnetisation when $B$ is zero, so one moves smoothly from the up state to the down state. But below the critical temperature there is a magnetisation in the absence of a magnetic

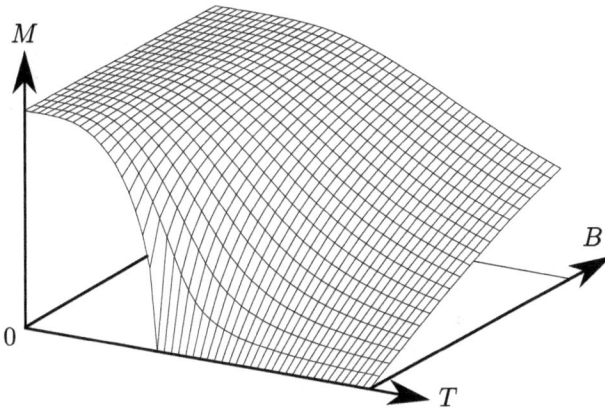

Fig. 4.3. Thermodynamic configuration diagram of a magnetic system.

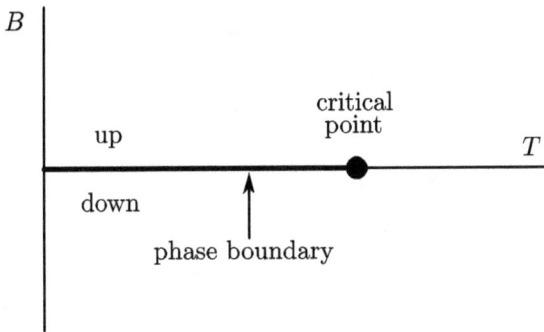

Fig. 4.4. Phase diagram for a magnetic system.

field. Then as $B$ decreases through zero, the magnetisation flips from up to down.

### 4.1.3 *Symmetry*

At the highest temperatures, a system will be at its most disordered. It will have the highest symmetry possible, and this will be the symmetry of the system's hamiltonian. As the temperature is lowered, when there is a phase transition, that symmetry is frequently broken; the resultant system will have a lower symmetry. Consider an isotropic magnet in zero magnetic field. The hamiltonian contains $S_1 \cdot S_2$, which is rotationally invariant. Above the transition temperature, there is no magnetisation and the system has *rotational symmetry*. But when cooled through the ferromagnetic transition, a magnetisation appears spontaneously. This defines a direction and the rotational symmetry is broken. Other systems have other symmetries broken:

| System | Symmetry broken |
|--------|-----------------|
| Crystal | Translational symmetry |
| Ferromagnet | Rotational symmetry |
| Ferroelectric | Inversion symmetry |
| Superfluid | Gauge symmetry |

In general, the ordered phase will possess a symmetry *lower* (i.e. less symmetric) than that of the system hamiltonian.

Note, however, that the liquid–gas transition does not involve a change of symmetry; not all transitions involve symmetry breaking. But we might say that in this case "homogeneity" is broken.

### 4.1.4 *Order of phase transitions*

Historically, the first stage in discussing phase transitions from a general perspective was the introduction, by Ehrenfest, of the idea of the *order* of the transition [6]; this was the first step in classifying the nature of the non-analytic behaviour at the transition. When two phases coexist, they have a common temperature and a common pressure (magnetic field). Thus, the phases will each have the same Gibbs free energy.

In Fig. 4.5, we show the variation of the Gibbs free energy $G$ for two different phases of a system — say solid and fluid, where the curves intersect. The equilibrium state will correspond to the lower $G$, so we see that the phase transition occurs at the point where the Gibbs free energy is the same for both phases. The observed $G$ will thus display a "kink" at the phase transition.

Traditionally, phase transitions were characterised, by Paul Ehrenfest, on the basis of the nature of the kink in $G$. If the $n$th derivative of $G$ with respect to $T$ (keeping other intensive variables constant) is the first discontinuous one, then it was said that the transition was of $n$th order. Now, since $\partial G/\partial T|_p = -S$, we see that the discontinuity in $\partial G/\partial T$ is the change in entropy between the two phases. Thus, at a first-order transition the

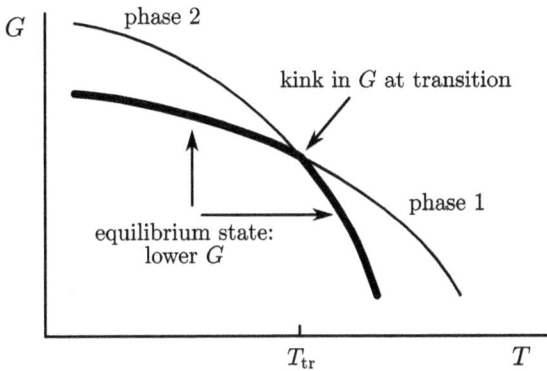

Fig. 4.5. Variation of Gibbs free energy for two phases.

entropy changes discontinuously, while at second and higher order transitions the entropy varies continuously. And since the latent heat, the heat absorbed, is given by $T\Delta S$, it follows that there *is* latent heat involved with a first-order transition, but not with higher orders.

Nowadays, the classification scheme is not used in quite this way; there are only first- and second-order transitions. First-order transitions are defined as above. And all transitions which are not first-order are called second-order. So, if there is latent heat involved in a phase transition, then it is first order, otherwise it is second order. Following the entropy considerations above, first-order transitions are also called discontinuous transitions, while so-called second-order transitions are also referred to as continuous transitions. A different ways of specifying the distinction will be seen in the following section.

### 4.1.5 *The order parameter*

The modern general theory of phase transitions started with the work of Landau and others in the 1930's. In developing a *general* treatment of phase transition phenomena, we must use a language applicable to all systems. On warming a ferromagnet (in zero field), the magnetisation goes to zero at the transition point. On warming a fluid along the coexistence curve, the difference between the liquid and gas densities goes to zero at the transition point. And on warming a superfluid, the superfluid density goes to zero at the transition point. The general feature is that there is *some quantity* which goes to zero at the transition point. When there is a symmetry broken on cooling through the transition, this special quantity will be related to that symmetry.

The special quantity is a measure of the order present in the system; it is called the *order parameter*. Felix Bloch introduced the concept of the order parameter in 1932 and it was developed subsequently by Landau. In the general case, we shall use $\varphi$ to denote the order parameter. For convenience, we will sometimes normalise $\varphi$ to be unity in the fully ordered state.

The nature of the order parameter is important; the magnetisation of a ferromagnet is a vector; the fluid density is a scalar. The order parameter for superfluid $^4$He is a complex variable; those for liquid crystals and for superfluid $^3$He are tensors. We use $n$ to indicate the dimensionality of the order parameter.

| System | Order parameter | $\varphi$ | | $n$ |
|--------|-----------------|-----------|--|-----|
| Ferromagnet | Magnetisation | $\mathbf{M}$ | vector | 3 |
| Ferroelectric | Polarisation along axis | $P$ | real scalar | 1 |
| Fluid | Density difference | $n - n_c$ | real scalar | 1 |
| Superfluid $^4$He | Ground state wavefunction | $\Psi_0$ | complex scalar | 2 |
| Superconductor | Pair wavefunction | $\Psi_s$ | complex scalar | 2 |
| Ising | Ising "magnetisation" | $m$ | real scalar | 1 |

Note that a complex scalar may, equivalently, be represented as a two-component (real) vector.

The ferromagnet will be treated in Section 4.3 and again in Section 4.5.2. The ferroelectric is treated in Section 4.6. The fluid is covered in Section 4.2. Superfluids and superconductors are beyond the scope of this book. The Ising model is treated in Section 4.4. This is a magnet model where the interaction is restricted to the $z$ direction. Its order parameter is a scalar ($n = 1$). And there is a further magnet model where the interaction is restricted to the $x$–$y$ plane. This is called the XY model; its order parameter is two-dimensional (or a complex scalar) so $n = 2$. It is mentioned briefly in Section 4.4.6.

The behaviour of the order parameter at the transition point gives a way to distinguish the order of the phase transition. As with the entropy, the order parameter changes discontinuously in a first-order transition, but continuously in a second-order transition; this reinforces the designation *discontinuous* and *continuous* transition. This behaviour is indicated in Fig. 4.6.

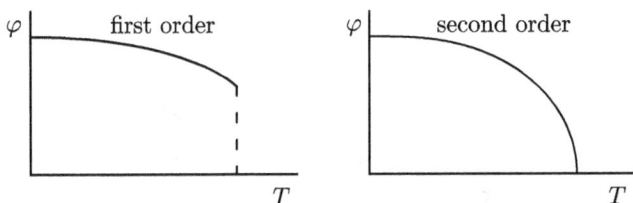

Fig. 4.6. Order parameter in first and second order transitions.

Note that the apparent order of the transition may depend on the way the phase boundary is crossed. Thus, while in general the liquid–gas transition is first-order, it is second order when crossing the critical point from along the coexistence curve. Similarly, in the ferromagnet below $T_c$, if the field is varied through zero, the magnetisation changes discontinuously, while it varies continuously when the field is zero and the transition is effected by varying the temperature. Furthermore, in the ferromagnet the magnetisation can vary continuously from a positive to a negative value by going *around* the critical point — just as in a fluid.

Returning to the first question posed in Section 4.1.1 about predicting transitions from the system hamiltonian, we stated in that section that the difficulty was in knowing the *nature* of the transition. This can now be reinterpreted as saying the difficulty is in identifying what the order parameter might be in the ordered phase.

Sometimes the order parameter will be one of the usual thermodynamic variables: magnetisation for the ferromagnet, polarisation for the ferroelectric and density for the fluid. However, it may be that the order parameter is some other — possibly unfamiliar — quantity such as the ground state wave function for superfluid $^4$He or the pair wave function in superconductors.

A further important quantity is the thermodynamic variable *conjugate* to the order parameter. For the ferromagnet, this is the magnetic field, while for the ferroelectric, it is the electric field. In the case of the fluid, the conjugate variable is the chemical potential. The conjugate variable is important since it is by varying this quantity that we can couple to the order parameter in order to study the transition experimentally.

### 4.1.6 *Conserved and non-conserved order parameters*

A new concept we must introduce at this stage is that of *conserved* and *non-conserved* order parameters. In the magnetic transition, as we have seen, below the transition there is a spontaneously occurring magnetisation. This is the order parameter for that system. The value of the order parameter is determined from thermodynamic arguments by minimising the appropriate free energy.

In the case of the gas–liquid transition, the order parameter is essentially the density (more precisely the difference between the density and its value at the critical point). Above the transition, the density of the system is determined by external constraints: the volume and the number of atoms. Below the transition, the system separates into liquid and gas components. The liquid has a high density, while the gas has a low density. However, the *mean* density is still a fixed quantity; it is determined by the external constraints and not by minimising a free energy. The fluid system is thus said to have a *conserved order parameter*. We shall see that the binary mixture is another such system. The magnet, by contrast, has a *non-conserved* order parameter.

For a full treatment of systems with a conserved order parameter, we must allow the order parameter to vary from place to place by introducing a position-dependent *order parameter density* [7]. It is then the integral of the order parameter density over the system's volume, which is the conserved quantity. The solution to this problem then gives the spatial dependence of the order parameter — in particular, it gives the variation at the interface between the different phases. There is a simpler approach, due to Gibbs [8], that ignores the energy associated with the interface and simply considers coexisting regions with two different (uniform) values of the order parameter.

### 4.1.7   Critical exponents

In Section 4.1.1, we stated that many properties of systems in the vicinity of phase transitions do not depend on microscopic details, but are shared by dissimilar systems. In using a common description of phase transition phenomena, it is customary and convenient to specify the nature of the singular behaviour of quantities at the critical point. This is achieved through the introduction of *critical exponents* (and critical amplitudes). One of the successes of the modern theories of critical phenomena is in finding relations between the various critical exponents; this is achieved using scaling arguments, as we shall see in the following section.

The first members of the critical exponent family are denoted by $\alpha$, $\beta$, $\gamma$ and $\delta$. These describe the singularity of the heat capacity, order parameter, susceptibility and equation of state, respectively. In terms of the reduced temperature

$$t = \frac{T - T_c}{T_c} \tag{4.1.1}$$

they are defined (using the ferromagnet variables, for example) through

$$\text{heat capacity } C \sim |t|^{-\alpha}$$
$$\text{order parameter } m \sim |t|^{\beta}$$
$$\text{susceptibility } \chi \sim |t|^{-\gamma}$$
$$\text{equation of state at } T_c \ m \sim |B|^{1/\delta}.$$

Here, the order parameter is the magnetisation density, the magnetic moment per unit volume.

There are two more critical exponents, which are connected with the spatial variation of fluctuations in the order as the critical point is approached. For this, we need to introduce the spatial correlation function for the order parameter. We define the correlation function

$$g(\mathbf{r}) = \langle m(\mathbf{r})m(0) \rangle \tag{4.1.2}$$

where $m(\mathbf{r})$ is the magnetisation density at position $\mathbf{r}$. The physical significance of correlation functions such as these is explained in Chapter 5 in Sections 5.1.2 and 5.1.3.

In the vicinity of the critical point the behaviour of the order parameter, spatial correlation function is written as

$$g(\mathbf{r}) \sim r^{-p} e^{-r/l}. \tag{4.1.3}$$

And from this we obtain two more exponents, $\nu$ and $\eta$. These describe the divergence in the correlation length $l$ and the power law decay $p$ that remains at $t = 0$, when $l$ has diverged. The exponents are defined through

$$\text{correlation length } l \sim |t|^{-\nu}$$
$$\text{power law decay at } T_c \ p = d - 2 + \eta$$

where $d$ is the dimensionality of the system (it will turn out to be of interest to consider systems of dimensions other than three).

A comprehensive theory of critical phenomena will give values for the critical exponents. However, it turns out that there are fundamental relations between the critical exponents, so that of the six exponents only two are independent. This is understood from *scaling theory*, which we will encounter in the following section.

### 4.1.8 *The scaling hypothesis*

Scaling theory involves the application of *dimensional analysis* to the study of the critical point. Near the critical point, there are large fluctuations in the order parameter of the system. A dramatic example of this is the critical opalescence observed in fluids. This is a consequence of the density fluctuations that become very large at the critical point. As the critical point is approached, the fluctuations occur over longer and longer distances; this is the correlation length $l$ referred to in the previous section.

Now, $l$ is an important length parameter in the system, and it becomes of macroscopic magnitude near the critical point. Since this length is macroscopic, it implies that microscopic details of the system become unimportant close to the critical point. *Thus,* the universality in critical phenomena.

Scaling theory relies on the hypothesis that close to the critical point the anomalous part of all quantities with the dimension of length will be proportional to the characteristic length $l$. And quantities of the dimension [length]$^n$ will correspondingly be proportional to $l^n$. We use the notation [...] to denote the dimensions of a quantity. We consider the dimensions of some of the quantities that appear in, or lead to, the critical exponent definitions. A clear account of scaling theory is given in Huang's *Statistical Mechanics* [3]; we have drawn heavily on that discussion.

**(a) Heat capacity critical exponent $\alpha$:** Let us start with the Gibbs free energy $G$, which, by differentiating twice with respect to temperature, will give the heat capacity. Now, $G/kT$ is dimensionless, although extensive, so that $g = G/kTV$, which is finite in the thermodynamic limit, has the dimensions of inverse volume:

$$[g] = L^{-d}. \tag{4.1.4}$$

The scaling hypothesis is that $L$ is proportional to the critical length $l$, whose critical exponent is defined to be $v$:

$$l \sim |t|^{-v} \tag{4.1.5}$$

so that in the vicinity of the critical point we expect

$$g \sim |t|^{vd}. \tag{4.1.6}$$

Now, heat capacity is the second derivative of the Gibbs free energy with respect to temperature. So, the definition of the exponent $\alpha$

$$C \sim |t|^{-\alpha} \tag{4.1.7}$$

implies the critical behaviour of $g$ can also be written as

$$g \sim |t|^{2-\alpha}. \tag{4.1.8}$$

Equating these two forms for the exponent of $g$ then gives

$$\alpha = 2 - vd, \tag{4.1.9}$$

showing how the heat capacity critical exponent relates to the critical length exponent and the system dimensionality.

**(b) Order parameter critical exponent $\beta$:** Now, consider the order parameter — in the ferromagnet case this is the magnetisation per unit volume. The critical exponent $\beta$ is defined through

$$m \sim |t|^{\beta}. \tag{4.1.10}$$

But we can also express the dimensionality of the order parameter from the spatial correlation function (correlation functions and their significance will be discussed in Chapter 5)

$$g(\mathbf{r}) = \langle m(\mathbf{r})m(0)\rangle \sim r^{-p}e^{-r/l}. \tag{4.1.11}$$

Thus,

$$\begin{aligned}[m^2] &= L^{-p} \\ &= L^{-(d-2+\eta)},\end{aligned} \tag{4.1.12}$$

so that

$$[m] = L^{(2-d-\eta)/2}. \tag{4.1.13}$$

Now, according to the scaling hypothesis, that $L \sim |t|^{-v}$, we then have

$$m \sim |t|^{-v(2-d-\eta)/2}. \tag{4.1.14}$$

Equating these two forms for the exponent of $m$ then gives

$$\beta = -v(2 - d - \eta)/2, \tag{4.1.15}$$

showing how the order parameter critical exponent relates to the critical length exponent, the power law decay exponent and the system dimensionality.

**(c) Susceptibility critical exponent $\gamma$:** Next, we consider the susceptibility, which gives the critical exponent $\gamma$ defined by

$$\chi \sim |t|^{-\gamma}. \tag{4.1.16}$$

Now, the susceptibility is related to the order parameter's spatial correlation function $g(\mathbf{r})$ defined above. This may be seen from the fluctuation expression for the susceptibility obtained at the end of Section 2.10.2, Eq. (2.10.19):

$$\chi = \mu_0 \frac{1}{VkT} \langle M^2 \rangle. \tag{4.1.17}$$

The magnetisation $M$ is given by the integral over the magnetisation density $m$. Thus, in $n$ dimensions

$$M = \int m(\mathbf{r}) \, d^n r \tag{4.1.18}$$

and then

$$\langle M^2 \rangle = \int d^n r_1 \int d^n r_2 \, \langle m(\mathbf{r}_1) m(\mathbf{r}_2) \rangle. \tag{4.1.19}$$

On the assumption of translational invariance, we can arbitrarily fix $\mathbf{r}_1$ to be at the origin and remove the first integral by simply multiplying by $V$, the $n$-dimensional volume

$$\langle M^2 \rangle = V \int \langle m(0) m(\mathbf{r}) \rangle \, d^n r. \tag{4.1.20}$$

And then the susceptibility is given by

$$\chi = \mu_0 \frac{1}{kT} \int \langle m(0) m(\mathbf{r}) \rangle \, d^n r. \tag{4.1.21}$$

This is a most important result. Our "derivation" relies on results of Chapter 2 for non-interacting particles. But the result is more general than that; thus our treatment should be regarded as no more than a plausibility argument.

The critical behaviour of the correlation function is given by

$$\langle m(\mathbf{r})m(0)\rangle \sim r^{-p}e^{-r/l} \qquad (4.1.22)$$

then it follows that

$$[kT\chi] = L^{-p} \times L^d$$
$$= L^{2-\eta} \qquad (4.1.23)$$

from the definition of $p$: $p = d - 2 + \eta$. So, using the scaling hypothesis $L \sim |t|^{-\nu}$, we then have

$$kT\chi \sim |t|^{\nu(2-\eta)}. \qquad (4.1.24)$$

Equating these two forms for the exponent of $\gamma$ then gives

$$\gamma = \nu(2 - \eta), \qquad (4.1.25)$$

showing how the susceptibility critical exponent relates to the critical length exponent and the power law decay exponent.

**(d) Equation of state critical exponent $\delta$:** Finally, we look at the equation of state. From the definitions of $\beta$ and $\delta$ we can write

$$m \sim |t|^{-\beta} \quad \text{and} \quad m \sim B^{1/\delta} \qquad (4.1.26)$$

so then we have

$$B \sim |t|^{\beta\delta}. \qquad (4.1.27)$$

Now, $B$ is related to the Gibbs free energy through

$$m = -\frac{\partial g}{\partial B} \qquad (4.1.28)$$

so that

$$[B] = [g]/[m]. \qquad (4.1.29)$$

We have already found that $[g] = L^{-d}$ and $[m] = L^{(2-d-\eta)/2}$ so that

$$[B] = [g]/[m] = L^{-(2+d-\eta)/2}. \qquad (4.1.30)$$

Using the scaling hypothesis $L \sim |t|^{-\nu}$, we then have

$$B \sim |t|^{\nu(2+d-\eta)}. \qquad (4.1.31)$$

Equating these two forms for the exponent of $B$ then gives

$$\delta = \nu(2 + d - \eta)/2\beta, \qquad (4.1.32)$$

for the equation of state critical exponent.

By taking linear combinations of the above results, we express the relations in the traditional manner as

$$
\begin{aligned}
\gamma &= v(2 - \eta) & \text{Fisher law} \\
\alpha + 2\beta + \gamma &= 2 & \text{Rushbrooke law} \\
\gamma &= \beta(\delta - 1) & \text{Widom law} \\
vd &= 2 - \alpha & \text{Josephson law.}
\end{aligned}
\tag{4.1.33}
$$

The experimental demonstration of these results is strong evidence in favour of the scaling hypothesis. Thus, it would appear that the correlation length *is* the only length of importance in the vicinity of the critical point. A consequence of the hypothesis is that only two critical exponents need be calculated for a specific system. Note the Josephson law is the only one to make explicit mention of the spatial dimensionality $d$.

### 4.1.9 Scaling of the free energy

The scaling hypothesis has important consequences for the free energy of systems in the vicinity of the critical point. In particular, it has important consequences for the *singular* part of the free energy. Following Fisher [9], we shall define the reduced free energy $f(T, B)$ as

$$
f = \frac{-\Delta F}{kTV}
\tag{4.1.34}
$$

where $\Delta F$ is the deviation of the singular part of the free energy from its value at the critical point. We have divided by $kT$ to make the quotient dimensionless. However, note that is still extensive, so we divide by the volume to produce an intensive quantity, with dimensions of inverse volume.

The scaling hypothesis is equivalent to the assumption that in the vicinity of the critical point the reduced free energy has the following functional structure:

$$
f(t, B) = A\, |t|^{2-\alpha}\, Y\left( D\frac{B}{|t|^{\Delta}} \right).
\tag{4.1.35}
$$

Here, $A$ and $D$ are non-universal parameters that depend on the particular system. Essentially, $A$ sets the energy scale and $D$ sets the magnetic field scale. The quantities $\alpha$ and $\Delta$ are the two universal exponents — recall the discussion of the previous section that there are only two independent exponents. Here, $\alpha$ is the heat capacity exponent and $\Delta$ is related to the familiar critical exponents through $\Delta = 2 - \alpha - \beta$. The universal function $Y(y)$ is defined so that $Y(0) = 1$; thus, the need for the $A$ prefactor.

There are two branches to $Y$, one for $t > 0$ and one for $t < 0$. And as $y \to \infty$, the two branches must meet as $t$ goes through zero.

From the above free energy function, all the scaling laws of the previous section may be derived; Problems 4.1 and 4.2 treat two examples. Thus, we may regard the assumption that the free energy has the above form to be equivalent to the scaling hypothesis of the previous section.

## 4.2   First-Order Transition — An Example

In this section, we shall consider the liquid–gas transition in a fluid. This is an example of a first-order transition with a conserved order parameter. The order parameter for this system is the density. The system has a conserved order parameter since its mean density is fixed (the volume is regarded as fixed).

### 4.2.1   *Coexistence*

In transitions involving a conserved order parameter, the ordered phase can only evolve through spatial variation of the order parameter density — so that its integral over all space remains constant. Now, a full solution of such a system would determine this spatial variation. As discussed in Section 4.1.6, we shall use the approximation due to Gibbs [8] to sidestep this complexity. The approximation must be justified *a posteriori*, but its basic assumption is that the system evolves into regions with two distinct values for the order parameter: two coexisting phases. At the boundary between the phases, the order parameter will vary between that of the two distinct phases. The change will not be abrupt, as that is energetically costly, but the Gibbs approach is a "thin wall" approximation where the fraction of particles in the intermediate regions is assumed to be extremely small.

When two phases coexist, then adding a small quantity of heat energy will result in the conversion of a small amount of the ordered phase to the disordered phase; there will be latent heat involved. Similarly, changing the volume at constant temperature will alter the proportions of the two phases. The coexistence region for a liquid–gas system is shown in Fig. 4.7. Clearly, during coexistence the pressure remains constant; the pressure depends only on the temperature.

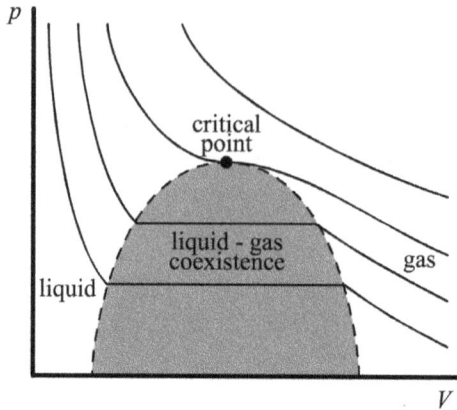

Fig. 4.7.   Coexistence region, showing $p - V$ isotherms for liquid–gas system.

A system will separate into two phases if it is energetically favourable to do so. Let us consider a fluid system held at constant temperature and volume. Then the appropriate thermodynamic potential is the Helmholtz free energy $F(T, V)$. This free energy will be minimised when a constraint on the system is removed. In this case, the "constraint" is the requirement that the system be uniform or homogeneous. We now remove this constraint and ask whether the free energy could be reduced through the system becoming inhomogeneous.

Figure 4.8 shows the Helmholtz free energy of a system at a given temperature as a function of *volume per particle* or specific volume $v$. The solid curve in this figure shows the free energy that a homogeneous system would have. The system comprises $N$ atoms or molecules occupying a volume $V$. The volume per particle of the homogeneous system is $v_0 = V/N$. We now ask if the system could lower its free energy by becoming inhomogeneous. In particular, we explore the possibility that it segregates into regions of specific volume $v_1$ and $v_2$.

If this happens, it will, nevertheless, be subject to the constraint that the total number of particles is fixed. Thus, if a fraction $\alpha_1$ of the particles is in regions of specific volume $v_1$ and a fraction $\alpha_2 = 1 - \alpha_1$ in regions of specific volume $v_2$, then

$$v_1 \alpha_1 + v_2 \alpha_2 = v_0 \tag{4.2.1}$$

so that the fractions are given by (the lever rule)

$$\alpha_1 = \frac{v_2 - v_0}{v_2 - v_1}, \quad \alpha_2 = \frac{v_0 - v_1}{v_2 - v_1}. \tag{4.2.2}$$

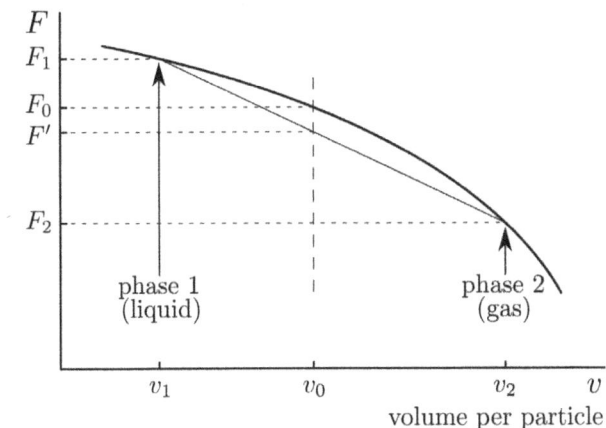

Fig. 4.8. Helmholtz free energy curve.

Then the free energy of the inhomogeneous system will be given by

$$F = \alpha_1 F_1 + \alpha_2 F_2$$

$$= \frac{v_2 F_1 - v_1 F_2}{v_2 - v_1} - \frac{F_1 - F_2}{v_2 - v_1} v_0. \tag{4.2.3}$$

This expression is linear in $v_0$. It is a linear interpolation between points $(v_1, F_1)$ and $(v_2, F_2)$. Thus, the free energy of this inhomogeneous system will be given by the chord in Fig. 4.8, where it intersects the line $v = v_0$; this is indicated as $F'$ in the figure.

We now have the condition for the system to remain homogeneous or to segregate. This depends upon whether $F'$ is below $F_0$ or not; whenever $F'$ falls below $F_0$, it will be favourable to separate into regions of different specific volume, or density. Thus, if the free energy curve is concave, as in Fig. 4.2, phase separation will occur, while if the curve is convex, then it is favourable to remain homogeneous.

The lowest free energy is achieved when the straight line takes its lowest possible position, as shown in Fig. 4.9. This determines the equilibrium state: of having the lowest possible free energy for a given overall particle density. For obvious reasons, it is referred to as the double-tangent construction.

The two coexisting phases have a common tangent in the $F$–$V$ plane. In other words, $\partial F / \partial V |_T$ is the same for both phases. Now, the differential

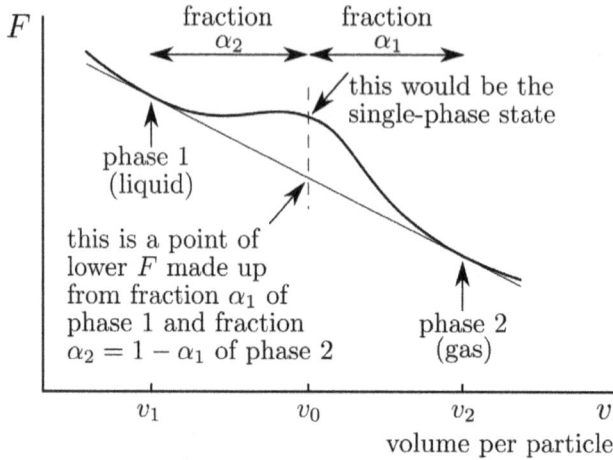

Fig. 4.9. Double-tangent construction for phase coexistence.

relation for the Helmholtz free energy is

$$dF = -SdT - pdV \tag{4.2.4}$$

so that

$$p = -\left.\frac{\partial F}{\partial V}\right|_T ; \tag{4.2.5}$$

the derivative is simply (minus) the pressure. In equilibrium, the coexisting phases will have a common pressure — as expected.

### 4.2.2 *Van der Waals fluid*

We encountered the van der Waals equation of state in the previous chapter. We saw there how it provided a way of approximating the effects of inter-particle interactions in a mean-field manner. At low densities, it provided a good description of gas-like behaviour, while at high densities it provided a good description of liquid-like behaviour. We shall now see how this model, when correctly interpreted, is also capable of providing a simple description for the gas–liquid transition. If we plot the equation of state

$$\left(p + \frac{aN^2}{V^2}\right)(V - Nb) = NkT \tag{4.2.6}$$

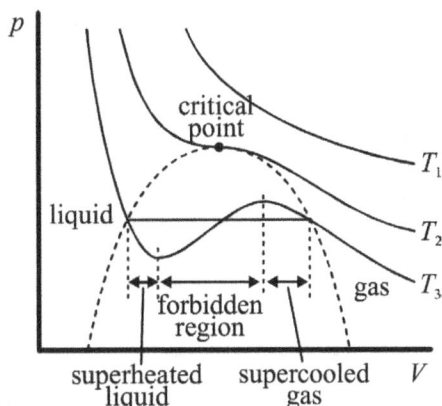

Fig. 4.10. Isotherms of the van der Waals equation.

we obtain the following curves for different temperatures $T_1 > T_2 > T_3$ in Fig. 4.10.

Below the critical point, the curve in the gas–liquid region appears strange. We know from the considerations of the previous section that it is preferable for phase separation to occur and the flat line is the usual behaviour. However, the curve represents the behaviour which would occur *if* the system remained homogeneous. There is a region of superheating and a region of supercooling. Although energetically unfavourable, it is possible in very clean systems that the new phase is not immediately nucleated. Then one can move down an isotherm from the liquid phase into the superheated region. Similarly, one can move from the gas, up the isotherm, into the supercooled region. But when the curve changes direction, you *have* to go to the two-phase state since the homogeneous phase is unstable if $\partial p/\partial V$ is positive. That would mean the pressure increasing when the volume increased!

### 4.2.3 *The Maxwell construction*

A van der Waals *isotherm* is a $p - V$ curve, given by the van der Waals equation at a given temperature. However, in the coexistence region the real isotherm must be a horizontal line: the pressure is a constant. Thus, between points A and B of Fig. 4.11 the van der Waals curve is replaced by a straight line. The question is where to position the line; what is the constant pressure during coexistence?

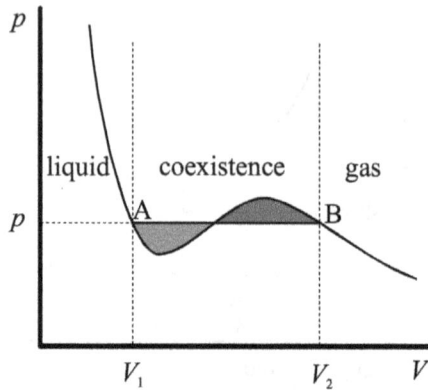

Fig. 4.11.   Maxwell construction on an isotherm.

Since $p = -\partial F/\partial V|_T$, it follows that the $p$–$V$ curve is essentially (minus) the derivative of the free energy graph of Fig. 4.9. In particular the curved $p$–$V$ isotherm comes from the derivative of the homogeneous system free energy curve, while the constant pressure coexistence isotherm comes from the derivative of the double-tangent line. From this the coexistence, pressure may be determined in the following way.

The difference in the free energy between the pure liquid and the pure gas, $F_1 - F_2$, may be found by integrating $p$ with respect to $V$ along the isotherm:

$$F_1 - F_2 = \int_{V_2}^{V_1} p \, dV. \tag{4.2.7}$$

Now, we may integrate either along the straight line or along the curve between points A and B of Fig. 4.11. If we integrate along the straight line, the free energy difference is the area under the straight line joining A and B. And if we integrate along the curve, the free energy difference is then the area under the van der Waals curve joining A and B. The free energy must be the same either way and thus the area of the curve above the straight line must be equal to the area below the line. This is known as the *Maxwell construction*. It permits determination of the pressure of the coexisting phases.

When the two phases coexist they have equal temperature, pressure and chemical potential. We have accommodated temperature equality since we are talking about an isotherm. We have also accommodated pressure equality since this followed directly from the double-tangent construction. It remains to consider the equality of chemical potentials of the two phases; this means that particles do not wish to flow from one phase to the other. We shall examine the consequence of requiring that the chemical potential at point A be equal to that at point B, in Fig. 4.11. This will give a further insight into the Maxwell construction.

The chemical potential is the Gibbs free energy per particle, $\mu = G/N$. So, let us find the change in chemical potential between points A and B and set this to zero. The differential expression for G is given by

$$dG = -SdT + Vdp. \tag{4.2.8}$$

So at constant temperature, we then have

$$\Delta G = \int_{\text{point A}}^{\text{point B}} V dp. \tag{4.2.9}$$

It is convenient to re-express this through an integration by parts:

$$\Delta G = pV \Big|_{p, V_A}^{p, V_B} - \int_{V_A}^{V_B} p \, dV$$

$$\tag{4.2.10}$$

$$= p(V_A - V_B) - \int_{V_A}^{V_B} p \, dV.$$

The first term is the area below the straight line joining A and B. The second term is the area below the van der Waals curve joining A and B. Thus, $\Delta G$ is the difference between these two areas. And so the requirement that the chemical potentials are equal at points A and B again gives the requirement that the area of the shaded part of the curve must be zero; we recover the Maxwell construction.

The book by Mazenko [10] treats the Maxwell construction for the van der Waals fluid quite extensively.

### 4.2.4  *The critical point*

At the critical point, the distinction between the liquid and the gas disappears; the two phases become equivalent. This means that at the critical point the discontinuity in $\partial G/\partial T$ vanishes and thus the transition becomes second-order. In other words, the first-order transition becomes second-order at the critical point. For the liquid–gas case, if the density is fixed to be equal to that at the critical point, then as the system is cooled it will pass through the critical point to a state of two-phase coexistence.

Professionals in the business of phase transitions usually restrict their attention to what is happening in the vicinity of the critical point — and to the universality of the behaviour that emerges there. Thus, the concentration on critical exponents, etc.

The critical point for the liquid–gas system is the point of inflection, where

$$\left.\frac{\partial p}{\partial V}\right|_T = 0 \quad \text{and} \quad \left.\frac{\partial^2 p}{\partial V^2}\right|_V = 0. \tag{4.2.11}$$

The volume, pressure and temperature at this point are found, in terms of the van der Waals parameters $a$ and $b$, to be

$$V_c = 3Nb, \quad p_c = \frac{a}{27b^2}, \quad kT_c = \frac{8a}{27b}. \tag{4.2.12}$$

(There is a beautiful calculus-free derivation of this result in Stanley's book [11].)

### 4.2.5  *Corresponding states*

If we define reduced dimensionless variables $v$, $\pi$ and $t^1$ by

$$v = \frac{V}{V_c}, \quad \pi = \frac{p}{p_c}, \quad t = \frac{T}{T_c} \tag{4.2.13}$$

then the van der Waals equation takes on the *universal* form

$$\left(\pi + \frac{3}{v^2}\right)\left(v - \frac{1}{3}\right) = \frac{8t}{3}. \tag{4.2.14}$$

This is universal in that there are no system-specific quantities. In other words, when the critical volume, temperature and pressure of a system

---

[1]Note that the $t$ defined here is different from the reduced temperature defined in Section 4.1.7.

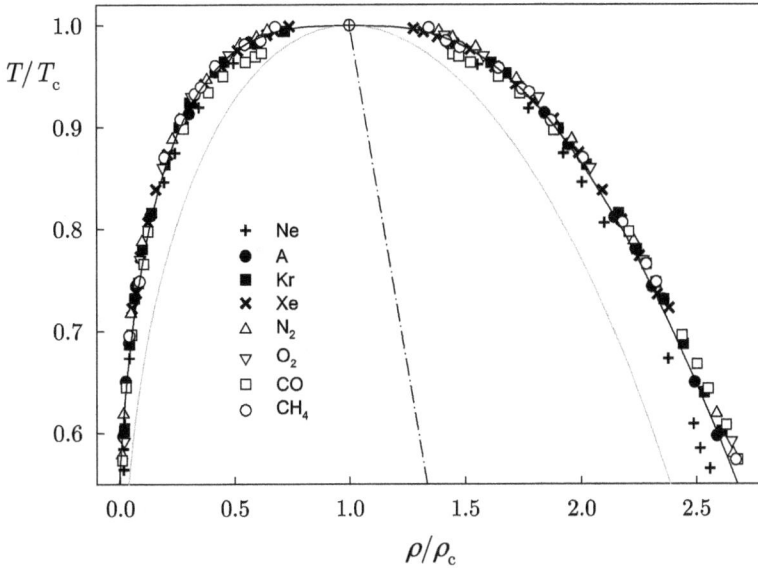

Fig. 4.12.    Guggenheim plot of liquid–gas coexistence.

are known, then in terms of the reduced variables all fluids should obey the same equation. This is known as the *Law of corresponding states*. Furthermore, the quantity $p_c V_c / N k T_c$ is predicted to have the universal value $3/8 = 0.375$ for all liquid–gas systems.

As a demonstration of Corresponding States, Fig. 4.12 shows liquid–gas coexistence data for a number of substances plotted in reduced form, originally by Guggenheim [12] in 1945; see also [13, 14]. The points do indeed fall reasonably well onto a universal curve. However, this is not the curve predicted from the van der Waals equation, the grey curve in the figure.

We conclude that the law of corresponding states *does* seem to be followed, but the van der Waals equation does not give a good description of the universal behaviour.

**Explanation:**    The explanation of this is that the law of corresponding states is not reliant on the precise details of the van der Waals equation. It follows solely from the assumption that the energy of interaction between the particles is of the form

$$U(r) = \varepsilon\, u(r/\sigma) \tag{4.2.15}$$

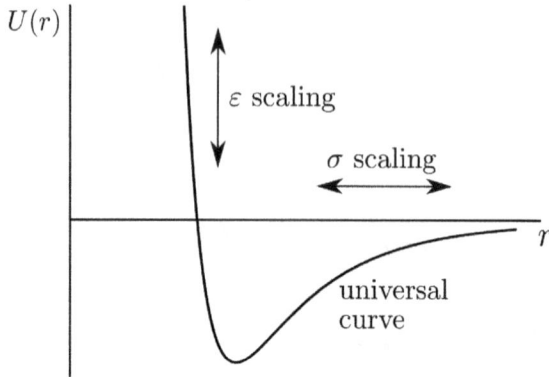

Fig. 4.13.   Scaling of universal interaction function.

where the energy parameter $\varepsilon$ and the distance parameter $\sigma$ are different for different substances, but the functional form of $u$ is the same. In other words, the assumption is that the interaction is of a universal form with a scaling *energy* and a scaling *length* characterising the substance. This is indicated in Fig. 4.13.

So, everything which follows from this interaction, such as the equation of state, the critical quantities, etc., must be functions only of these two scaling parameters. Thus, the universality when working in terms of the reduced variables, even if this is not precisely in accordance with the predictions of the van der Waals equation. Furthermore, since $p_c$, $V_c$ and $T_c$ are simply functions of $\varepsilon$ and $\sigma$, we can eliminate the two variables from the three equations so that the critical compressibility factor $z_c = p_c V_c / N k T_c$ is independent of $\varepsilon$ and $\sigma$. While the van der Waals equation gives 0.375 for this quantity, experimental measurements give a value of $0.292 \pm 0.002$ for many substances, see Table 4.1.

We note that a calculation of $z_c$ for the Lennard-Jones 6–12 potential, gives gives $z_c = 0.281$ [15].

Table 4.1.   Critical compressibility factor $z_c = p_c V_c / N k T_c$.

|       | $^4$He | Ne    | A     | Kr    | Xe    | $N_2$ | $O_2$ | CO    | $CH_4$ |
|-------|--------|-------|-------|-------|-------|-------|-------|-------|--------|
| $z_c$ | 0.308  | 0.305 | 0.290 | 0.291 | 0.290 | 0.291 | 0.292 | 0.294 | 0.290  |

The curve drawn by Guggenheim plotting his data, Fig. 4.12, has a remarkably simple form:

$$\frac{\rho - \rho_c}{\rho_c} = \frac{3}{4}\left(1 - \frac{T}{T_c}\right) \pm \frac{7}{4}\left(1 - \frac{T}{T_c}\right)^{1/3}. \tag{4.2.16}$$

Here, the plus sign applies to the liquid (right side) and the minus sign to the gas (left side). Since $(\rho - \rho_c)/\rho_c$ is the order parameter for this transition, we see Guggenheim's formula corresponds to the order parameter critical exponent $\beta = 1/3$. By contrast, the van der Waals coexistence curve in Fig. 4.12 (the grey curve), obtained by applying the Maxwell construction to the van der Waals equation of state, gives

$$\frac{\rho_1 - \rho_c}{\rho_c} = \frac{2}{5}\left(1 - \frac{T}{T_c}\right) \pm 2\left(1 - \frac{T}{T_c}\right)^{1/2} \tag{4.2.17}$$

in the vicinity of the critical point. This would correspond to the order parameter critical exponent being $\beta = 1/2$.

### 4.2.6 Dieterici's equation

We encountered the Dieterici equation in Section 3.5.1. This is a phenomenological equation with far less microscopic justification than the van der Waals equation. However, as stated in Chapter 3, the Dieterici equation gives a better description of behaviour near the critical point than does the van der Waals equation. We write the equation in the form

$$p(V - Nb) = NkTe^{-Na/kTV} \tag{4.2.18}$$

where, as we saw, the parameters $a$ and $b$ have precisely the same interpretation as in the van der Waals equation.

The critical point for the liquid–gas system is the point of inflection, where

$$\left.\frac{\partial p}{\partial V}\right|_T = 0 \quad \text{and} \quad \left.\frac{\partial^2 p}{\partial V^2}\right|_T = 0. \tag{4.2.19}$$

The volume, pressure and temperature at this point are found to be

$$V_c = 2Nb, \quad p_c = \frac{a}{4b^2}e^{-2}, \quad kT_c = \frac{a}{4b}. \tag{4.2.20}$$

Then in terms of the reduced dimensionless variables $v$, $\pi$ and $t$

$$v = \frac{V}{V_c}, \quad \pi = \frac{p}{p_c}, \quad t = \frac{T}{T_c} \tag{4.2.21}$$

the Dieterici equation takes on the universal form

$$\pi\left(v - \frac{1}{2}\right) = \frac{t}{2}e^2 e^{-2/tv}. \tag{4.2.22}$$

As with the reduced van der Waals equation, this is universal in that there are no system-specific quantities. In other words, when the critical volume, temperature and pressure of a system are known, then in terms of the reduced variables all fluids should obey the same equation.

For the Dieterici equation, the critical compressibility factor is given by

$$z_c = \frac{p_c V_c}{NkT_c} = 2e^{-2}$$

$$= 0.271. \tag{4.2.23}$$

This is closer to the experimental value around 0.29 than is the van der Waals value of 0.375. For this reason, it is claimed the Dieterici equation gives a better description of a fluid's behaviour in the vicinity of the critical point. However, there has been a revival of interest in the Dieterici equation [16] with the claim that it can give a better description of the entire liquid–gas coexistence region than can the van der Waals equation.

### 4.2.7   *Quantum mechanical effects*

The observant reader will notice in Fig. 4.12 that the data points for *liquid* neon show a small but consistent deviation from the universal line. Neon is the lightest molecule shown in the figure and its mass is the key. This may be seen even more dramatically in the behaviour of helium, which is significantly lighter. In Fig. 4.14, we show the Guggenheim plot coexistence data again, this time augmented by points from the two helium isotopes [17, 18].

The breakdown of Corresponding States occurs as a consequence of quantum effects; this happens when the de Broglie wavelength $\lambda$ becomes larger than the size of the particles. Then it is no longer appropriate to treat the particles as classical objects since their "extent" is determined by quantum mechanics.

A particle of mass $m$ and energy $\varepsilon$ has a momentum $p = \sqrt{2m\varepsilon}$ and a de Broglie wavelength $\lambda = h/\sqrt{2m\varepsilon}$. This is compared with the particle's

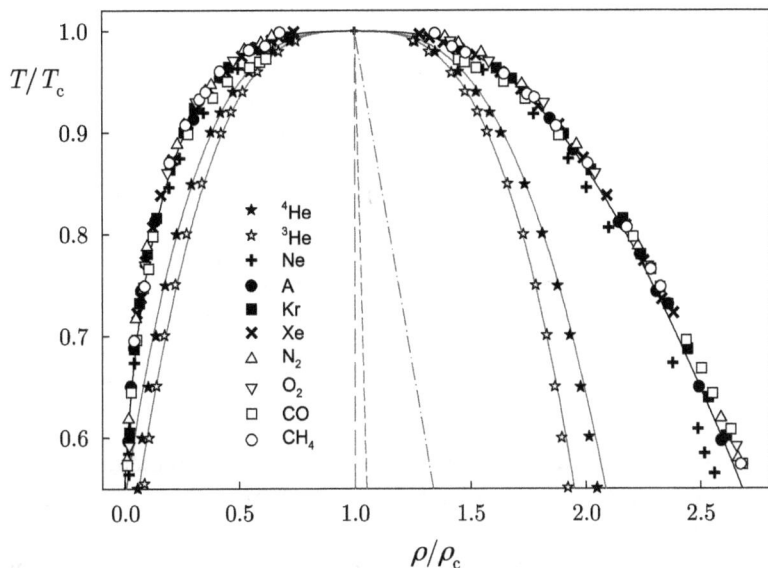

Fig. 4.14. Liquid–gas coexistence including data for helium.

size $\sigma$. The ratio (apart from a factor of $\sqrt{2})^2$ is known as the de Boer parameter $\Lambda^*$:

$$\Lambda^* = \frac{h}{\sigma\sqrt{m\varepsilon}}. \tag{4.2.24}$$

So whenever $\Lambda^*$ is less than one, we may regard the particles as being classical. But if $\Lambda^*$ is greater than one, then quantum effects will be important. In a general view of this, we may take the energy and size parameters as the scaling parameters of an un-specified interaction function, in the spirit of Section 3.2.7. However, for numerical calculation, we may take $\varepsilon$ and $\sigma$ to be those of the Lennard-Jones interaction. These [23] and the corresponding values of $\Lambda^*$ for the inert gases are shown in Table 4.2.

For most substances, we see that $\Lambda^*$ is small; the de Broglie wavelength is significantly less than the particle size and the particles therefore behave classically. Neon is marginal; here the de Broglie wavelength is beginning to become significant. And the classical picture is entirely inappropriate in the case of helium; there quantum effects become crucial.

---

[2] Some authors use different numerical factors in the definition of $\Lambda^*$; we follow de Boer's original expression.

Table 4.2. Lennard-Jones and de Boer parameters for the inert gases.

| | $\varepsilon/k(K)$ | $\sigma(\text{Å})$ | $m(\text{AMU})$ | $m(\text{kg})$ | $\Lambda^*$ |
|---|---|---|---|---|---|
| $^3$He | 10.14 | 2.56 | 3.016 | $5.01 \times 10^{-27}$ | 3.092 |
| $^4$He | 10.14 | 2.56 | 4.003 | $6.65 \times 10^{-27}$ | 2.685 |
| Ne | 36.21 | 2.74 | 20.18 | $3.35 \times 10^{-26}$ | 0.591 |
| Ar | 120.93 | 3.40 | 39.95 | $6.63 \times 10^{-26}$ | 0.185 |
| Kr | 162.93 | 3.65 | 83.80 | $1.39 \times 10^{-25}$ | 0.103 |
| Xe | 231.72 | 3.98 | 131.29 | $2.18 \times 10^{-25}$ | 0.063 |
| Ra | 283.00 | 4.36 | 222.00 | $3.69 \times 10^{-25}$ | 0.040 |

We should note also that a full quantum treatment would include the effect of statistics; recall that $^4$He obeys Bose–Einstein statistics, whereas $^3$He obeys Fermi–Dirac statistics.

A particle localised within a distance $\sigma$ will have a momentum uncertainty $\sim h/\sigma$ corresponding to a zero-point kinetic energy of $\sim h^2/2m\sigma^2$. This can be compared with the attractive energy $\varepsilon$. We see that $\Lambda^{*2}$ is approximately the ratio of the zero-point energy to the attractive interaction.

There is a quantum extension of the law of corresponding states, introduced by de Boer [19] in 1948.

One should note, however, that while the helium points do not collapse onto the other points of the Guggenheim plot, the behaviour in the vicinity of the critical point still corresponds to an order parameter critical exponent $\beta \approx 1/3$ [20].

## 4.3 Second-Order Transition — An Example

### 4.3.1 *The ferromagnet*

The essential phenomenon associated with the ferromagnet is that below a certain temperature a magnetisation will spontaneously appear in the absence of an applied magnetic field. It is understood that the interaction responsible for ferromagnetism is the exchange interaction between electron spins. The origin of the exchange interaction is the necessity to antisymmetrise the electronic wavefunction, together with the Coulomb repulsion between electrons. Then the symmetric and the antisymmetric wavefunctions have different energies and this may be written as an *effective* spin-dependent hamiltonian:

$$\mathscr{H}_x = -\hbar J \sum_{i,j}^{nn} \mathbf{S}_i \cdot \mathbf{S}_j \qquad (4.3.1)$$

called the *Heisenberg* hamiltonian. You should be familiar with this from your Atomic Physics studies. The key feature of this interaction is that when $J$ is positive, the energy is minimised when the spins are parallel; ferromagnetism is the energetically favourable state. And when $J$ is negative, the favourable state occurs when neighbouring spins are antiparallel; this is an antiferromagnet. Note that the exchange interaction is rotationally invariant. The sum is over nearest neighbours and $J$ is called the exchange frequency.

As written above, the sum counts each pair twice since both $i$ and $j$ vary freely. For this reason, we will sometimes use the alternate way of expressing the same thing:

$$\mathscr{H}_x = -2\hbar J \sum_{i<j}^{nn} \mathbf{S}_i \cdot \mathbf{S}_j \qquad (4.3.2)$$

since in this case the restriction $i < j$ avoids the double counting.

The phase diagram for a ferromagnet is shown in Fig. 4.15. The system is restricted to the surface, and there is a second, lower surface at negative $B$ and $M$. The surface is symmetric under $M \to -M$, $B \to -B$. Note that there is a smooth variation between positive and negative magnetisation only at $B = 0$, for $T > T_c$. Otherwise the change is discontinuous.

For temperatures below the critical temperature, a magnetisation isotherm is shown in Fig. 4.16. Since $B = 0$ when the magnetisation inverts, there is no cost in energy.

### 4.3.2 *The Weiss model*

The behaviour of the ferromagnet is contained in its hamiltonian. In the presence of a magnetic field **B**, assumed to point in the $z$ direction, the hamiltonian is

$$\mathscr{H} = -\hbar\gamma\mathbf{B} \cdot \sum_{i} \mathbf{S}_i - \hbar J \sum_{i,j}^{nn} \mathbf{S}_i \cdot \mathbf{S}_j \qquad (4.3.3)$$

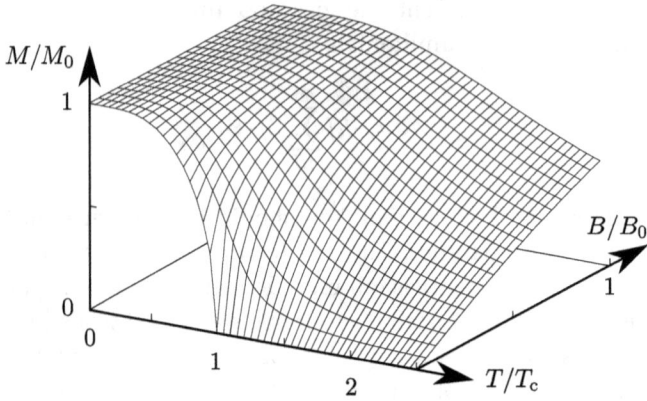

Fig. 4.15.   Phase diagram for a magnetic system.

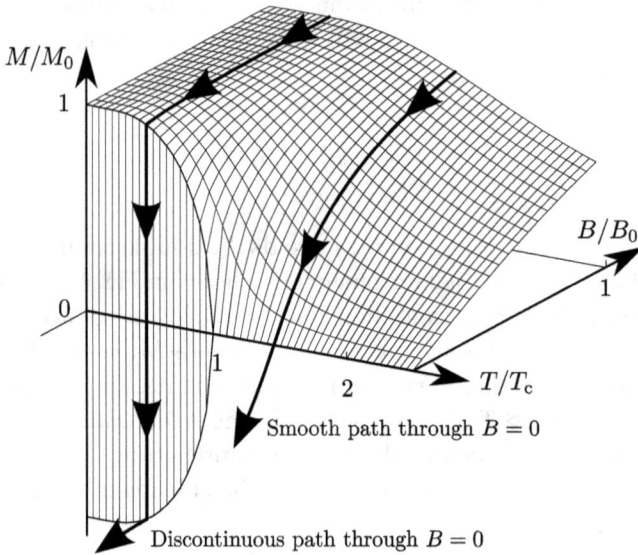

Fig. 4.16.   Variation of $M$ as $B$ goes through zero.

where the first term is the usual $\mathbf{M} \cdot \mathbf{B}$ term for the Zeeman interaction between magnetic moments and a magnetic field; $\gamma$ is the magnetogyric (gyromagnetic) ratio of the electron. Writing this hamiltonian as

$$\mathscr{H} = -\hbar\gamma \left\{ \mathbf{B} + \frac{J}{\gamma} \sum_{j}^{\mathrm{nn}} \mathbf{S}_j \right\} \cdot \sum_i \mathbf{S}_i \qquad (4.3.4)$$

we see that the effect of the exchange term is to contribute some extra magnetic field at the site of each spin due to its neighbours.

The extra field at each site is different; it is a result of the particular disposition of the site's neighbours. Furthermore, the field at a site will vary because of the time evolution of the neighbouring moments induced by the hamiltonian. However, the time-averaged field at different sites will be expected to be the same. We shall take the extra field to be the same, and constant, throughout the specimen. This is thus a mean field treatment. The mean extra magnetic field is then

$$
\mathbf{b} = \frac{J}{\gamma} \left\langle \sum_{j}^{\mathrm{nn}} \mathbf{S}_j \right\rangle
$$

$$
= \frac{nJ}{\gamma} \langle \mathbf{S} \rangle ,
$$

(4.3.5)

where $n$ is the number of nearest neighbours of a site. Now, the magnetisation $\mathbf{M}$ is given by

$$
\mathbf{M} = N\gamma\hbar \langle \mathbf{S} \rangle
$$

(4.3.6)

so that

$$
\langle \mathbf{S} \rangle = \mathbf{M}/N\gamma\hbar
$$

(4.3.7)

and the mean extra magnetic field can then be written as

$$
\mathbf{b} = \frac{nJ}{N\gamma^2\hbar}\mathbf{M}.
$$

(4.3.8)

This field is proportional to the magnetisation. This was the basis of the model introduced by Pierre Weiss in 1907 to explain ferromagnetism. He didn't know about quantum mechanics; he postulated such an internal magnetic field, but with no knowledge of its origin.

We have, then, a single-particle effective hamiltonian. The properties of this system may be found from the paramagnet model presented in Chapter 2. There we found the magnetisation (total magnetic moment) of an assembly of $N$ spins of magnitude $\hbar/2$ is given by

$$
M = N\frac{\gamma\hbar}{2} \tanh \frac{\gamma\hbar B}{2kT},
$$

(4.3.9)

pointing in the direction of $\mathbf{B}$.

Now, we must add the mean extra field **b**, Eq. (4.3.8), so that the magnetisation is then

$$M = N\frac{\gamma\hbar}{2} \tanh \frac{\gamma\hbar}{2kT} \left\{ B + \frac{nJ}{N\gamma^2\hbar}M \right\}. \qquad (4.3.10)$$

This is an implicit equation relating magnetisation, magnetic field and temperature. It provides a mathematical representation of the phase diagram shown in the previous section. The equation is, however, nonlinear and impossible to solve analytically. But we can find the spontaneous magnetisation and the critical temperature; we do this in the following section.

### 4.3.3 *Spontaneous magnetisation*

In zero-applied magnetic field, a magnetisation will spontaneously appear when the temperature falls below the critical temperature. This may be seen from the equation for magnetisation. When $B = 0$, this becomes

$$M = N\frac{\gamma\hbar}{2} \tanh \left\{ \frac{nJ}{2kTN\gamma}M \right\}. \qquad (4.3.11)$$

This is still a nonlinear and implicit equation that is difficult to solve analytically. But we can adopt a graphical method, which is highly instructive. We define the auxiliary quantity $X$ by

$$X = \frac{nJ}{2kTN\gamma}M \qquad (4.3.12)$$

and then we have two simultaneous equations in two unknowns. The solution corresponds to the intersection of the curves representing the two equations:

$$M = N\frac{\gamma\hbar}{2} \tanh X$$
$$M = \frac{2kTN\gamma}{nJ}X. \qquad (4.3.13)$$

These equations are plotted for three different temperatures in Fig. 4.17.

We observe there is always a solution at $M = 0$. At high temperatures, as expected, this is the only solution. At low temperatures, however, there is a second solution where the line intersects the curve at a non-zero value of $M$. We shall see in Section 4.5.2 that this corresponds to the energetically favourable solution. Thus, for temperatures below a critical value, there

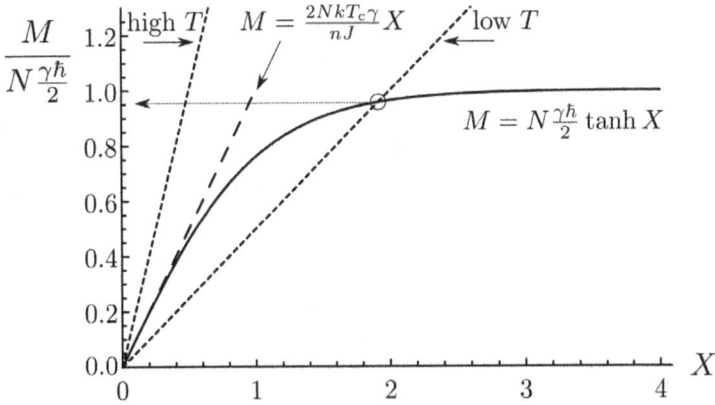

Fig. 4.17. Graphical solution for spontaneous magnetisation.

will be a non-zero magnetisation. The critical temperature is that for which the straight line is tangential with the tanh at the origin, when

$$N\frac{\gamma\hbar}{2} = \frac{2kT_c N\gamma}{nJ},\tag{4.3.14}$$

that is,

$$T_c = \frac{\hbar n}{4k}J.\tag{4.3.15}$$

In the case of a 2d square lattice, for which $n = 4$, one then has precisely

$$kT_c = \hbar J\tag{4.3.16}$$

an example of our "rule of thumb" for the transition temperature mentioned in Section 4.1.1.

Although an analytic solution for $M$ in terms of $T$ is not possible, we can solve for $T$ in terms of $M$. We note that the implicit equation for the spontaneous magnetisation, Eq. (4.3.11), can be written in the elegant form

$$\frac{M}{M_0} = \tanh\left\{\frac{M}{M_0}\frac{T_c}{T}\right\}\tag{4.3.17}$$

relating the two (reduced) variables $M/M_0$ and $T/T_c$, where $M_0 = N\gamma\hbar/2$, the saturation magnetisation, and $T_c = \hbar nJ/4k$, the critical temperature. This equation may be inverted and solved in the following way. We write it as

$$\frac{M}{M_0}\frac{T_c}{T} = \tanh^{-1}\left(\frac{M}{M_0}\right)$$

$$= \frac{1}{2}\ln\left(\frac{1 + M/M_0}{1 - M/M_0}\right),$$

(4.3.18)

so that

$$\frac{T}{T_c} = \frac{M}{M_0}\left/\tanh^{-1}\left(\frac{M}{M_0}\right)\right.$$

$$= \frac{2M/M_0}{\ln[(1 + M/M_0)/(1 - M/M_0)]}.$$

(4.3.19)

Admittedly, this gives the temperature in terms of the magnetisation rather than *vice versa*, but it is an explicit expression. When plotted, it gives the form for the variation of spontaneous magnetisation with temperature. This solution is shown in Fig. 4.18.

### 4.3.4  *Critical behaviour*

In the vicinity of the critical point, the magnetisation will be very small. In that case, we may expand Eq. (4.3.19) as

$$1 - \frac{T}{T_c} = \frac{1}{3}\left(\frac{M}{M_0}\right)^2 + \frac{4}{45}\left(\frac{M}{M_0}\right)^4 + \frac{44}{945}\left(\frac{M}{M_0}\right)^6 + \cdots$$

(4.3.20)

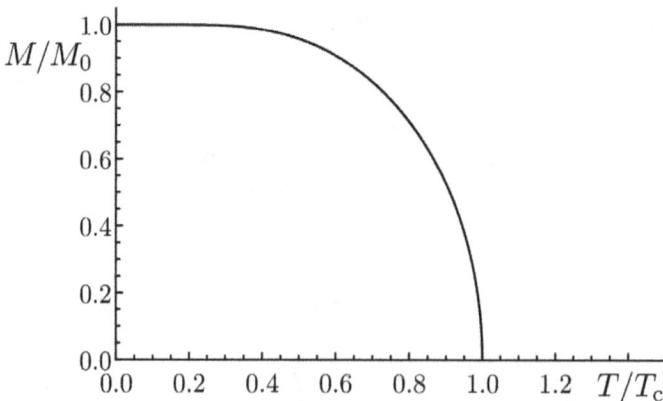

Fig. 4.18.   Spontaneous magnetisation of a ferromagnet.

and this series may be inverted, giving

$$\frac{M}{M_0} = \sqrt{3}\left(1 - \frac{T}{T_c}\right)^{1/2} - \frac{2\sqrt{3}}{5}\left(1 - \frac{T}{T_c}\right)^{3/2} + \cdots. \tag{4.3.21}$$

So in the vicinity of the critical point

$$\frac{M}{M_0} \sim \sqrt{3}\left(1 - \frac{T}{T_c}\right)^{1/2}. \tag{4.3.22}$$

This shows the initial growth of the magnetisation as the temperature is decreased through the critical point.

The dominant, singular, part of this behaviour is contained in the factor $(1 - T/T_c)^{1/2}$. In particular, it is the exponent ½ which characterises how $M$ "takes off" from zero. This is the critical exponent $\beta$ introduced in Section 4.1.7. Our conclusion is thus that the Weiss mean field model gives a value of ½ for $\beta$.

The coefficient of the dominant singular temperature behaviour, here the coefficient of $(1 - T/T_c)^{1/2}$, is called the *critical amplitude*. For the Weiss mean field model, the magnetisation critical amplitude is thus seen to be $\sqrt{3}$.

### 4.3.5  *Magnetic susceptibility*

The equation of state for the non-interacting paramagnet is given, from Chapter 2, by

$$M = N\frac{\gamma\hbar}{2}\tanh\frac{\gamma\hbar B}{2kT} \tag{4.3.23}$$

or, in terms of the saturation magnetisation $M_0 = N\gamma\hbar/2$:

$$M = M_0\tanh\frac{M_0 B}{NkT}. \tag{4.3.24}$$

At high temperatures/low polarisation, the tanh is expanded to leading order, resulting in a linear relation between $M$ and $B$

$$M = \frac{M_0^2}{NkT}B = \frac{C}{T}B \tag{4.3.25}$$

in terms of the Curie constant $C$. And in this case, we defined the *magnetic susceptibility* $\chi$

$$\chi = \frac{\mu_0}{V}\frac{M}{B}. \tag{4.3.26}$$

The susceptibility of the non-interacting paramagnet is then

$$\chi = \frac{\mu_0 M_0^2}{VNk}\frac{1}{T} = \frac{\mu_0}{V}\frac{C}{T} \tag{4.3.27}$$

noting, particularly, the $1/T$ dependence of the susceptibility — Curie's law.

We now include the effect of the mean field, which can be written (in terms of $T_c$) as

$$b = \frac{Nk}{M_0^2}T_c M \tag{4.3.28}$$

so that now the magnetisation is

$$M = \frac{M_0^2}{NkT}\left(B + \frac{Nk}{M_0^2}T_c M\right)$$
$$= \frac{M_0^2}{NkT}B + M\frac{T_c}{T}. \tag{4.3.29}$$

This may be solved to give

$$M = \frac{M_0^2}{Nk(T - T_c)}B = \frac{C}{T - T_c}B \tag{4.3.30}$$

so that the susceptibility is now

$$\chi = \frac{M_0^2}{\mu_0 VNk(T - T_c)} = \frac{\mu_0}{V}\frac{C}{T - T_c}. \tag{4.3.31}$$

This is shown in Fig. 4.19; it is similar to Curie's law, except that the temperature $T$ is replaced by $T - T_c$. By contrast to Curie's law, this is referred to as the Curie–Weiss law.

The Curie–Weiss susceptibility exhibits divergent behaviour as the critical point is approached. The exponent of $(T - T_c)^{-1}$ is 1. This is the critical exponent $\gamma$. Thus, we conclude that the Weiss mean field model gives a value of 1 for the exponent $\gamma$.

We see that by plotting the inverse susceptibility as a function of temperature, the transition temperature may be estimated by extrapolating high temperature measurements to zero. Linear extrapolation corresponds to mean field behaviour. In Problem 4.11, you will see that for realistic values of $\gamma$ above unity, the actual transition temperature will be slightly lower than the mean field estimate.

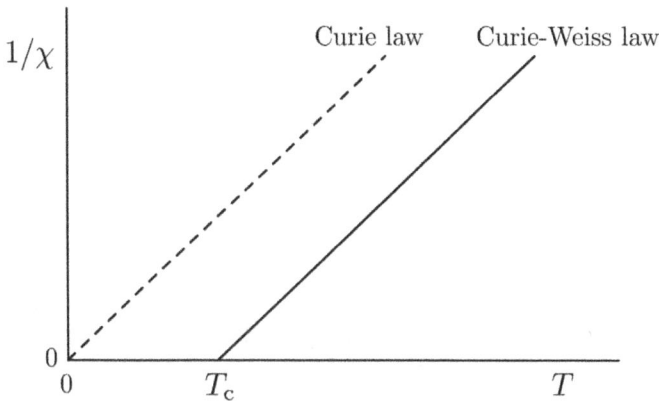

Fig. 4.19. Curie and Curie–Weiss laws.

### 4.3.6 *The ground state and Goldstone modes*

The essential feature of the ferromagnetic transition is that below a critical temperature a spontaneous magnetisation appears. This happens in the absence of an applied magnetic field. The magnetisation appears in a completely arbitrary direction; the interaction responsible for the transition is the Heisenberg exchange hamiltonian and this is rotationally invariant. Nevertheless, the vector magnetisation must point in some direction. And thus, the transition breaks the symmetry of the hamiltonian. There is, of course, also the $M = 0$ solution and this *does* respect the symmetry of the hamiltonian. But the non-zero $M$ solution, the symmetry-breaking solution, is energetically favourable. This means that the ground state of the system is highly degenerate, Fig. 4.20.

We may denote the set of degenerate ground states by $|\hat{r}\rangle$, where $\hat{r}$ is the unit vector pointing in the direction of the magnetisation. It is an observed fact that the ground state of this system is always one of the $|\hat{r}\rangle$ states; one never observes a linear superposition of such states, even though this is allowed by the laws of quantum mechanics. This puzzle is essentially the paradox of Schrödinger's cat, which goes to the very heart of quantum theory. In the measurement process, it appears that some quantum states "are more equal than others". Zurek [21] has argued that the favoured states occur as a consequence of interaction with the environment — a fundamental feature of the measurement process.

The ground state corresponds to a uniform order parameter; the magnetisation points in the same direction throughout the specimen. Now, it

Fig. 4.20.   Different possible ground states.

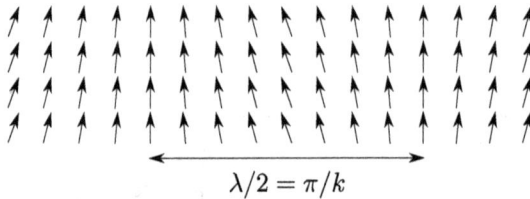

$$\lambda/2 = \pi/k$$

Fig. 4.21.   Spatial variation of the order parameter.

costs energy to deform the order parameter: a system with a deformed order parameter would no longer be in the ground state. So, if we consider a sinusoidal spatially varying order parameter $\mathbf{M(r)}$, Fig. 4.21,

$$\mathbf{M(r)} = M_0\big\{ \cos(\mathbf{k}\cdot\mathbf{r})\,\hat{\mathbf{x}} + \sin(\mathbf{k}\cdot\mathbf{r})\hat{\mathbf{y}} \big\} \qquad (4.3.32)$$

then this will have an energy above that of the ground state. This expression for $\mathbf{M(r)}$ is the spatial part of a spin wave with wave-vector $\mathbf{k}$. In the limit $\mathbf{k} \to 0$, the expression reduces to the uniform ground state with the magnetisation pointing in the $x$ direction. Thus, the energy of the spin wave goes to zero continuously as $k$ goes to zero.

This is the physical content of Goldstone's theorem, which states that when a continuous symmetry is broken, there will be excitations involving variations in the order parameter and the dispersion relation for these excitations, $\varepsilon(k)$, satisfies

$$\varepsilon(k) \to 0 \quad \text{as} \quad k \to 0. \qquad (4.3.33)$$

These excitations are known as Goldstone modes or, when quantised, as Goldstone bosons. They are the low energy excitations and thus they determine the low temperature thermal properties of the system.

The fact that $\varepsilon(k)$ goes to zero continuously means that there is no energy gap between the ground state and the excitations. This may also be interpreted as saying that the Goldstone bosons have zero mass. And this property follows only when the interaction responsible for the symmetry breaking, here the exchange hamiltonian, is of short range. We may contrast this with the superconducting transition. There the long range of the Coulomb force results in a gap in the plasmon excitation spectrum; then the bosons have mass. This is the condensed matter analogue of the Higgs mechanism of quantum field theory, whereby elementary particles acquire mass.

Note that Goldstone modes only appear when a *continuous* symmetry is broken. The order parameter cannot vary continuously in space when it is a discrete symmetry that is broken, such as that for the Ising model treated in the Section 4.4.

## 4.4 The Ising and Other Models

### 4.4.1 *Ubiquity of the Ising model*

The Ising model is a simple hamiltonian model introduced to treat ferromgnetism. It was originally proposed in the early 1920's by Wilhelm Lenz in an attempt to explain ferromagnetism from microscopic first principles. This was after Weiss had introduced the mean field model of ferromagnetism, but before the advent of the quantum Heisenberg exchange hamiltonian (which we had used to motivate the Weiss mean field). This pre-quantum model comprises an assembly of magnetic moments, each of which can point either parallel or antiparallel to a given direction, with a nearest neighbour interaction. It was Lenz's hope that this very simplest of interacting systems would exhibit a ferromagnetic transition. Lenz set this as a problem to his student Ernst Ising. Ising was able to solve only the one-dimensional case and he was disappointed to discover that there was no phase transition. It was then not until the 1940's that the two-dimensional case was solved by Lars Onsager, giving a ferromagnetic transition at a finite temperature. However, the three-dimensional Ising model is still not solved; probably it does not admit an analytic solution. A very readable account of the history of the Ising model is given in the review by Brush [22].

Note that the (Weiss) mean field model and its predictions make no mention of the spatial dimension. Within that context, the discovery of the difference between the one- and two-dimensional case of the Ising model was rather surprising.

The Ising model is one of the simplest descriptions of a system that leads to a phase transition. It is important furthermore because, notwithstanding its magnetic origin, it can be applied to a variety of dissimilar physical systems. This is an example of the phenomenon of *universality* in phase transitions. Perhaps of even more importance is the fact that the Ising model in two dimensions has been solved analytically. This is one of the very few microscopic models of interacting systems that have been solved exactly, to exhibit a phase transition.

The Ising model is specified in terms of a hamiltonian, so it is a complete microscopic description — but it keeps this description as simple as possible. It is a lattice model; the moments are fixed on sites. Often interactions are permitted only between nearest neighbours. The order parameter is a scalar: the expectation value of the Ising "spin". The insight of Lenz was in stripping away all superfluous aspects of the ferromagnet while keeping the essence of the system that leads to a transition.

We will write the Ising hamiltonian down in the following section. For the present, we simply note that this involves the energies of neighbouring pairs of the magnetic moments. We shall denote the directions in which the moments can point as up and down. Parallel neighbours have one energy, $\varepsilon_{\uparrow\uparrow}$, while antiparallel neighbours have another, $\varepsilon_{\uparrow\downarrow}$. In the case that

$$\varepsilon_{\uparrow\uparrow} < \varepsilon_{\uparrow\downarrow}, \tag{4.4.1}$$

the parallel state is energetically favourable and the ordered phase will thus be ferromagnetic. But when

$$\varepsilon_{\uparrow\uparrow} > \varepsilon_{\uparrow\downarrow}, \tag{4.4.2}$$

the antiparallel state is favoured and the ordered phase will be antiferromagnetic.

The Ising description need not, however, be restricted to real magnetic moments. For example, consider a binary mixture of A atoms and B atoms, as we shall do in Section 4.7. Here, again the interaction energies depend upon neighbouring particles. We can imagine that an A atom is represented by an "up moment" and a B atom by a "down moment". The analogue of the ferromagnetic state is phase-separation, when the A atoms coalesce together as do the B atoms. However, if it is favourable for unlike atoms to be neighbours, then we have the analogue of the antiferromagnet and the ordered phase will be a superlattice structure.

Finally, consider a fluid system modelled as a lattice gas. We can interpret an up moment as a molecule on the site, while a down moment indicates an empty site: a hole. Adsorption on a surface is an example of this model in two dimensions.

We see how, by interpreting the parameters appropriately, the Ising model may be used to describe a variety of physical systems. In most of the following, we will use the language of the magnetic case, but the generalisation to other systems is straightforward. However, one should note the distinction between conserved and non-conserved order parameters. For the binary mixture, the order parameter is conserved; the numbers of A and B atoms are fixed. However, in the magnetic case the order parameter is not conserved as spins may flip. The lattice gas model has a fixed number of molecules. If the number of available sites is fixed, then the order parameter is conserved. But if the number of available sites is unlimited, then the order parameter is not conserved. In the magnetic description, we may imagine applying a magnetic field to the system; this will determine the fraction of up and down moments. In the case of the non-conserved order parameter, this field is a free parameter. But for the conserved order parameter case, this field is a "chemical potential" whose value is determined by the conserved quantity; it fixes the number of A and B atoms in the binary mixture or the number of molecules in the lattice gas model.

### 4.4.2 *Magnetic case of the Ising model*

In the magnetic Ising model, we have a magnetic moment on each lattice site which may point either "up" or "down". As there are only two states, this is reminiscent of a spin ½ problem. It is thus appropriate to utilise a spin operator description, taking the up direction as pointing along the $z$ axis. We then assign to each site a spin with eigenvalue $S^z = +1/2$ if it carries an "up" moment and $S^z = -1/2$ if it carries a "down" moment.[3] We must count the number of parallel and antiparallel neighbouring pairs. This may be effected through consideration of the product of two neighbouring spins' eigenvalues; it will have magnitude $+1/4$ if the neighbours are parallel and $-1/4$ if they are antiparallel:

$$S_i^z S_j^z = +1/4 \quad \text{for parallel spins}$$
$$= -1/4 \quad \text{for antiparallel spins.}$$

---

[3]Traditional treatment of the Ising model ascribes the values +1 and −1 to the two values of the Ising "spin". We adopt a different convention that makes better connection with the spin ½ quantum mechanical description of the Heisenberg magnet.

It then follows that

$$2S_i^z S_j^z - 1/2 = 0 \quad \text{for parallel spins}$$
$$= -1 \quad \text{for antiparallel spins,}$$

so that the quantity $-(2S_i^z S_j^z - 1/2)$ can be used to count the number of antiparallel spins. Similarly,

$$2S_i^z S_j^z + 1/2 = 1 \quad \text{for parallel spins}$$
$$= 0 \quad \text{for antiparallel spins,}$$

so the quantity $2S_i^z S_j^z + 1/2$ can be used to count the number of parallel spins.

Thus, we have

$$N_{\uparrow\downarrow} = \sum_{i>j\,\text{nn}} (2S_i^z S_j^z - 1/2),$$

$$N_{\uparrow\uparrow} = \sum_{i>j\,\text{nn}} (2S_i^z S_j^z + 1/2),$$

(4.4.3)

where $\uparrow\uparrow$ indicates parallel spins of either orientation. In terms of the interaction energies $\varepsilon_{\uparrow\uparrow}$ and $\varepsilon_{\uparrow\downarrow}$ introduced in the previous section, the energy of the system is then

$$E = N_{\uparrow\uparrow}\varepsilon_{\uparrow\uparrow} + N_{\uparrow\downarrow}\varepsilon_{\uparrow\downarrow}$$

$$= -2(\varepsilon_{\uparrow\downarrow} - \varepsilon_{\uparrow\uparrow}) \sum_{i>j\,\text{nn}} 2S_i^z S_j^z + \text{const.}$$

(4.4.4)

The constant term may be ignored. And then the expression for the Ising hamiltonian may be written as

$$\mathcal{H} = -2\hbar J \sum_{i>j\,\text{nn}} S_i^z S_j^z,$$

(4.4.5)

where $\hbar J = \varepsilon_{\uparrow\downarrow} - \varepsilon_{\uparrow\uparrow}$. Note that the definition of $J$ in our treatment of the Ising model is a factor of four greater than that used in conventional treatments. This has been done purposely to maintain consistency with the $J$ used in the Heisenberg model (footnote 3 on page 245).

We can see from the expression for $\mathcal{H}$ that this is just a *part of* the Heisenberg hamiltonian considered previously. The Heisenberg interaction $\mathbf{S}_i \cdot \mathbf{S}_j$ is here replaced by the product $S_i S_j$; essentially, we are taking only the $z - z$ part of the dot product.

If we include an external magnetic field $B$ parallel to the up direction, then the hamiltonian is

$$\mathscr{H} = -2\hbar J \sum_{i>j}^{nn} S_i^z S_j^z - \gamma \hbar B \sum_i S_i^z. \tag{4.4.6}$$

The absence of any $x$ or $y$ spin operators is of vital importance. There are only $z$ spin operators and these will therefore all commute. This means that while we have used the quantum-mechanical language of spin ½ and spin operators, this is simply a convenient mathematical device and the model is really classical; there are no quantum aspects to the system.

We should also point out that the Weiss mean field treatment of the Heisenberg ferromagnet is equally appropriate for the Ising model. Thus, most of the results of Section 4.3.1 carry over to the Ising case. The only difference is that the order parameter for the Heisenberg magnet can point in any direction, while the Ising magnetisation can only point "up" or "down". The Heisenberg magnet breaks a continuous symmetry while the Ising transition breaks a discrete symmetry.

### 4.4.3 *Ising model in one dimension*

As Ising discovered, in one dimension the model exhibits no phase transition. This is an important deviation from the mean field prediction, indicating that the spatial dimension is important. The absence of a transition in 1d may be seen in the following elegant argument, due to Landau [28].

The essence of Landau's argument is to show that at any non-zero temperature the ordered phase is energetically unfavourable. Let us assume that there is an ordered phase below some critical temperature and let us examine the stability of this phase. We are considering a chain of $N$-ordered spins. Now, let us introduce a small element of disorder by reversing all spins from the $n$th site onwards, Fig. 4.22.

This will increase the energy by $\hbar J$ (there is no applied magnetic field), the contribution coming from the opposite-pointing pair. The entropy of the system will also increase. Since there is a choice of $N$ sites, after which spins are flipped, the reversal will increase the entropy by $k \ln N$. The effect

Fig. 4.22. Chain with spins reversed from one point onwards.

of the reversal is thus to increase the free energy $F = E - TS$ by

$$\Delta F = \hbar J - kT \ln N. \tag{4.4.7}$$

The equilibrium state of the system is that for which $F$ is a minimum. In the thermodynamic limit ($N \to \infty$), the second term of $\Delta F$ will always dominate so that the reversal operation results in a reduction of the free energy. And further reversal operations will further reduce the free energy. Thus, for all non-zero temperatures, the entropy term wins out over the energy term and the equilibrium configuration for the system is the disordered state. So, only at $T = 0$ can the ordered state exist; at all finite temperatures, it is unstable. There is no finite-temperature phase transition in an infinite one-dimensional lattice.

This behaviour is shown from a full microscopic calculation. In the presence of a magnetic field $B$, the fractional magnetisation of the one-dimensional Ising system is given (see Plische and Bergersen [24] for details) by

$$m = \frac{\sinh \gamma \hbar B / kT}{\left( \sinh^2 \gamma \hbar B / kT + e^{-\hbar J/kT} \right)^{1/2}}. \tag{4.4.8}$$

As expected, we see that when $B = 0$ there is no spontaneous magnetisation at any temperature except $T = 0$. Nevertheless, at low temperatures such that

$$\sinh^2 \gamma \hbar B / kT \gg e^{-\hbar J/kT}, \tag{4.4.9}$$

for any field $B$, however small, there will occur the saturation magnetisation.

### 4.4.4   Ising model in two dimensions

The literature on the 2d Ising model is vast; there are entire books devoted to the subject. There are various methods for calculating the partition function and the transition temperature. However, all are complicated, requiring more space than we can reasonably devote in this book. We thus content ourselves with quoting the main results. For further details, we refer the reader to Landau and Lifshitz [28].

In two dimensions, the Ising model exhibits a phase transition. The analytic expression for the transition temperature was derived by Onsager in 1944. The free energy was found, somewhat later, to be

$$F = -Nk\ln\left(2\cosh\hbar J/2kT\right) - \frac{Nk}{2\pi}\int_0^\pi \ln\frac{1}{2}\left(1 + \sqrt{1 - \kappa^2\sin^2\varphi}\right)d\varphi$$

(4.4.10)

where

$$\kappa = \frac{2}{\cosh\hbar J/2kT \coth\hbar J/2kT}.$$

(4.4.11)

The transition occurs at the singularity in $F$, at a critical temperature of

$$T_c = 0.567\hbar J/k,$$

(4.4.12)

the solution of the equation

$$2\tanh^2\hbar J/2kT_c = 1 \quad \text{or} \quad \sinh\hbar J/2kT_c = 1.$$

(4.4.13)

The spontaneous (reduced) magnetisation, the order parameter, is given by

$$m = \left\{1 - (\sinh\hbar J/2kT)^{-4}\right\}^{1/8}.$$

(4.4.14)

This is shown in Fig. 4.23.

The critical exponents for the 2d Ising model may be calculated. We compare these with the mean field values we obtained already, together with others, from Section 4.4.5:

| 2d Ising model | $\alpha$ | $\beta$ | $\gamma$ | $\delta$ | $\nu$ | $\eta$ |
|---|---|---|---|---|---|---|
| mean field | 0 | ½ | 1 | 3 | ½ | 0 |
| exact calc. | 0 | ⅛ | ¾ | 15 | 1 | 0 |

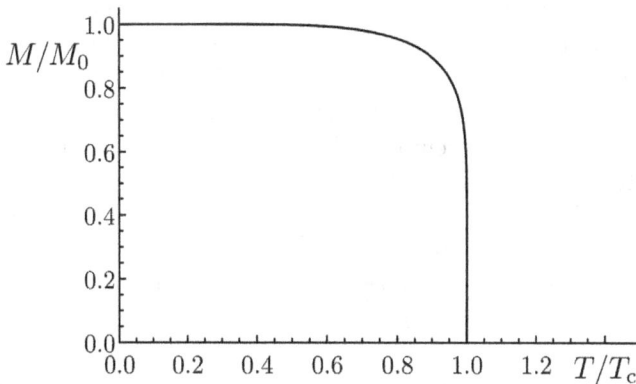

Fig. 4.23. Order parameter for 2d Ising model.

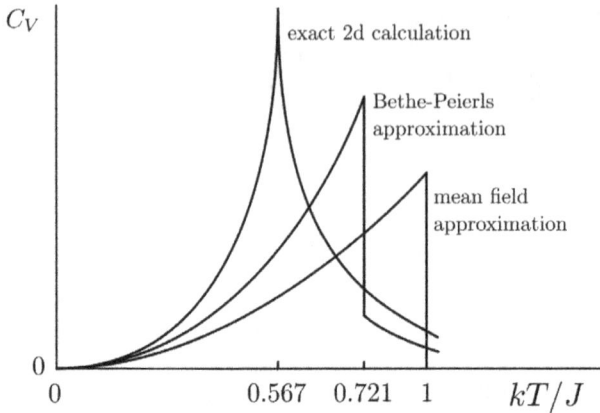

Fig. 4.24.   Heat capacity of 2d Ising model.

The heat capacity of the Ising system in the vicinity of the phase transition is shown in Fig. 4.24, together with that calculated from two approximate models.

The mean field calculation gives the simplest behaviour. Note that as the temperature is reduced from above the transition, there is no "precursor" of the impending critical behaviour. The Bethe-Peierls approximation is an attempt to do rather better than the mean field calculation. It considers the effects of nearest neighbour interactions exactly, while the spins further away are treated in mean field. Here, the predicted critical temperature is somewhat better and there is a small increase in the heat capacity before the transition occurs.

The precursor in the heat capacity is an indication that something is "going on" before the transition actually occurs. This is evidence of the importance of fluctuations in the vicinity of the critical point. In Section 4.5.2, we shall see how the mean field/Landau approach indicates, through the dramatic flattening of the free energy at the critical point, that large fluctuations in the order parameter can occur at negligible cost. However, mean field theories take no account of the effects of such fluctuations. The initial assumption that all particles experience the same constant "mean field" precludes the incorporation of such fluctuations.

Fisher [25] has commented that the fluctuations close to the critical point may be regarded as the system "trying-out" regions of the other phase (but finding it energetically unfavourable). Thus, the action on *both* sides of the transition.

### 4.4.5 *Mean field critical exponents*

In treating the Weiss model of the ferromagnet, we obtained values for two of the critical exponents. In that mean field treatment, no mention was made of the dimensionality of the order parameter or of the dimensionality of space. Thus, mean field arguments will apply equally to the ferromagnet and the Ising model. Note, however, that in contrast to the results of the previous two sections, the mean field treatment gives results independent of the spatial dimension.

Recall we found $\beta = 1/2$ for the order parameter exponent and $\gamma = 1$ for the susceptibility exponent. In this section, we will use general mean field arguments to extend that discussion, to obtain $\alpha$, the heat capacity exponent and $\delta$ the equation of state exponent.

(a) **Heat capacity critical exponent $\alpha$:**  The heat capacity is found from the internal energy of the system by differentiating with respect to temperature. Now, the internal energy is given by

$$E = -MB \qquad (4.4.15)$$

where $B$ is the sum of the external and internal fields.

So, in the absence of an external field, we have

$$\begin{aligned} E &\sim M^2 & T < T_c \\ E &= 0 & T > T_c. \end{aligned} \qquad (4.4.16)$$

But we know that $M \sim (T - T_c)^{1/2}$; that is the order parameter critical behaviour. Thus, we have

$$\begin{aligned} E &\sim T - T_c & T < T_c \\ E &= 0 & T > T_c. \end{aligned} \qquad (4.4.17)$$

And differentiating to find the heat capacity, this gives

$$\begin{aligned} C &= \text{const} & T < T_c \\ &= 0 & T > T_c. \end{aligned} \qquad (4.4.18)$$

The result is that the heat capacity is *discontinuous* at the critical point. Observe this in the mean field curve of Fig. 4.24. Such behaviour is observed in the superconducting transition, but not for real ferromagnets.

In terms of a critical exponent, a discontinuity is understood as a power-law behaviour

$$C \sim \frac{1}{x}(T - T_c)^x \qquad (4.4.19)$$

as the quantity $x$ tends to zero. Thus, we have the conclusion that the heat capacity critical exponent $\alpha$ is zero in the mean field model.

**(b) Equation of state critical exponent $\delta$:** We now turn to the critical exponent connected with the equation of state at $T = T_c$. Recall that $\delta$ is defined by

$$M \sim B^{1/\delta} \qquad (4.4.20)$$

evaluated at the critical point. We have to consider here the behaviour of the magnet in the presence of a magnetic field. The equation of state is conveniently written as

$$\frac{M}{M_0} = \tanh\left\{\frac{M_0 B}{NkT} + \frac{M}{M_0}\frac{T_c}{T}\right\}. \qquad (4.4.21)$$

Now, in the vicinity of the critical point, $M$ is small and if we restrict the discussion to small values of $B$, then we may expand the tanh. Thus, we expand as far as

$$\frac{M}{M_0} = \frac{M_0 B}{NkT} + \frac{M}{M_0}\frac{T_c}{T} - \frac{1}{3}\left\{\frac{M_0 B}{NkT} + \frac{M}{M_0}\frac{T_c}{T}\right\}^3 + \cdots. \qquad (4.4.22)$$

Since we are considering the relation between $M$ and $B$ *at* the critical point, we set $T = T_c$, giving

$$\frac{M}{M_0} = \frac{M_0 B}{NkT_c} + \frac{M}{M_0} - \frac{1}{3}\left\{\frac{M_0 B}{NkT_c} + \frac{M}{M_0}\right\}^3, \qquad (4.4.23)$$

or

$$\frac{M_0 B}{NkT_c} = \frac{1}{3}\left\{\frac{M_0 B}{NkT_c} + \frac{M}{M_0}\right\}^3. \qquad (4.4.24)$$

This can be rearranged as

$$\frac{M}{M_0} = \left\{\frac{3M_0 B}{NkT_c}\right\}^{1/3} - \frac{M_0 B}{NkT_c}. \qquad (4.4.25)$$

Now, when $B$ is small, it is the first term on the right-hand side that will dominate. Thus, we find that at the critical point

$$M \sim B^{1/3}, \qquad (4.4.26)$$

and so we identify the critical exponent $\delta$ as 3 in the mean field approximation.

We are unable to derive the mean field results $\nu = \frac{1}{2}$ and $\eta = 0$ here as the calculations are rather more complicated. See Huang [3] for details.

We give the critical exponents for the three-dimensional Heisenberg ferromagnet as follows, comparing values calculated from the mean field theory with those measured experimentally [11].

| 3d ferromagnet | $\alpha$ | $\beta$ | $\gamma$ | $\delta$ | $\nu$ | $\eta$ |
|---|---|---|---|---|---|---|
| mean field | 0 | ½ | 1 | 3 | ½ | 0 |
| experiment | −0.14 | 0.3 | 1.4 | 4.8 | 0.7 | 0.04 |

The corresponding exact values for the two-dimensional Ising model were given in a table in Section 4.4.4.

From these results, we must conclude that while the mean field theory provides a good *qualitative* model for behaviour at the critical point, quantitatively it leaves something to be desired. This is because of the limitations imposed by the fundamental assumption of the mean field model; using a *mean* field involved neglecting the variation in the actual field from site to site. It is precisely these fluctuations, which are neglected, that determine the behaviour in the vicinity of the critical point.

### 4.4.6  *The XY model*

The interaction hamiltonian for the XY model is given by

$$\mathscr{H} = -2\hbar J \sum_{\substack{i>j}}^{nn} \left( S_i^x S_j^x + S_i^y S_j^y \right). \tag{4.4.27}$$

This is an extension of the Ising hamiltonian, but not the full Heisenberg hamiltonian. The order parameter for this model is the magnetisation in the $x$–$y$ plane; it is a vector of dimension $n = 2$. Thus, the order parameter is two-dimensional. Since the magnetisation can point in any direction in the $x$–$y$ plane, it can vary continuously; the XY transition thus breaks a continuous symmetry. This model is regarded as a good description for the conventional superconducting transition and the superfluid transition in liquid $^4$He, since the order parameter in this case is a two-component vector (or a complex scalar). Estimates for critical exponents for the XY model in three spatial dimensions are given in Table 4.3 in Section 4.9.3.

### 4.4.7 *The spherical model*

The spherical model was proposed by Mark Kac (pronounced Cats), in 1952. The specification of the original model is probably not so intuitively appealing! But subsequently it was realised that it was equivalent to the limiting case of the $n$−vector model as $n \to \infty$. In other words, one takes spin variables of dimension $n$, with an interaction of the form

$$\mathscr{H} = -2\hbar J \sum_{i>j}^{nn} \left( S_i^\alpha S_j^\alpha + S_i^\beta S_j^\beta + S_i^\gamma S_j^\gamma + \cdots + S_i^n S_j^n \right), \qquad (4.4.28)$$

in the limit of infinite $n$. Written in this way we can interpret the model as an extension of the Heisenberg model in a direction opposite to that of the Ising (or XY) model.

In 1952, Theodore Berlin obtained the analytic solution of this model for spatial dimension $d = 1, 2$ and 3 and this was subsequently extended to *continuous n* [26]. Berlin found that in one and two dimensions there was no transition to an ordered phase, but there was a transition at finite temperature in three dimensions. For a three-dimensional simple cubic lattice, the critical temperature is found to be

$$kT_c = 0.989\hbar J. \qquad (4.4.29)$$

So unlike the Ising model, there is no transition in two dimensions.

Another important result is that for $d > 4$, the behaviour is independent of $d$ and equivalent to that of the mean field model.

Estimates for critical exponents for the spherical model in three spatial dimensions are given in Table 4.3 in Section 4.9.3.

## 4.5 Landau Theory of Phase Transitions

### 4.5.1 *Landau free energy*

In order to develop a general theory of phase transitions, it is necessary to extend the concept of the free energy. For definiteness we will, in this section, consider the case of a ferromagnet, but it should be appreciated that the ideas introduced apply more generally.

The (magnetic) Helmholtz free energy has proper variables $T$ and $B$. In differential form

$$dF = -SdT - MdB \qquad (4.5.1)$$

and the entropy and magnetisation are thus given by

$$S = -\frac{\partial F}{\partial T}\bigg|_B, \quad M = -\frac{\partial F}{\partial B}\bigg|_T. \tag{4.5.2}$$

When the temperature and magnetic field are specified the magnetisation (the order parameter in this system) is determined — from the second of the above differential relations. It is possible however, perhaps because of some constraint, that the system may be in a *quasi-equilibrium* state with a different value for its magnetisation. Then the Helmholtz free energy corresponding to that state will be greater than its equilibrium value; it will be a minimum when $M$ takes its equilibrium value. This is equivalent to the law of entropy increase discussed in Chapter 1 and extended in Appendix B on thermodynamic potentials.

The conventional Helmholtz free energy applies to systems in thermal equilibrium, as do all thermodynamic potentials. The above discussion leads us to the introduction of a "constrained" Helmholtz free energy for quasi-equilibrium states, which we shall write as

$$F(T, B : M). \tag{4.5.3}$$

The conventional Helmholtz free energy is a mathematical function of $T$ and $B$; when $T$ and $B$ are specified, then the equilibrium state of the system has Helmholtz free energy $F(T, B)$. The system we are considering here is prevented from achieving its full equilibrium state since its magnetisation is constrained to take a certain value. The full equilibrium state of the system is that for which $F(T, B : M)$ is minimised with respect to variations in $M$, i.e.

$$\frac{\partial F(T, B : M)}{\partial M} = 0. \tag{4.5.4}$$

This constrained free energy is often called the Landau free energy (or *clamped* free energy). In a "normal" system, one would expect the Landau free energy to possess a simple minimum at the equilibrium point, as shown in Fig. 4.25.

### 4.5.2 *Landau free energy for the ferromagnet*

In order to visualise what underlies the Landau theory of phase transitions, we shall review the Weiss model of the ferromagnetic transition from the point of view of the constrained free energy. Thus, we shall consider the Landau free energy of this system.

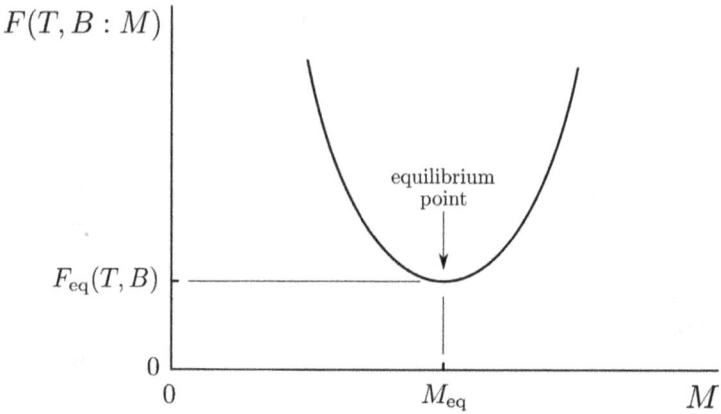

Fig. 4.25.   Variation of Landau free energy with magnetisation.

The Helmholtz free energy is defined as

$$F = E - TS. \tag{4.5.5}$$

We will evaluate the internal energy and the entropy separately. Since we require the *constrained* free energy, we must be sure to keep $M$ as an explicit variable. The internal energy is given by

$$E = - \int B \, dM \tag{4.5.6}$$

(assuming the vectors $\mathbf{B}$ and $\mathbf{M}$ in the same direction). The magnetic field, in the Weiss model, is the sum of the applied field $B_0$ and the local (mean) field $b$:

$$B = B_0 + b. \tag{4.5.7}$$

We shall write the local field in terms of the critical temperature:

$$b = \frac{Nk}{M_0^2} T_c M. \tag{4.5.8}$$

Integrating up the internal energy, we obtain

$$E = -B_0 M - \frac{NkT_c}{2} \left( \frac{M}{M_0} \right)^2. \tag{4.5.9}$$

For the present, we will consider the case where there is no external applied field. Then $B_0 = 0$, and in terms of the reduced magnetisation

$m = M/M_0$ (the order parameter), the internal energy is

$$E = -\frac{NkT_c}{2}m^2. \tag{4.5.10}$$

Now, we turn to the entropy. This is most easily obtained from the expression[4]

$$S = -Nk \sum_j p_j \ln p_j \tag{4.5.11}$$

where $p_j$ are the probabilities of the single-particle states. It is simplest to treat spin one half, which is appropriate for electrons. Then there are two states to sum over:

$$S = -Nk \left[ p_\uparrow \ln p_\uparrow + p_\downarrow \ln p_\downarrow \right]. \tag{4.5.12}$$

Now, these probabilities are simply expressed in terms of $m$, the fractional magnetisation

$$p_\uparrow = \frac{1+m}{2} \quad \text{and} \quad p_\downarrow = \frac{1-m}{2} \tag{4.5.13}$$

so that the entropy becomes

$$S = \frac{Nk}{2} \left[ 2\ln 2 - (1+m)\ln(1+m) - (1-m)\ln(1-m) \right]. \tag{4.5.14}$$

We now assemble the free energy $F = E - TS$, to obtain

$$F = -\frac{Nk}{2} \left\{ T_c m^2 + T \left[ 2\ln 2 - (1+m)\ln(1+m) - (1-m)\ln(1-m) \right] \right\}. \tag{4.5.15}$$

This is plotted, in Fig. 4.26, for temperatures less than, equal to and greater than the critical temperature.

The occurrence of the ferromagnetic phase transition can be understood quite clearly from this figure. For temperatures above $T_c$, we see there is a single minimum in the Landau free energy at $m = 0$, while for temperatures below $T_c$, there are two minima either side of the origin. The symmetry changes precisely at $T_c$. There the free energy has flattened, meaning that $m$ may make excursions around $m = 0$ with negligible cost of free energy — hence the large fluctuations at the critical point.

Strictly speaking, Fig. 4.26 applies to the Ising case where the order parameter is a scalar; the order parameter must choose between the values

---

[4]This is the *Gibbs* entropy evaluated in a *Boltzmann* ensemble.

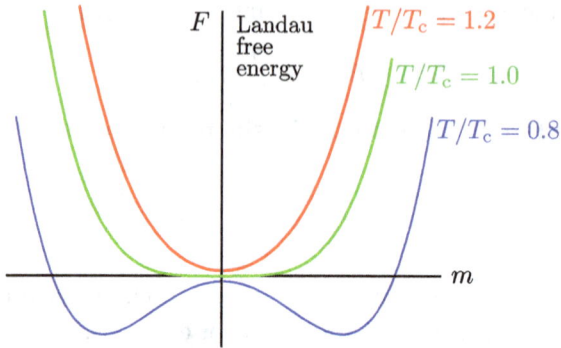

Fig. 4.26.   Landau free energy.

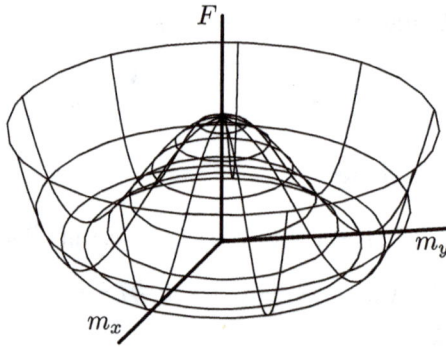

Fig. 4.27.   Landau free energy for XY model.

$\pm m$. The XY model is represented by Fig. 4.27 where, below the transition, the magnetisation can point anywhere in the $x$–$y$ plane in the minimum of the "wine bottle" potential. Similarly, for the Heisenberg model, Fig. 4.27 would apply where the $m_x$ and $m_y$ axes really refer to the three axes of the order parameter $m_x$, $m_y$ and $m_z$.

The behaviour in the vicinity of the critical point is even clearer if we expand the free energy in powers of the order parameter. Straightforward evaluation gives, neglecting the constant term,

$$F = \frac{Nk}{2}\left\{(T - T_c)m^2 + \frac{T_c}{6}m^4 + \cdots\right\}. \tag{4.5.16}$$

We see that the essential information is contained in the *quartic* form for the free energy in the vicinity of the critical point. This will have either one or two minima, depending on the values of the coefficients; the critical point occurs where the coefficient of $m^2$ changes sign. This abstraction of the *essence* of critical phenomena in terms of the general behaviour of the free energy was formalised by Landau. We have replaced the variable $T$ in the second term by the constant $T_c$ since we are restricting discussion to the vicinity of the critical point.

### 4.5.3  *Landau theory — Second-order transitions*

Landau's theory concerns itself with what is happening in the *vicinity* of the phase transition. In this region, the magnitude of the order parameter will be small and the temperature will be close to the transition temperature. When the problem is stated in this way, the approach is clear: The Landau free energy should be expanded as a power series in these small parameters. The fundamental assumption of the Landau theory of phase transitions is that in the vicinity of the critical point, the free energy is an *analytic* function of the order parameter so that it may indeed be thus expanded. In fact, this assumption is, in general, invalid. The Landau theory is equivalent to mean field theory (in the vicinity of the transition) and, as with mean field theory, it is the fluctuations in the order parameter at the transition point which are neglected in this approach. Nevertheless, Landau theory provides a good insight to the general features of phase transitions, describing many of the qualitative features of systems even if the values of the critical exponents are not quite correct.

The free energy is written as a power series in the order parameter $\varphi$

$$F = F_0 + F_1\varphi + F_2\varphi^2 + F_3\varphi^3 + F_4\varphi^4 + \cdots. \tag{4.5.17}$$

First, we note that $F_0$ can be ignored here since the origin of the energy is entirely arbitrary. Considerations of symmetry and the fact that $F$ is a scalar may determine that other terms should be discarded. For instance, if the order parameter is a vector (magnetisation), then odd terms must be discarded since only $\mathbf{M} \cdot \mathbf{M}$ will give a scalar. In this case the free energy is symmetric in $\varphi$, so that

$$F = F_2\varphi^2 + F_4\varphi^4 + \cdots. \tag{4.5.18}$$

There is a minimum in $F$ when $\mathrm{d}F/\mathrm{d}\varphi = 0$. Thus, the equilibrium state may be found from

$$\frac{\mathrm{d}F}{\mathrm{d}\varphi} = 2F_2\varphi + 4F_4\varphi^3 = 0. \tag{4.5.19}$$

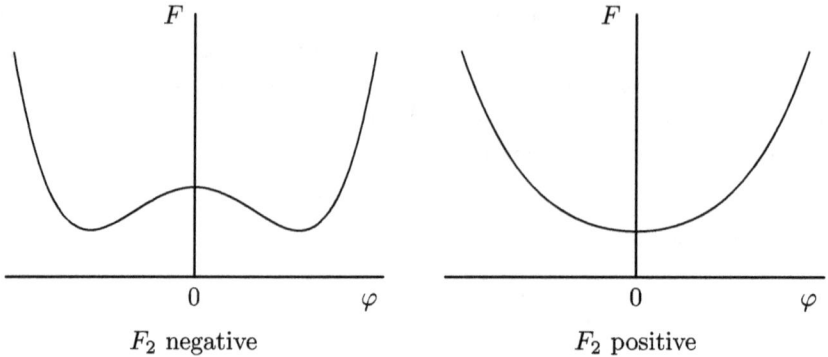

Fig. 4.28.   Minima in free energy for positive and negative $F_2$.

This has solutions

$$\varphi = 0 \quad \text{and} \quad \varphi = \pm\sqrt{\frac{-F_2}{2F_4}}. \tag{4.5.20}$$

We must have $F_4$ positive so that $F$ increases far away from the minima to ensure stability: $\varphi$ must be bounded. Then if $F_2 < 0$, there are three stationary points, while if $F_2 > 0$, there is only one. Thus, the nature of the solution depends crucially on the sign of $F_2$ as can be seen in Fig. 4.28.

It is clear from the figures that the critical point corresponds to the point where $F_2$ changes sign. Thus, expanding $F_2$ and $F_4$ in powers of $T - T_c$ we require, to leading order

$$F_2 = a(T - T_c) \quad \text{and} \quad F_4 = b \tag{4.5.21}$$

so that

$$\varphi = 0 \qquad\qquad T > T_c$$
$$\varphi = \pm\sqrt{\frac{a(T_c - T)}{2b}} \quad T < T_c. \tag{4.5.22}$$

This behaviour reproduces the critical behaviour of the order parameter, as we have previously calculated for the Weiss model.

We see that when cooling through the critical point, the equilibrium magnetisation starts to grow. But precisely at the transition point the magnetisation must decide in which way it is going to point. This is a *bifurcation*, Fig. 4.29. It is referred to as a pitchfork bifurcation because of its shape. This is only for a broken *discrete* symmetry.

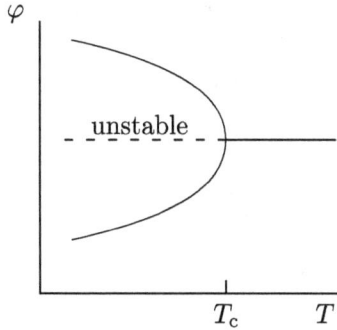

Fig. 4.29. Pitchfork bifurcation.

This model gives the critical exponent for the order parameter, $\beta$, to be ½. This is expected, since in the vicinity of the critical point the description is equivalent to the Weiss model.

When $\varphi$ is small (close to the transition), it follows that the leading terms in the expansion are the dominant ones and we can ignore the higher-order terms. We have seen in this section that only terms up to fourth order in $\varphi$ were necessary to give a Landau free energy leading to a second-order transition: a free energy that can vary between having a single and a double minimum depending on the values of the coefficients. Landau's key insight here was to appreciate that it is not that the higher order terms *may* be discarded, it is that they *must* be discarded in order to distil the *generic* properties of the transition. When terms above $\varphi^4$ are discarded, this is known as the $\varphi^4$ model.

### 4.5.4  *Heat capacity in the Landau model*

We have already seen that the Weiss model leads to a discontinuity in the heat capacity (critical exponent $\alpha = 0$). We can now examine this from the perspective of the Landau theory. The free energy is given by

$$F = F_0 + a(T - T_c)\varphi^2 + b\varphi^4. \tag{4.5.23}$$

We have included the $F_0$ term here and we shall permit it to have some smooth temperature variation, which is nothing to do with the transition.

The entropy is found by differentiating the free energy

$$S = -\frac{\partial F}{\partial T} = S_0 - a\varphi^2. \tag{4.5.24}$$

(Note, we do not need to consider the temperature dependence of $\varphi$ here since $\partial F / \partial \varphi = 0$.) This shows the way the entropy drops as the ordered phase is entered. We see that the entropy is continuous at the transition

$$T > T_c, \quad \varphi = 0 \qquad\qquad S = S_0(T)$$

$$T < T_c, \quad \varphi = \pm\sqrt{\frac{a(T_c - T)}{2b}} \quad S = S_0(T) + \frac{a^2}{2b}(T - T_c). \tag{4.5.25}$$

The heat capacity is given by

$$C = \frac{\partial Q}{\partial T} = T\frac{\partial S}{\partial T}. \tag{4.5.26}$$

Thus, we find

$$T > T_c, \quad C = C_0$$

$$T < T_c, \quad C = C_0 + \frac{a^2 T}{2b}. \tag{4.5.27}$$

At the transition, there is a discontinuity in the heat capacity given by

$$\Delta C = \frac{a^2 T_c}{2b}. \tag{4.5.28}$$

This is in accordance with the discussion of the Weiss model; here we find the magnitude of the discontinuity in terms of the Landau parameters.

### 4.5.5  *Ferromagnet in a magnetic field*

In the presence of a magnetic field, a ferromagnet exhibits certain features characteristic of a first-order transition. There is no latent heat involved, but, in the Ising magnet, the transition can exhibit hysteresis. In the case of a first-order transition, the free energy may no longer be symmetric in the order parameter; odd powers of $\varphi$ might appear in $F$. And the energy of a ferromagnet in the presence of a magnetic field has an additional term $(-\mathbf{M} \cdot \mathbf{B})$ linear in the magnetisation.

For temperatures above $T_c$, the effect of a free energy term linear in $\varphi$ is to shift the position of the minimum to the left or right. Thus, there is an equilibrium magnetisation proportional to the applied magnetic field.

To start with, let us consider the Ising magnet ($n = 1$, discrete symmetry breaking) in the presence of a magnetic field. For temperatures below $T_c$, the effect of the linear term is to raise one minimum and to lower the other. We plot the form of $F$ as a function of the order parameter for a number of different applied fields in Fig. 4.30.

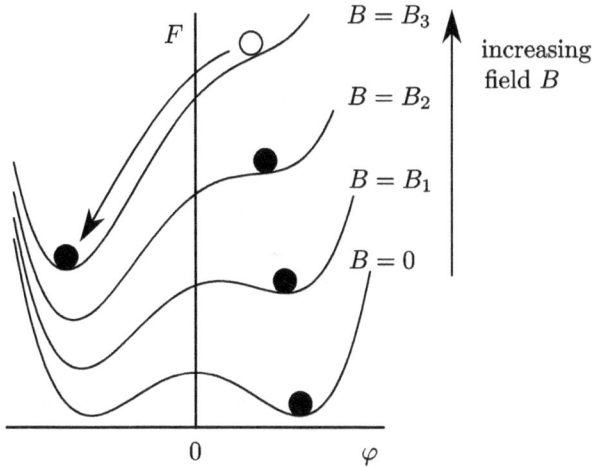

Fig. 4.30.    Variation in free energy for different magnetic fields — Ising case.

We see that in zero external field, since then $F$ is symmetric in $\varphi$, there are two possible values for the order parameter either side of the origin. Once a field is applied, the symmetry is broken. Then one of the states is the absolute minimum. But the system will remain in the metastable equilibrium. As the field is increased, the left-hand side of the right-hand minimum gets shallower and shallower. When it flattens, it is then possible for the system to move to the absolute minimum on the other side.

These considerations indicate that the behaviour of the magnetisation discussed in Section 4.3.1 is not quite correct for the Ising magnet. There we stated that the magnetisation inverts as $B$ goes through zero. But in reality there will be *hysteresis*, a specific characteristic of first-order transitions, Fig. 4.31. That discussion gave the equilibrium behaviour, whereas hysteresis is a quasi-equilibrium phenomenon.

This hysteretic behaviour we have discussed only applies to the Ising magnet, not to the XY magnet nor the Heisenberg magnet, both of which have broken *continuous* symmetries. Then the magnetisation can always shift in the distorted "wine bottle" potential to move to the lowest free energy, Fig. 4.32.

Real (Heisenberg) magnets do, however, exhibit hysteresis. This happens through the formation of *domains*: regions where the magnetisation points in different directions. Domain formation provides extra entropy from the domain walls at some energy cost. This can lower the free energy at finite temperatures. When one encounters un-magnetised iron at room

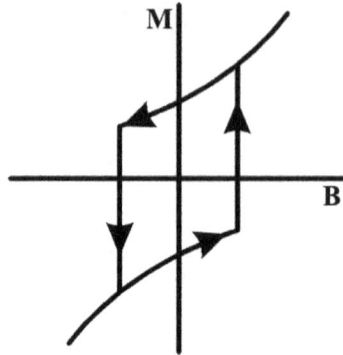

Fig. 4.31.   Hysteretic variation of magnetisation below $T_c$.

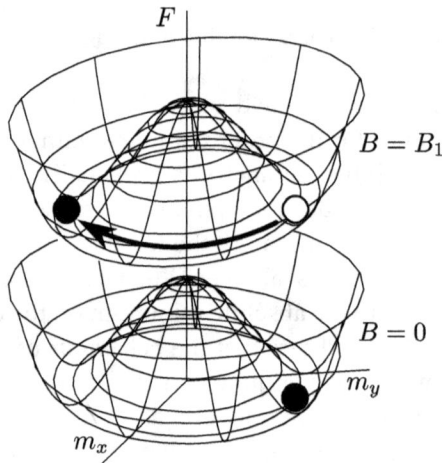

Fig. 4.32.   Variation in free energy for different magnetic fields — XY case.

temperatures, which is far below its critical temperature of 1044 K, this is because of the cancellation of the magnetisation of the domains.

## 4.6   Ferroelectricity

### 4.6.1   *Description of the phenomenon*

In certain ionic solids, a spontaneous electric polarisation can appear below a particular temperature. A typical example of this is barium titanate, which has a transition temperature of about 140 °C. There is a

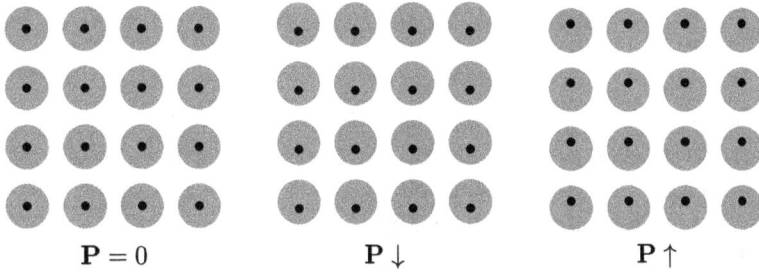

Fig. 4.33. Unpolarised state and ferroelectric state showing two polarisations.

good description of ferroelectricity in Kittel's *Introduction to Solid State Physics* [23].

The essential feature of the ferroelectric transition is that the "centre of charge" of the positive ion core becomes displaced from that of the negative electron cloud, Fig. 4.33. In this way, a macroscopic polarisation appears. The polarised state is called a *ferroelectric*. The transition is associated with a structural change in the solid.

In describing this transition, we see that the order parameter is the electric polarisation **P**. However, the polarisation is restricted in the directions it may point. In the figure, it may point either up or down. Thus, it is a discrete symmetry that is broken; in reality, the effective order parameter is a scalar. We note that this system has a non-conserved order parameter.

At the phenomenological level, the transition has many things in common with the Ising model of the magnetic transition of the ferromagnet, although the description is very different at the microscopic level. One special feature of the ferroelectric case is that the transition may be first- or second-order. We will use the Landau approach to phase transitions to examine this system and we shall be particularly interested in what determines the order of the transition.

Our previous example of a first-order transition, the liquid–gas system, was one with a conserved order parameter. Similarly, phase separation in a binary alloy, to be treated in Section 4.7, is a first-order system with a conserved order parameter. Here, however, we will encounter a first-order transition where the order parameter is not conserved.

### 4.6.2 *Landau free energy*

The order parameter is the electric polarisation, but in keeping with our general approach we shall denote it by $\varphi$ in the following. In the spirit of

Landau, we expand the (appropriate) free energy in powers of $\varphi$:

$$F = F_0 + F_1\varphi + F_2\varphi^2 + F_3\varphi^3 + F_4\varphi^4 + F_5\varphi^5 + F_6\varphi^6 + \cdots. \quad (4.6.1)$$

On the assumption that the crystal lattice has a centre of symmetry, the odd terms of the expansion will vanish. Furthermore, in the absence of an applied electric field, there will be no $F_1\varphi$ term. To determine the equilibrium state, we can ignore the constant term $F_0$ and so the free energy simplifies to

$$F = F_2\varphi^2 + F_4\varphi^4 + F_6\varphi^6 + \cdots. \quad (4.6.2)$$

Here, we have allowed for the possibility that terms higher than the fourth power may be required (unlike the ferromagnetic case).

When we considered the ferromagnet, we truncated the series at the fourth power. A necessary condition to do this was that the final coefficient, $F_4$, was positive so that the order parameter remained bounded. If we were to have a system where $F_4$ were negative, then we would have to include higher order terms until a positive (even) one were found. The simplest example is where $F_6$ is positive and we would truncate there. We will see that in this case the *order* of the transition is determined by the *sign* of the $F_4$ term.

If $F_4$ is positive, then in the spirit of the Taylor expansion we can ignore the $F_6$ term and the general behaviour, in the vicinity of the transition, is just as the (Ising) ferromagnet considered previously, as shown in Fig. 4.34. As the temperature is cooled below the transition point, the order parameter grows continuously from its zero value, a second order transition.

If $F_4$ is negative, then in the simplest case we need a positive $F_6$ term to terminate the series. The general form of the free energy curve then has

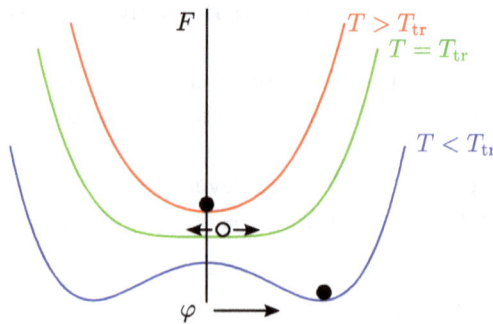

Fig. 4.34.   Conventional second-order transition.

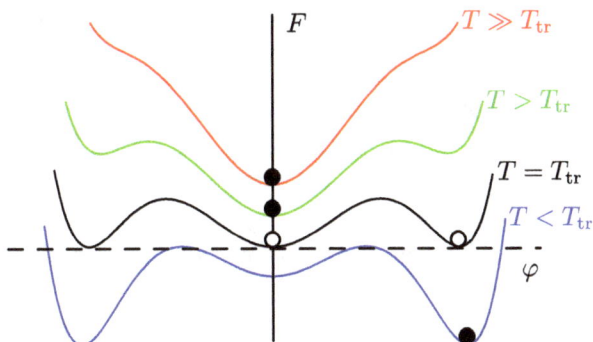

Fig. 4.35.   Inclusion of sixth-order terms.

three minima, one at $\varphi = 0$ and one equally spaced at either side, as shown in Fig. 4.35.

At high temperatures, the minimum at $\varphi = 0$ is lower, while at low temperatures, the minima either side will be lower. The transition point corresponds to the case where all three minima occur at the same value of $F$, since then $\varphi$ can jump from one minimum to another (at no energy cost).

Here, we see that the transition is characterised by a jump in the order parameter and there is the possibility of hysteresis in the transition since the barrier to the lower minimum must be surmounted. These are the characteristics of a first-order transition.

### 4.6.3   *Second-order case*

The analysis follows that previously carried out for the ferromagnet case since $F_4$ is positive and, following the Landau prescription, $F_6$ is then discarded. One thus concludes

$$\varphi = 0 \qquad\qquad T > T_c$$

$$\varphi = \sqrt{\frac{a(T_c - T)}{2b}} \qquad T < T_c, \tag{4.6.3}$$

the conventional Landau result for a second-order transition.

Problem 4.15 explores the inclusion of the $F_6$ term in the second-order case, demonstrating that this does not change the order parameter critical exponent $\beta$ from its value of ½.

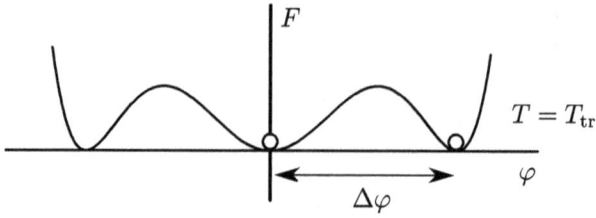

Fig. 4.36.  Jump in the order parameter at the transition.

### 4.6.4  *First-order case*

In a first-order transition, we *must* include the sixth-order term in the free energy expansion. In this case, we have

$$F = F_2\varphi^2 + F_4\varphi^4 + F_6\varphi^6, \tag{4.6.4}$$

where we note that now $F_4 < 0$ and $F_6 > 0$.

*Order parameter jump at the transition*

At the transition point, there are three minima of the free energy curve and they have the same value of $F$, namely zero (since, conveniently, we have taken $F_0$ to be zero) (Fig. 4.36). Thus, the transition point is characterised by the conditions

$$F(\varphi) = 0 \quad \text{and} \quad \frac{\mathrm{d}F}{\mathrm{d}\varphi} = 0 \tag{4.6.5}$$

or

$$F_2\varphi^2 + F_4\varphi^4 + F_6\varphi^6 = 0$$
$$2F_2\varphi + 4F_4\varphi^3 + 6F_6\varphi^5 = 0. \tag{4.6.6}$$

We know we have the $\varphi = 0$ root; this can be factored out. The others are found from solving the simultaneous equations

$$F_2 + F_4\varphi^2 + F_6\varphi^4 = 0$$
$$F_2 + 2F_4\varphi^2 + 3F_6\varphi^4 = 0. \tag{4.6.7}$$

The solution to these is

$$\varphi^2 = -\frac{F_4}{2F_6} \tag{4.6.8}$$

or

$$\varphi = \pm\sqrt{\frac{-F_4}{2F_6}} \qquad (4.6.9)$$

as we know that $F_4$ is negative.

This gives the discontinuity in the order parameter at the transition since $\varphi$ will jump between zero and this value at the transition. Thus, we can write

$$\Delta\varphi = \sqrt{\frac{-F_4}{2F_6}}. \qquad (4.6.10)$$

This vanishes as $F_4$ goes to zero; in other words, the transition becomes second-order. Thus, we have shown that when $F_4$ changes sign, the transition becomes second-order. The point of changeover from first- to second-order is referred to as the *tricritical point*.

## Transition temperature

At the transition point, the value of $F_2$ may be found as the second solution to the simultaneous equation pair, Eq. (4.6.7). The result is

$$F_2 = \frac{F_4^2}{4F_6}. \qquad (4.6.11)$$

This gives the relation which must hold between $F_2$, $F_4$ and $F_6$ at the transition. In the second-order case, we saw that the transition point corresponded to the vanishing of $F_2$. We see from Eq. (4.6.11) that this cannot be so in the first-order case. Nevertheless, we continue to call the point at which $F_0$ vanishes the critical point. In other words, for first order, the transition point is different from the critical point.

We make the assumption that $F_2$ varies with temperature as it does in the second-order case

$$F_2 = a(T - T_c) \qquad (4.6.12)$$

since, at the microscopic level, the system doesn't know if it is first-order or second-order. Moreover, we assume the coefficients $F_4$ and $F_6$ are constant or they vary but slowly in the vicinity of the transition. The "serious" temperature variation is in $F_2$ and we take $F_4$ and $F_6$ to be temperature independent.

Denote $T_{tr}$ the temperature at the transition; there $F_2$ takes the value

$$F_2 = a(T_{tr} - T_c). \tag{4.6.13}$$

Combining this with Eq. (4.6.11) gives

$$a(T_{tr} - T_c) = \frac{F_4^2}{4F_6} \tag{4.6.14}$$

or

$$T_{tr} = T_c + \frac{1}{4a}\frac{F_4^2}{F_6}. \tag{4.6.15}$$

The value of $F_4$ may vary with some external parameter such as strain, (as in Problem 4.17). Then if $F_4$ increases from a negative value to zero, $T_{tr}$ reduces to $T_c$. As $F_4$ increases through zero, the first-order transition becomes second-order and the transition temperature remains equal to $T_c$.

$$F_4 > 0, \quad T_{tr} = T_c, \qquad\qquad \text{second-order}$$
$$F_4 < 0, \quad T_{tr} = T_c + \frac{1}{4a}\frac{F_4^2}{F_6}, \quad \text{first-order.} \tag{4.6.16}$$

This is plotted in Fig. 4.37.

**Terminology:** In the literature on ferroelectricity it is common to use a different designation for the various temperature parameters. The transition temperature, our $T_{tr}$, is commonly called the *Curie* temperature with symbol $T_C$, and what we have called the critical temperature, our $T_c$, is commonly designated $T_0$. Our choice makes for consistency with the other phase transition systems treated.

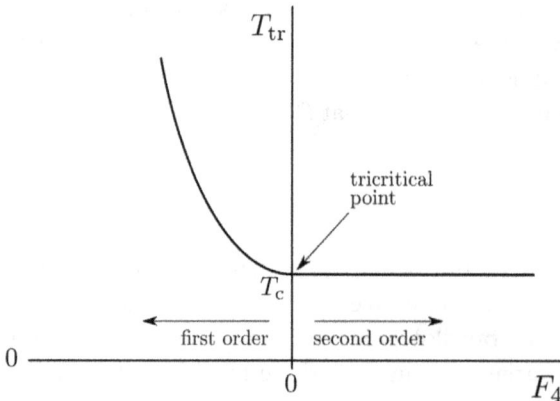

Fig. 4.37.  Variation of transition temperature with $F_4$.

*Temperature dependence of the order parameter*

In the ordered phase, the value of the order parameter is that corresponding to the lowest minimum of the free energy. We thus require

$$\frac{dF}{d\varphi} = 0, \quad \frac{d^2F}{d\varphi^2} > 0 \qquad (4.6.17)$$

where the second condition ensures the stationary point is a minimum. Using the sixth-order expansion for the free energy, differentiating it and discarding the $\varphi = 0$ solution gives

$$\varphi^2 = \frac{|F_4|}{3F_6}\left\{1 \pm \sqrt{1 - \frac{3F_2F_6}{F_4^2}}\right\}. \qquad (4.6.18)$$

Remember that $F_4$ is negative here. We have to choose the positive root to obtain the *minima*; the negative root gives the maxima. Using the conventional temperature variation for $F_2$ together with the expression for the transition temperature $T_{tr}$ in terms of the critical temperature $T_c$ gives the temperature variation of the order parameter for the first-order transition:

$$\varphi^2 = \frac{|F_4|}{3F_6}\left\{1 + \sqrt{1 - \frac{3}{4}\frac{(T - T_c)}{T_{tr} - T_c}}\right\}. \qquad (4.6.19)$$

Figure 4.38 shows the temperature variation of the order parameter. This shows clearly the discontinuity in the order parameter at $T_{tr}$, characteristic of a first-order transition.

Problem 4.18 considers the extension of Fig. 4.38 to distinguish the stable and metastable regions, together with the spinodal points for the transition.

*Caveat*

There is a contradiction in the use of the Landau expansion for treating first-order transitions. The justification for expanding the free energy in

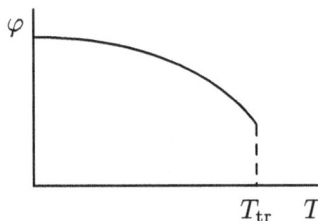

Fig. 4.38. Order parameter in a first-order transition.

powers of the order parameter and the truncation of this expansion is based on the assumption that the order parameter is very small. Another way of expressing this is to say that the Landau expansion is applicable to the vicinity of the critical point. This is fine in the case of second-order transitions where, when the ordered phase is entered, the order parameter grows continuously from zero. However, the characteristic of a first-order transition is the discontinuity in the order parameter. In this case, the assumption of a vanishingly small order parameter is invalid. When the ordered phase is entered, the order parameter jumps to a finite value. There is then no justification for a small–$\varphi$ expansion of the free energy. There are two ways of accommodating this difficulty.

Notwithstanding the invalidity of terminating the free energy expansion, the *qualitative* discussion of Section 4.6.2 still holds. In particular, it should be apparent that the general features of a first-order transition can be accounted for by a free energy which is a sixth-order polynomial in the order parameter. From this perspective we may regard the Landau expansion as putting the simplest of mathematical flesh on the bones of a qualitative model.

It is clear, however, that the free energy expansion becomes more respectable when the discontinuity in the order parameter becomes smaller, particularly as it goes to zero. Thus the theory is expected to describe well the changeover between first order and second order at the tricritical point. Transitions where the discontinuity in the order parameter is small are referred to as being *weakly first order*. We conclude that the Landau expansion should be appropriate for weakly first-order transitions.

### 4.6.5   *Entropy and latent heat at the transition*

The temperature dependence of the free energy in the vicinity of the transition is given by

$$F = F_0(T) + a(T - T_c)\varphi^2 + F_4\varphi^4 + F_6\varphi^6. \qquad (4.6.20)$$

Here, $F_0(T)$ gives the temperature dependence in the disordered, high temperature phase, and the coefficient of $\varphi^2$ gives the dominant part of the temperature dependence in the vicinity of the transition. The temperature dependence of the free energy here is similar to that of the second-order case, treated in the context of the ferromagnet.

The entropy is given by the temperature derivative of $F$ according to the standard thermodynamic prescription

$$S = -\frac{\partial F}{\partial T},\tag{4.6.21}$$

in this case

$$S = -\frac{\partial F_0}{\partial T} - a\varphi^2.\tag{4.6.22}$$

The second term indicates the reduction in entropy that follows as the order grows in the ordered phase. (As in the previous section, we do not need to consider the temperature dependence of $\varphi$ since $\partial F/\partial\varphi = 0$.) We see that if there is a discontinuity in the order parameter, then there will be a corresponding discontinuity in the entropy of the system.

In the first-order transition, there is a discontinuity in the order parameter at $T = T_{tr}$. The latent heat is then given by

$$
\begin{aligned}
L &= -T_{tr}\Delta S\\
&= aT_{tr}\Delta\varphi^2.
\end{aligned}\tag{4.6.23}
$$

This shows how the latent heat is directly related to the discontinuity in the order parameter — both characteristics of a first-order transition. The discontinuity in $\varphi$ is given by

$$\Delta\varphi = \sqrt{\frac{-F_4}{2F_6}}\tag{4.6.24}$$

so we can write the latent heat as

$$L = aT_{tr}\frac{|F_4|}{2F_6}.\tag{4.6.25}$$

Now, $F_4$ is negative for a first-order transition. This expression shows how the latent heat vanishes as $F_4$ vanishes when the transition becomes second-order.

### 4.6.6 Soft modes

In the ferroelectric, the excitations of the order parameter are optical phonons: the positive and negative charges oscillate in opposite directions. In other words, the polarisation oscillates. These are not Goldstone bosons since it is not a continuous symmetry that is broken; indeed the excitations have a finite energy (frequency) in the $p \to 0$ ($k \to 0$) limit. This is a characteristic of optical phonons.

The frequency of the optical phonons depends on the restoring force of the interatomic interaction and the mass of the positive and negative ions. Now, let us consider the destruction of the polarisation as a ferroelectric is warmed through its critical temperature. We are interested here in the case of a second-order transition. At the critical point, the Landau free energy exhibits anomalous broadening; the restoring force vanishes and the crystal becomes unstable. And this means that the frequency of the optical phonon modes will go to zero: they become "soft".

For further details of soft modes, see the books by Burns [27] and by Kittel [23].

## 4.7 Binary Mixtures

We have already seen the gas–liquid transition of a fluid as an example of a system exhibiting a first-order transition with conserved order parameter. And in the previous section the ferroelectric had a first-order transition with non-conserved order parameter. In this section, we shall examine phase separation in a binary mixture. This is a first-order transition with a conserved order parameter. By comparing and contrasting these different systems we should be able to identify their similarities and differences. We should also note that the binary alloy model treated in this section is equivalent to the Ising model (with a conserved order parameter). And the procedures described in Section 4.7.2 and beyond correspond to the mean field treatment of this Ising model.

### 4.7.1  *Basic ideas*

Consider a mixture of two atomic species A and B, with relative proportions $x$ and $1 - x$. For clarity, we shall consider a solid system, but many of the arguments will also apply to liquid mixtures. We thus imagine an alloy whose composition is specified by $A_x B_{1-x}$. We assume that the energy of the system is determined by the nearest neighbour interactions and we will denote

$$\varepsilon_{aa} \quad \text{energy of a single A–A bond}$$
$$\varepsilon_{bb} \quad \text{energy of a single B–B bond}$$
$$\varepsilon_{ab} \quad \text{energy of a single A–B bond.}$$

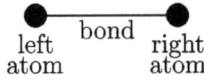

Fig. 4.39.  Bond joining two atoms.

A microstate of this system is specified by indicating the occupancy of each lattice site as an A or a B atom. Each microstate will have a given energy, found from a consideration of what neighbours each site has, by summing over the bond energies. And from these energies the partition function could be found. This would then give all the thermodynamic properties of the system. Unfortunately, such a procedure would be prohibitively complicated. Instead, we shall use an approximation method.

### 4.7.2  *Model calculation*

We make the assumption that the A and B atoms are distributed randomly: the occupancy of a given site is independent of the occupancy of its neighbours. Thus, we are considering a homogeneous mixture. We take the system to be at a specified temperature and volume so we calculate the *Helmholtz* free energy for this system. So, we need to know the energy and the entropy.

The key point is to start by considering a given *bond* joining two neighbouring atoms. We label one atom as the *left* atom and the other as the *right* (Fig. 4.39).

Each of the atoms may be an A atom or a B atom. Thus, there are four different configurations for the bond: A–A, A–B, B–A, B–B.

| left atom \ right atom | A | B |
|---|---|---|
| A | $\varepsilon_{aa}$ | $\varepsilon_{ab}$ |
| B | $\varepsilon_{ab}$ | $\varepsilon_{bb}$ |

Configuration energies

The concentration of the A atoms is taken to be $x$; then the concentration of the B atoms is $1 - x$. In the general case, we could allow the concentration to vary with position. Then the probability the left atom is A is $x_l$; the probability it is B is $1 - x_l$; the probability the right atom is A is $x_r$; the probability it is B is $1 - x_r$. The four bond configurations A–A, A–B, B–A, B–B, then have the probabilities: $x_l x_r$, $x_l(1 - x_r)$, $(1 - x_l)x_r$, $(1 - x_l)(1 - x_r)$.

| left atom \ right atom | A | B |
|---|---|---|
| A | $x_l x_r$ | $x_l(1 - x_r)$ |
| B | $(1 - x_l)x_r$ | $(1 - x_l)(1 - x_r)$ |

Configuration probabilities

### 4.7.3 *System energy*

We have defined the energies $\{\varepsilon_{aa}, \varepsilon_{bb}, \varepsilon_{ab}\}$ of the three types of bonds above (of course $\varepsilon_{ab} = \varepsilon_{ba}$) and we now know the probability of occurrence of each configuration. The mean energy for the bonds will be the sum of the energy of each state multiplied by its probability

$$\bar{e}_{lr} = \varepsilon_{aa}x_l x_r + \varepsilon_{ab}x_l(1 - x_r) + \varepsilon_{ab}(1 - x_l)x_r + \varepsilon_{bb}(1 - x_l)(1 - x_r). \quad (4.7.1)$$

Now, we assume that the composition of the system is homogeneous so that the concentration is independent of position; $x_l = x_r = \text{constant} = x$. In this case, the mean energy per bond reduces to

$$\bar{e} = x^2 \varepsilon_{aa} + (1 - x)^2 \varepsilon_{bb} + 2x(1 - x)\varepsilon_{ab}, \quad (4.7.2)$$

which may be rearranged as

$$\bar{e} = x\varepsilon_{aa} + (1 - x)\varepsilon_{bb} + x(1 - x)\{2\varepsilon_{ab} - (\varepsilon_{aa} + \varepsilon_{bb})\}. \quad (4.7.3)$$

The expression for the internal energy is found by considering a system containing $N$ atomic sites in a lattice where each atom has $s$ neighbours. Then the number of neighbouring bonds will be $Ns/2$; the divisor of 2 removes double counting. The internal energy of the system is then $E = Ns\bar{e}/2$ or

$$E = \frac{Ns}{2}\left[x^2\varepsilon_{aa} + (1 - x)^2\varepsilon_{bb} + 2x(1 - x)\varepsilon_{ab}\right]. \quad (4.7.4)$$

This may be written in a more suggestive form as

$$E = \frac{Ns}{2}\left\{x\varepsilon_{aa} + (1 - x)\varepsilon_{bb} + 2x(1 - x)\left[\varepsilon_{ab} - \frac{1}{2}(\varepsilon_{aa} + \varepsilon_{bb})\right]\right\}. \quad (4.7.5)$$

Here, the first two terms represent the energy of the separate pure phases, which we denote by $E_0$

$$E_0 = \frac{Ns}{2}\left[x\varepsilon_{aa} + (1 - x)\varepsilon_{bb}\right]. \quad (4.7.6)$$

The third term gives the extra energy upon mixing the species, $E_m$, as follows:

$$E_m = Nsx(1 - x)\left[\varepsilon_{ab} - \frac{1}{2}(\varepsilon_{aa} + \varepsilon_{bb})\right]. \tag{4.7.7}$$

It is then convenient to define the energy parameter

$$\varepsilon = \varepsilon_{ab} - \frac{1}{2}(\varepsilon_{aa} + \varepsilon_{bb}), \tag{4.7.8}$$

the difference between the "unlike" neighbour energy and the mean of the two "like" neighbour energies. This will turn out to be the characteristic energy of the system. Then the energy of mixing takes the simple form

$$E_m = Nsx(1 - x)\varepsilon. \tag{4.7.9}$$

Observe that the energy of mixing is invariant under the transformation $x \rightarrow 1 - x$. This is saying that $E_m$ is symmetric about the line $x = 1/2$. Any system whose energy satisfies this condition (regardless of the precise expression for its energy of mixing) is referred to as a *strictly regular solution*.

An *ideal solution* is one for which $\varepsilon = 0$. In this case, the energy of mixing is zero; the energy is independent of the microscopic structure. An ideal solution cannot lower its energy by changing its structure.

By contrast, when $\varepsilon \neq 0$, then the solution can lower its energy by rearranging itself in an ordered structure. When $\varepsilon < 0$, then A–B bonds are preferred and the ordered state will be a superlattice. (Clearly this would only happen in a solid.) And when $\varepsilon > 0$, then A–A and B–B bonds are preferred; the ordered state will comprise coexisting regions of phase-separated atoms. (This would happen in both solids and liquids.)

### 4.7.4 Entropy

Each site can be occupied by an A atom or a B atom; it has two states. The probability that it is an A atom is $x$ and the probability that it is a B atom is $1 - x$. The entropy of a given site is then $-k[x \ln x + (1 - x)\ln(1 - x)]$, so that for a solid of $N$ atomic sites the entropy will be

$$S = -Nk\left[x \ln x + (1 - x)\ln(1 - x)\right]. \tag{4.7.10}$$

You should recognise that this expression for the entropy is similar to that for the spin ½ magnet; there each site also had two states.

### 4.7.5   Free energy

As the system has its temperature and volume fixed, the Helmholtz free energy is the appropriate thermodynamic potential to use. Since

$$F = E - TS \tag{4.7.11}$$

we then have

$$F = \frac{Ns}{2}\left[x\varepsilon_{aa} + (1-x)\varepsilon_{bb} + 2x(1-x)\varepsilon\right]$$
$$+ NkT\left[x\ln x + (1-x)\ln(1-x)\right]. \tag{4.7.12}$$

This is plotted in Fig. 4.40. We see that at high temperatures the free energy curve is a convex function; it has a single minimum. However, at lower temperatures a concave region develops and the curve has a double minimum.

We shall see that the low temperature free energy leads to phase separation, Fig. 4.41.

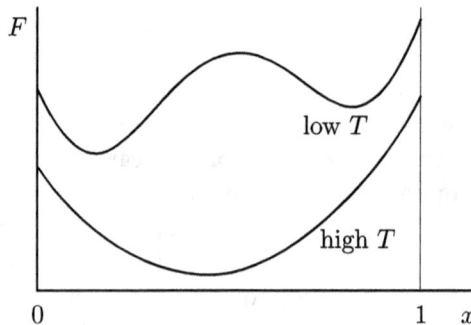

Fig. 4.40.   Free energy of binary alloy at high and low temperatures.

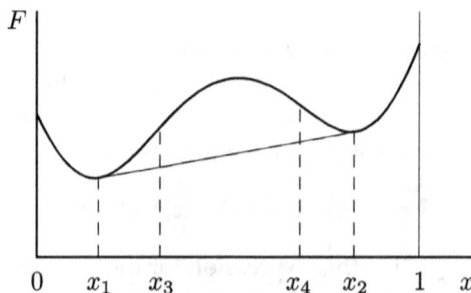

Fig. 4.41.   Free energy leading to phase separation.

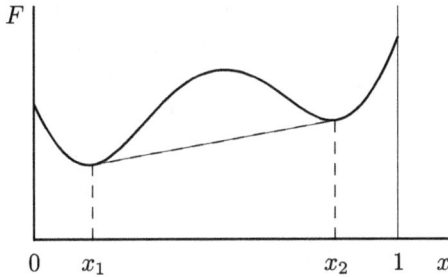

Fig. 4.42. Double-tangent construction.

Between $x = x_1$ and $x = x_2$, it is possible to lower the free energy by phase separation by dropping down from the curve to the double-tangent line, with a mixture of phases at densities $x_1$ and $x_2$, Fig. 4.42. This is analogous to the situation with the liquid–gas system treated in the van der Waals approach in Section 4.2.1. And as with the fluid case, in the phase separation region some parts are metastable, while some parts are unstable.

Instability occurs whenever $\partial F/\partial x$ is a *decreasing* function of $x$, that is, when $\partial^2 F/\partial x^2 < 0$. Then the system is unstable with respect to infinitesimal density fluctuations. This happens in the region $x_3$ to $x_4$. This is known as the *spinodal* region. If the temperature is quenched into this region, then one has spontaneous phase separation, referred to as *spinodal decomposition*.

By contrast, the regions $x_1$ to $x_3$ and $x_4$ to $x_2$ are *metastable*. Here, it is possible to remain in the homogeneous phase unless a density fluctuation of sufficient magnitude occurs. The system is unstable with respect to *finite* density fluctuations. Clearly, if one waits long enough, a fluctuation of sufficient magnitude will occur (but it might be a very long wait indeed).

In the phase diagram (the $x$–$T$ plane), the homogeneous phase and the metastable phase are separated by the phase separation or *binodal* line; this is the line of coexistence. The metastable phase and the unstable phase are separated by the spinodal line. We shall see that for the model considered here the locus of these two lines may be calculated.

### 4.7.6 *Phase separation — The lever rule*

The phase separation curve is found by the double-tangent construction, introduced in Section 4.2.1; this determines the region where the free energy may be reduced from its "homogeneous phase" value by separating into two phases.

The system will separate into regions of concentration $x_1$ and regions of concentration $x_2$. The *fractions* of the substance in the two phases are determined by the *lever rule*. Let us denote the fraction of the substance in the phase of concentration $x_1$ by $\alpha$; then the fraction in the other phase, of concentration $x_2$ will be $1 - \alpha$. The number of A atoms in the $x_1$ phase will be $N\alpha x_1$, while the number of A atoms in the $x_2$ phase will be $N(1 - \alpha)x_2$. The sum of these will give the total number of A atoms in the system, $Nx_0$, where $x_0$ is the concentration of A atoms in the homogeneous phase. Thus we have the equality

$$Nx_0 = N\alpha x_1 + N(1 - \alpha)x_2. \tag{4.7.13}$$

This may be solved for the fractions $\alpha$ and $1 - \alpha$ to give

$$\alpha = \frac{x_2 - x_0}{x_2 - x_1}$$
$$1 - \alpha = \frac{x_0 - x_1}{x_2 - x_1}. \tag{4.7.14}$$

This has a simple interpretation in terms of distances on the phase diagram. This is called the *lever rule*, Fig. 4.43.

### 4.7.7 *Phase separation curve — The binodal*

The concentrations $x_1$ and $x_2$ of the two separated phases are determined from the double-tangent construction. So, essentially, we must solve the

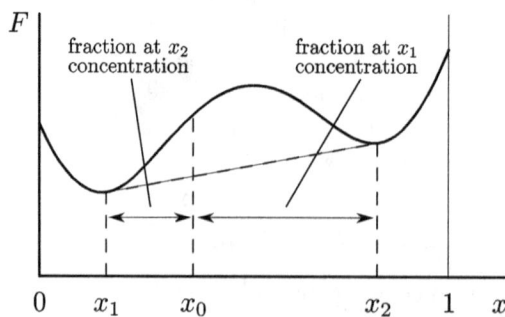

Fig. 4.43. The lever rule.

simultaneous equations

$$\frac{dF(x)}{dx}\bigg|_{x_1} = \frac{dF(x)}{dx}\bigg|_{x_2}$$

$$F(x_2) = F(x+1) + (x_2 - x_1)\frac{dF(x)}{dx}\bigg|_{x_1,x_2}$$

(4.7.15)

for $x_1$ and $x_2$.

In general, this could be a difficult problem. But there is an important simplification that helps in the case of a strictly regular solution. We can perform the double-tangent construction on the free energy of mixing $F_m$:

$$F_m = E_m - TS.$$

(4.7.16)

This is possible because the energy $E_0$ which we have neglected

$$E_0 = \frac{Ns}{2}[x\varepsilon_{aa} + (1-x)\varepsilon_{bb}]$$

is a linear function of $x$. This means that removing $E_0$ from the free energy is equivalent to a vertical shear of the free energy graph. The double-tangent construction is unchanged; the straight line still touches the free energy curve tangentially at the two points $x = x_1$ and $x = x_2$.

For a strictly regular solution, the free energy of mixing is symmetrical about $x = 1/2$. Thus, the two minima will be equally spaced either side of $x = 1/2$, and the symmetry now implies that the minima will be at the same height. In other words, on the free energy of mixing curve the double-tangent construction is a *horizontal* line joining the *minima* of the free energy, Fig. 4.44.

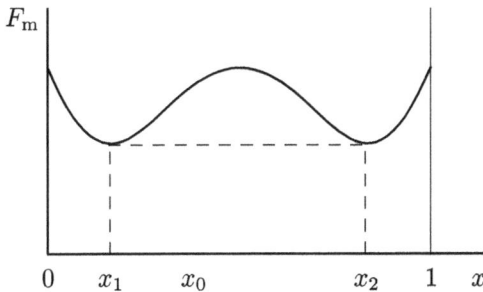

Fig. 4.44. Double-tangent construction for free energy of mixing.

The concentrations $x_1$ and $x_2$ of the two separated phases are determined from the condition

$$\frac{dF_m(x)}{dx} = 0 \qquad (4.7.17)$$

where

$$F_m(x) = Nsx(1 - x)\varepsilon + NkT\left[x\ln x + (1 - x)\ln(1 - x)\right]. \qquad (4.7.18)$$

The derivative of this is

$$\frac{dF_m}{dx} = Ns\varepsilon(1 - 2x) - NkT\ln\left(\frac{1 - x}{x}\right). \qquad (4.7.19)$$

We should set this expression equal to zero and then solve the equation for $x$, giving the two solutions $x_1$ and $x_2$ as a function of temperature. Unfortunately, an explicit expression cannot be obtained. However, it is possible to express $T$ as a function of $x$, which gives the locus of the phase separation line as follows:

$$T_{ps} = \frac{s\varepsilon(1 - 2x)}{k\ln[(1 - x)/x]}. \qquad (4.7.20)$$

The transformation $x \rightarrow 1 - x$ leaves the expression for $T_{ps}$ unchanged, so there are two solutions for $x$ corresponding to a given temperature.

The *critical temperature* $T_c$ corresponds to the maximum $T_{ps}$, occurring at $x = 1/2$ by symmetry. One must be careful in taking the limit correctly. Since

$$\lim_{x \rightarrow 1/2} \frac{1 - 2x}{\ln[(1 - x)/x]} = \frac{1}{2} \qquad (4.7.21)$$

it follows that the critical temperature is given by

$$T_c = \frac{s\varepsilon}{2k}. \qquad (4.7.22)$$

From this, we can write the phase separation temperature as

$$T_{ps} = \frac{2(1 - 2x)}{\ln[(1 - x)/x]}T_c. \qquad (4.7.23)$$

And from this, we plot the phase separation curve as shown in Fig. 4.45.

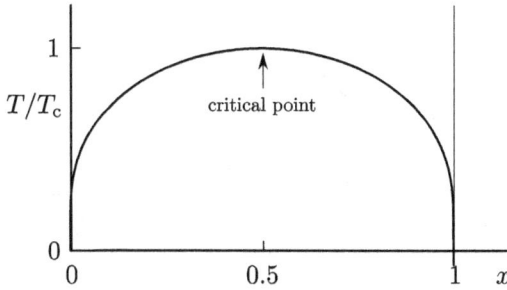

Fig. 4.45.  Phase separation of a binary alloy.

### 4.7.8  *The spinodal curve*

The spinodal curve traces out the region of instability in the $T$–$x$ plane. This is characterised by the vanishing of the second derivative of $F$ (or the free energy of mixing), points $x_3$ and $x_4$ of Fig. 4.41. Thus, we require

$$\frac{d^2F}{dx^2} = 0. \tag{4.7.24}$$

We found the first derivative in the previous section, Eq. (4.7.19):

$$\frac{dF_m}{dx} = Ns\varepsilon(1 - 2x) - NkT \ln\left(\frac{1-x}{x}\right)$$

and differentiating once more gives

$$\frac{d^2F_m}{dx^2} = -2Ns\varepsilon + NkT\left(\frac{1}{1-x} + \frac{1}{x}\right). \tag{4.7.25}$$

Setting this equal to zero gives the spinodal temperature as a function of $x$

$$T_{sp} = \frac{2s\varepsilon}{k}x(1-x) \tag{4.7.26}$$

or, in terms of the critical temperature

$$T_{sp} = 4x(1-x)T_c. \tag{4.7.27}$$

The spinodal line is below the phase separation line (Fig. 4.46). They meet at the critical point where $x = 1/2$ (for a strictly regular solution).

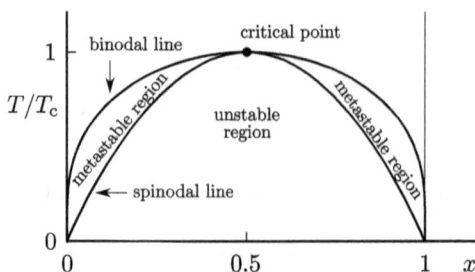

Fig. 4.46.   Phase separation and spinodal line.

### 4.7.9   *Entropy in the ordered phase*

The entropy of a homogeneous mixture, where the concentration of A atoms is $x$, is given by Eq. (4.7.10)

$$S = -Nk[x \ln x + (1 - x) \ln(1 - x)].$$

So, if $x_0$ is the concentration of A atoms in the high-temperature homogeneous phase, then the entropy of this phase is

$$S_0 = -Nk[x_0 \ln x_0 + (1 - x_0) \ln(1 - x_0)]. \tag{4.7.28}$$

This is a constant, determined by the chosen concentration $x_0$. Thus, in the homogeneous phase $S$ is independent of $T$, so that the heat capacity is zero:

$$C = 0 \text{ in the homogeneous phase.}$$

In the ordered low-temperature phase, there will be regions of low concentration $x_1$ and regions of high concentration $x_2$. Accordingly the entropy of these regions will be

$$S_1 = -N_1 k[x_1 \ln x_1 + (1 - x_1) \ln(1 - x_1)]$$
$$S_2 = -N_2 k[x_2 \ln x_2 + (1 - x_2) \ln(1 - x_2)]. \tag{4.7.29}$$

Now, for a *strictly regular solution*, as is our model, the phase diagram is symmetrical about $x = 1/2$ and we have

$$x_2 = 1 - x_1. \tag{4.7.30}$$

This has the important consequence that

$$S_1/N_1 = S_2/N_2; \tag{4.7.31}$$

the entropy per particle in the dilute phase and in the concentrated phase
is the same. In other words, each particle (regardless of the phase it is in)
"carries the same entropy".

So, the total entropy in the ordered low-temperature phase is

$$S = -Nk[x_1 \ln x_1 + (1 - x_1)\ln(1 - x_1)]. \tag{4.7.32}$$

The entropy is independent of the initial concentration $x_0$ and it depends
only on the concentration of the separated states; we can use $x_1$ or $x_2$.
This entropy depends on temperature because $x_1$ (or $x_2$) depends on
temperature

$$S(T) - Nk[x_1(T) \ln x_1(T) + (1 - x_1(T))\ln(1 - x_1(T))], \tag{4.7.33}$$

where $x(T)$ is given by the solution to Eq. (4.7.23)

$$T = \frac{2(1 - 2x)}{\ln[(1 - x)/x]} T_c. \tag{4.7.34}$$

Although an explicit expression for $S$ in terms of $T$ cannot be found, we
can make a parametric plot by varying $x_1$. This is shown in Fig. 4.47.

In the phase-separated region, the entropy (and therefore the heat
capacity) will be independent of the initial concentration. The temperature at which phase separation occurs will depend on $x_0$, but once the
separated region is entered, the entropy (and therefore the heat capacity)
will join a *universal curve*. This happens for a strictly regular solution.

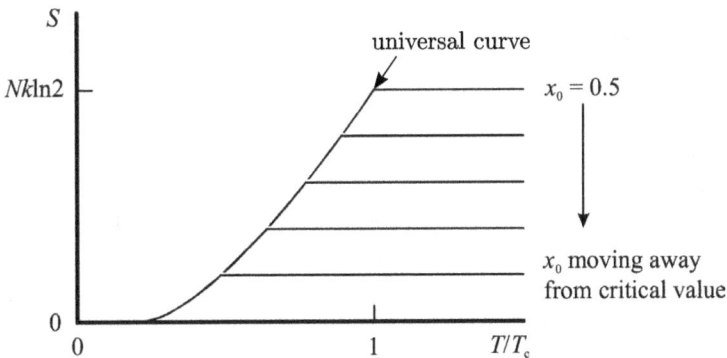

Fig. 4.47. Entropy of phase-separated binary mixture.

### 4.7.10   *Heat capacity in the ordered phase*

In the ordered phase, heat capacity is found from the entropy

$$C = \frac{dQ}{dT} = T\frac{dS}{dT}. \tag{4.7.35}$$

The differentiation is performed on a function of a function since entropy depends on $x_1$, and $x_1$ depends on $T$. Thus, we use

$$\frac{dS}{dT} = \frac{dS}{dx_1} \bigg/ \frac{dT}{dx_1}. \tag{4.7.36}$$

The derivatives are evaluated as

$$\frac{dS}{dx_1} = Nk\ln[(1-x_1)/x_1]$$

$$\frac{dT}{dx_1} = \frac{2T_c}{\ln[(1-x_1)/x_1]}\left\{\frac{1-2x_1}{x_1(1-x_1)\ln[(1-x_1)/x_1]} - 2\right\}. \tag{4.7.37}$$

Again, we cannot assemble these into an *explicit* function of $C$ in terms of $T$, but we can make use of a parametric plot since

$$T = \frac{2(1-2x_1)}{\ln[(1-x_1)/x_1]}T_c.$$

The heat capacity is $C = TdS/dT$, so that in this case

$$C = \frac{2(1-2x_1)}{\ln[(1-x_1)/x_1]}T_c\frac{Nk\ln[(1-x_1)/x_1]}{\frac{2T_c}{\ln[(1-x_1)/x_1]}\left\{\frac{1-2x_1}{x_1(1-x_1)\ln[(1-x_1)/x_1]} - 2\right\}}$$

$$= Nk\frac{(1-2x_1)\ln[(1-x_1)/x_1]}{\left\{\frac{1-2x_1}{x_1(1-x_1)\ln[(1-x_1)/x_1]} - 2\right\}}. \tag{4.7.38}$$

Thus, we have the implicit pair

$$\left. \begin{array}{l} C = Nk\dfrac{x_1(1-x_1)(1-2x_1)\ln^2[(1-x_1)/x_1]}{1-2x_1 - 2x_1(1-x_1)\ln[(1-x_1)/x_1]} \\[3mm] T = \dfrac{2(1-2x_1)}{\ln[(1-x_1)/x_1]}T_c. \end{array} \right\} \tag{4.7.39}$$

We plot this in Fig. 4.48.

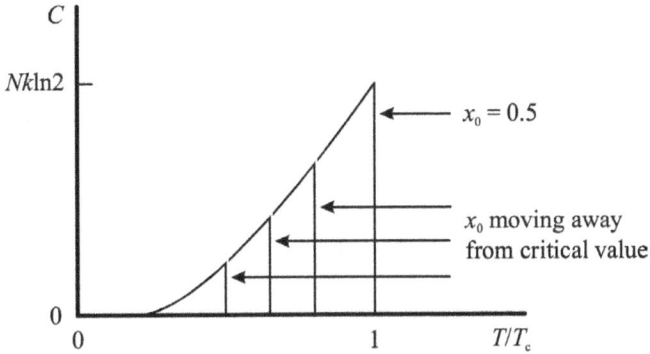

Fig. 4.48. Heat capacity of binary mixture.

If we eliminate $\ln[(1 - x_1)/x_1]$ from these equations, then one obtains the expression commonly quoted in the literature (e.g. Slater [29]):

$$\frac{C}{Nk} = \frac{2(1 - 2x_1)^2}{\frac{T}{T_c}\left(\frac{T/T_c}{2x_1(1-x_1)} - 2\right)},$$  (4.7.40)

but this is only one of an implicit pair.

The entropy is continuous at the transition and there is a simple discontinuity in the heat capacity. This means that there is *no latent heat* even though the transition is first order.

The absence of latent heat is due simply to the constraints under which we are studying the transition. The gas–liquid transition (and indeed the liquid–solid transition) are studied conventionally under the condition of constant pressure, often atmospheric pressure. If the transition were studied at constant volume, as in the above analysis, then there would be no latent heat. The fixed concentration $x_0$ of the binary mixture is analogous to the fluid system at constant density. To investigate the latent heat of the transition in a binary mixture requires the introduction of the *osmotic pressure* of the components and consideration of the transition under the (hypothetical) condition of constant osmotic pressure.

### 4.7.11  *Order of the transition and the critical point*

Below the transition, there are two coexisting phases. This is a general feature of any system with a conserved order parameter. Below the transition the mean value of the concentration $x$ is fixed $x_0$ and the proportions of the proportions of the two phases, of concentrations $x_1$ and $x_2$,

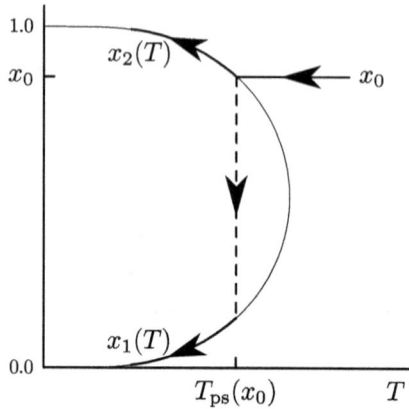

Fig. 4.49.   Order parameter values of the transition.

are determined by the requirement that the mean density be equal to $x_0$; this is the content of the lever rule.

So, the order parameter density for a system with conserved order parameter has two distinct values below the transition. Of course, this ignores the variation at the boundary between the two phases. For the binary mixture, these two values would be $x_1$ and $x_2$, Fig. 4.49.

Above the transition, the order parameter density is a constant $x_0$. If $x_0 > 1/2$, then on passing through the transition $x_2$ is continuous, but $x_1$ is discontinuous. It is the discontinuity that characterises the transition as being first order. But we reiterate that since $x_0$ is constant, the entropy is continuous, the heat capacity is not infinite, and there is no latent heat.

Things are different, however, at $x_0 = 1/2$. Here, the order parameter density values are both continuous at the transition so this qualifies as a *second-order* transition. We also observe that the phase separation point and the spinodal point coincide so that there is no metastable region, and so no hysteresis. And finally, since both the first and the second derivatives of the Helmholtz free energy (of mixing) vanish, the minimum is anomalously broad so that there can be large fluctuations in the order parameter (with negligible free energy cost). We recall these all as characteristics of a second-order transition. The point at $x = 1/2$, where the transition becomes second-order is thus called the *critical point*.

We have the general rule that a first-order transition becomes second-order at a critical point. It was the assumption of a strictly regular solution that placed the critical point at $x = 1/2$. In the general case, the critical point occurs at the concentration for which the transition temperature is a maximum (stationary).

### 4.7.12 The critical exponent β

The critical exponents refer to the critical point so we shall examine the binary mixture specifically in the $x_0 = 1/2$ case. It is convenient to utilise an order parameter that grows from zero at the critical point, so we shall define

$$\varphi = 2x - 1. \tag{4.7.41}$$

Then the phase separation curve

$$T = \frac{2(1-2x)}{\ln[(1-x)/x]} T_c \tag{4.7.42}$$

may be expressed in terms of $\varphi$ as

$$\frac{T}{T_c} = \frac{2\varphi}{\ln[(1+\varphi)/(1-\varphi)]}. \tag{4.7.43}$$

Observe this is identical to the corresponding equation for the Weiss ferromagnet, Eq. (4.3.19). It is plotted in Fig. 4.50.

We would like to expand $\varphi$ in powers of $1 - T/T_c$ to investigate the behaviour of the order parameter in the vicinity of the critical point, and to obtain the order parameter critical exponent $\beta$. This is most conveniently done by expanding $1 - T/T_c$ in powers of $\varphi$ and then inverting the series.

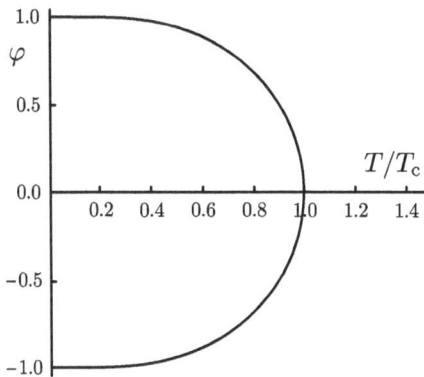

Fig. 4.50. Critical behaviour.

We find, from Eq. (4.7.43):

$$1 - \frac{T}{T_c} = \frac{1}{3}\varphi^2 + \frac{4}{45}\varphi^4 + \frac{44}{945}\varphi^6 + \cdots \tag{4.7.44}$$

which may be inverted to yield

$$\varphi = \sqrt{3}\left(1 - \frac{T}{T_c}\right)^{1/2} - \frac{2\sqrt{3}}{5}\left(1 - \frac{T}{T_c}\right)^{3/2} + \cdots. \tag{4.7.45}$$

In other words, we have found that

$$\varphi \sim \left(1 - \frac{T}{T_c}\right)^{1/2} \tag{4.7.46}$$

at the critical point, the usual mean field value of $\beta = 1/2$.

Finally, we must mention an important distinction between systems with conserved and non-conserved order parameters at the critical point. The diagram above shows the critical point may be similar in both cases, but the *interpretation* is different. For a non-conserved order parameter, when the critical point is approached, there is a *bifurcation* and the system breaks symmetry by choosing one particular branch. By contrast, for a conserved order parameter *both* possible branches are chosen; we have coexisting fractions determined by the lever rule.

## 4.8   Quantum Phase Transitions

### 4.8.1   *Introduction*

In a conventional second-order phase transition, there is an ordered phase existing at low temperatures. As the temperature is raised, this order is destroyed by thermal fluctuations. And at the critical point, the order disappears completely. The temperature of the critical point is determined by the competing requirements of energy minimisation, favouring order, and entropy maximisation, favouring disorder. Our consideration is restricted to second-order phase transitions where the "anomalous broadening" of the free energy minimum allows large excursions of the order parameter at negligible cost. These diverging fluctuations in the order parameter result in a collapse of the order and we can thus say that the transition is *driven* by the fluctuations.

Temperature is the control parameter in a conventional phase transition; as the temperature is varied, the critical point is traversed. By contrast, a *quantum phase transition* occurs at zero temperature. Here, the control parameter is some other variable such as pressure, magnetic field, alloy composition, and the transition occurs at $T = 0$ as the control parameter is varied. At $T = 0$, there are no thermal fluctuations present and the transition from the ordered phase to the disordered is driven purely by quantum-mechanical fluctuations. Such fluctuations are a consequence of the Uncertainty Principle.

The study of quantum phase transition thus concerns the different kinds of ground state a system can have for different values of the control parameter. There will be critical exponents associated with the quantum critical point and these display the same sorts of universality as do the conventional critical exponents.

As the temperature is increased from $T = 0$, the thermal fluctuations will make an increasing contribution to the critical behaviour. In the temperature–control parameter plane, there will thus be a region of ordered phase and a region of disordered phase separated by a phase boundary line. While at $T = 0$ the critical point is purely quantum, at the higher temperatures where the transition occurs the critical point will be classical. And the classical critical exponents are related rather simply to the quantum critical exponents – as we shall see.

Figure 4.51 shows an example of such a transition [30]. The material is LiHoF$_4$, where the electrons in Ho are coupled by an Ising-type interaction. The control parameter is a magnetic field applied transverse to the Ising spin direction. The dotted curve shows the result of calculation based on a mean field approximation of the Ising interaction. The solid line incorporates the effects of the nuclear hyperfine interaction as well.

### 4.8.2  *The transverse Ising model*

The transverse Ising model is a simple system exhibiting a quantum critical point. In this model, the Ising interaction is supplemented by a magnetic field transverse to the Ising direction. Thus, we write the hamiltonian as

$$\mathcal{H} = -2\hbar J \sum_{i>j\,\mathrm{nn}} S_z^i S_z^j - \gamma \hbar B_x \sum_i S_x^i. \qquad (4.8.1)$$

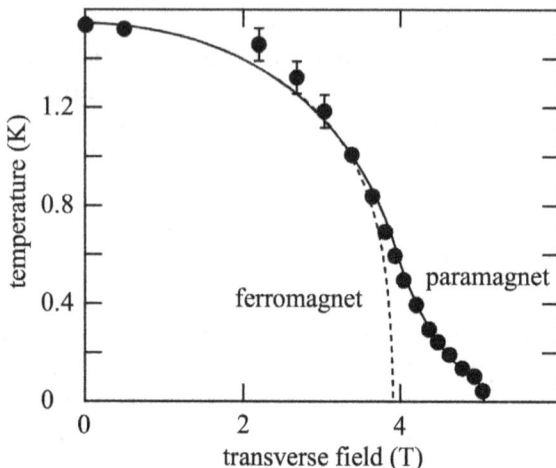

Fig. 4.51.   Phase boundary in a transverse Ising system. Reprinted with permission from Ref. [30].

The first term is the Ising interaction and the second term represents the interaction with an applied magnetic field in the $x$ direction, $B_x$.

In the absence of the external field, we have the conventional Ising model. The interaction favours parallel spins and the ground state then corresponds to aligned spins either in the $+z$ or the $-z$ direction. When the transverse field is applied, its effect is to induce transitions between the up and down spin states. This is because the hamiltonian no longer commutes with the spin $z$ component

$$\left[\mathscr{H}, S_z^i\right] \neq 0, \tag{4.8.2}$$

and using the Heisenberg equation of motion.

Thus, the effect of the transverse magnetic field is to cause fluctuations in the $z$ component of the spins. And these fluctuations will act to destroy the order of the Ising ground state. For high enough values of the transverse field, the order will be fully destroyed and then the ground state will be qualitatively different. Thus, the ground state is altered by the control parameter.

### 4.8.3   *Recap of mean field Ising model*

For clarity, let us recap the mean field treatment of the conventional Ising model, described now in a form suitable for the addition of a transverse field. In the absence of interactions, the magnetisation of an assembly of $N$ spin ½ magnetic moments $\mu$ is given by

$$M = N\mu \tanh\left(\frac{\mu B}{kT}\right) \qquad (4.8.3)$$

where $T$ is the temperature. Clearly, the magnetisation will point in the direction of the applied magnetic field. The saturation magnetisation $M_0$ is given by

$$M_0 = N\mu \qquad (4.8.4)$$

so we may write the magnetisation as

$$M = M_0 \tanh\left(\frac{M_0}{N}\frac{B}{kT}\right). \qquad (4.8.5)$$

In the mean field approach to the Ising hamiltonian, the $S_z^i S_z^j$ interaction is approximated by the average $nS_z^i \langle S_z \rangle$, where $n$ is the number of neighbours surrounding each particle. In terms of the magnetisation, this may be written as a magnetic field $b$ as follows:

$$b = \lambda M_z \hat{z}. \qquad (4.8.6)$$

(Since the magnetisation is defined as the total magnetic moment, the dimensions of $\lambda$ are $m_0$ over volume.) Spontaneous magnetisation then occurs, in the absence of $B$, as the solution to

$$M_z = M_0 \tanh\left(\frac{M_0}{N}\frac{\lambda M_z}{kT}\right) \qquad (4.8.7)$$

or

$$\frac{M_z}{M_0} = \tanh\left(\frac{M_z}{M_0}\frac{T_c}{T_0}\right) \qquad (4.8.8)$$

where $T_c$, the critical temperature, is given by

$$T_c = \frac{\lambda M_0^2}{Nk}. \qquad (4.8.9)$$

The tanh equation is solved by writing it as

$$\frac{M_z}{M_0}\frac{T_c}{T} = \tanh^{-1}\left(\frac{M_z}{M_0}\right)$$

$$= \frac{1}{2}\ln\left(\frac{1 + M_z/M_0}{1 - M_z/M_0}\right), \qquad (4.8.10)$$

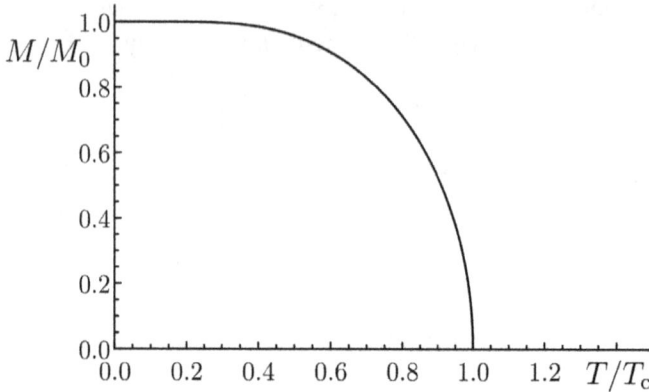

Fig. 4.52.   Spontaneous magnetisation of the Ising ferromagnet.

and then

$$\frac{T}{T_c} = \frac{2M_z/M_0}{\ln\left(\frac{1+M_z/M_0}{1-M_z/M_0}\right)}. \tag{4.8.11}$$

When plotted, this gives the usual form for the variation of spontaneous magnetisation with temperature, Fig. 4.52.

The critical behaviour may be found from a power series expansion, since $M_z$ is vanishingly small. The expansion is

$$\frac{T}{T_c} = 1 - \frac{1}{3}\left(\frac{M_z}{M_0}\right)^2 + \frac{4}{45}\left(\frac{M_z}{M_0}\right)^4 - \frac{44}{945}\left(\frac{M_z}{M_0}\right)^6 + \cdots \tag{4.8.12}$$

and this series may be inverted to yield

$$\frac{M_z}{M_0} = \sqrt{3}\left(1 - \frac{T}{T_c}\right)^{1/2} + \frac{2}{5}\sqrt{3}\left(1 - \frac{T}{T_c}\right)^{3/2} + \frac{12}{175}\sqrt{3}\left(1 - \frac{T}{T_c}\right)^{5/2} + \cdots . \tag{4.8.13}$$

Thus, we see that the magnetisation is continuous as $T \to T_c$; the transition is second order; the critical exponent $\beta$ has the mean field value of ½ and its critical amplitude is $\sqrt{3}$.

### 4.8.4   *Application of a transverse field*

We now have the applied magnetic field

$$\mathbf{B} = B_x \hat{\mathbf{x}} \tag{4.8.14}$$

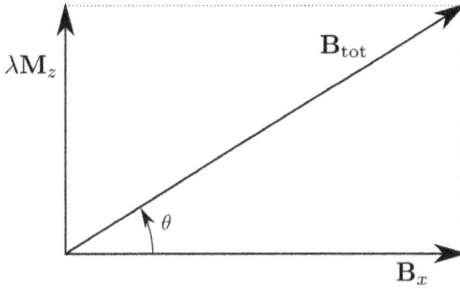

Fig. 4.53.   Resultant magnetic field.

to which must be added the Ising mean field

$$\mathbf{b} = \lambda M_z \hat{\mathbf{z}}. \tag{4.8.15}$$

Thus, the total magnetic field has magnitude

$$B_{\text{tot}} = \sqrt{B_x^2 + \lambda^2 M_z^2} \tag{4.8.16}$$

and it points in a direction $\theta$ from the $x$ axis, Fig. 4.53, where

$$\sin \theta = \frac{\lambda M_z}{\sqrt{B_x^2 + \lambda^2 M_z^2}}. \tag{4.8.17}$$

Now, the mean field recipe says that the magnetisation is given by

$$M = M_0 \tanh\left(\frac{M_0}{N} \frac{B_{\text{tot}}}{kT}\right) \tag{4.8.18}$$

and it points in the direction parallel to the field. Thus,

$$M = M_0 \tanh\left(\frac{M_0}{N} \frac{\sqrt{B_x^2 + \lambda^2 M_z^2}}{kT}\right). \tag{4.8.19}$$

However, what we are interested in is the magnetisation in the $z$ direction. We know the direction of $\mathbf{M}$: parallel to $\mathbf{B}_{\text{tot}}$. So, to find the component in the $\mathbf{z}$ direction, we require

$$M_z = M \sin \theta, \tag{4.8.20}$$

or

$$M_z = \frac{\lambda M_z}{\sqrt{B_x^2 + \lambda^2 M_z^2}} M_0 \tanh\left(\frac{M_0}{N} \frac{\sqrt{B_x^2 + \lambda^2 M_z^2}}{kT}\right). \tag{4.8.21}$$

This may be tidied up a little, but it is helpful first to re-express in terms of reduced variables. Since $M_0$ is the saturation magnetisation, it follows that $\lambda M_0$ is the "saturation" internal field. Let's measure the transverse field in multiples of this. That is, we define

$$b_x = B_x/\lambda M_0. \tag{4.8.22}$$

We also define the reduced variables

$$m_z = M_z/M_0 \qquad \text{and} \qquad t = T/T_c. \tag{4.8.23}$$

Then $m_z$ satisfies the implicit equation

$$\sqrt{b_x^2 + m_z^2} = \tanh \frac{\sqrt{b_x^2 + m_z^2}}{t}. \tag{4.8.24}$$

### 4.8.5  Transition temperature

The transition from the ordered phase to disordered phase corresponds to the vanishing of $M_z$. In general, the temperature at which the transition occurs will be a function of the transverse field. Setting $m_z = 0$ in the above equation then gives the ($b_x$ – dependent) critical temperature in the equation

$$b_x = \tanh \frac{b_x}{t_c(b_x)}. \tag{4.8.25}$$

This may be inverted to give

$$\frac{b_x}{t_c(b_x)} = \tanh^{-1} b_x$$
$$= \frac{1}{2} \ln \left( \frac{1 + b_x}{1 - b_x} \right) \tag{4.8.26}$$

so that

$$t_c(b_x) = \frac{2b_x}{\ln \left( \frac{1+b_x}{1-b_x} \right)}. \tag{4.8.27}$$

In terms of the full variables, this is

$$\frac{T_c(B_x)}{T_c(B_x = 0)} = \frac{2B_x/\lambda M_0}{\ln \left( \frac{\lambda M_0 + B_x}{\lambda M_0 - B_x} \right)} \tag{4.8.28}$$

and this is plotted in Fig. 4.54.

This curve corresponds to the dotted line plotted with the experimental data in Fig. 4.51 in the introduction.

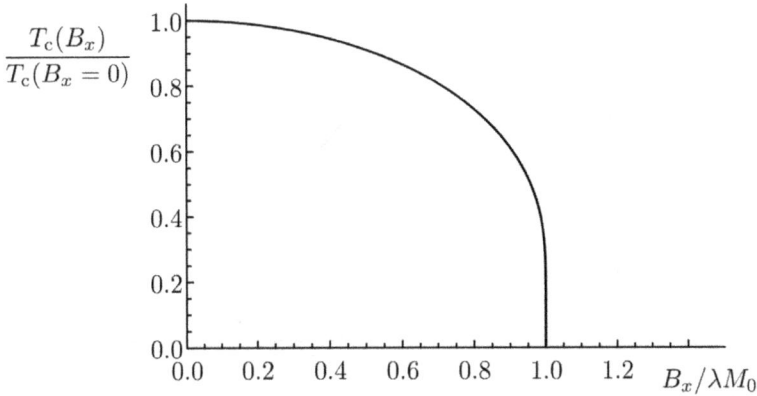

Fig. 4.54. Variation of critical temperature with transverse field.

### 4.8.6 *Quantum critical behaviour*

In terms of reduced variables, the equation of state for the system may be written as

$$t = \frac{2\sqrt{m_z^2 + b_x^2}}{\ln\left(\frac{1+\sqrt{m_z^2+b_x^2}}{1-\sqrt{m_z^2+b_x^2}}\right)}. \tag{4.8.29}$$

The quantum critical behaviour occurs at zero temperature. The control parameter is $b_x$; this is the analogue of temperature in the classical case. So, to study the zero-temperature phase diagram, we need the dependence of $m_x$ on $b_z$.

Since we are assuming that both $m_x$ and $b_z$ are finite, then in the limit of $t \to 0$ the logarithm must become infinite. And for this to be the case, its denominator will go to zero. In other words,

$$1 - \sqrt{m_z^2 + b_x^2} = 0 \tag{4.8.30}$$

or

$$m_z^2 = 1 - b_x^2. \tag{4.8.31}$$

This is plotted in Fig. 4.55.

To investigate the critical behaviour, we want to know what is happening in the vicinity of the quantum critical point, at $b_x = 1$. We write $m_z$ as

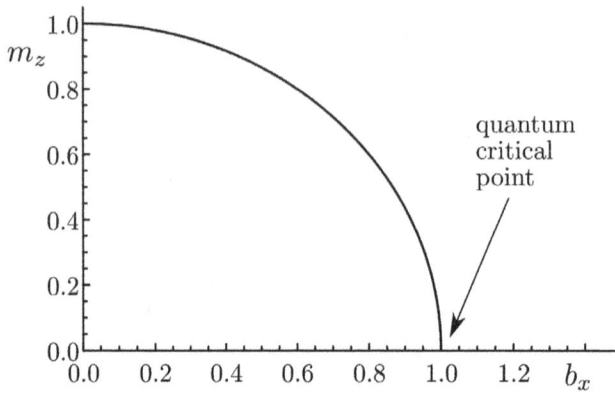

Fig. 4.55.   Zero-temperature phase diagram.

$$m_z = (1 - b_z^2)^{1/2} \tag{4.8.32}$$

and a series expansion then gives

$$m_z = \sqrt{2}(1 - b_x)^{1/2} - \frac{1}{2\sqrt{2}}(1 - b_x)^{3/2} - \frac{1}{16\sqrt{2}}(1 - b_x)^{5/2} + \cdots . \tag{4.8.33}$$

We see that the order parameter goes continuously to zero at the critical point so the transition is second order. And the above expansion gives $\beta$, the order parameter critical exponent, to be ½. The classical and the quantum critical behaviour can be compared. We have

$$\text{classical case } m_z = \sqrt{3}(1 - t)^{1/2}$$
$$\text{quantum case } m_z = \sqrt{2}(1 - b_x)^{1/2}.$$

We see that the critical *exponents* are the same in the quantum and the classical case, but the critical *amplitudes* are different.

### 4.8.7   *Dimensionality and critical exponents*

There is a general rule relating classical and quantum critical exponents and the dimensionality of the system. The scaling theory arguments outlined in Section 4.1.8 and in particular the Josephson critical exponent law demonstrate the importance of the system's dimensionality to the critical exponents.

Let us write the Boltzmann factor appearing in the system's partition function as $e^{-E\beta}$ where $\beta = 1/kT$. This has a certain similarity to the quantum generator of time evolution $e^{iEt/\hbar}$. We can thus regard the Boltzmann factor as a generator of an (imaginary) spatial displacement. And as the temperature goes to zero, the spatial displacement goes to infinity – what one understands as the thermodynamic limit. The sum or integral over Boltzmann factors in the partition covers the spatial extent of the system. And in the limit of zero temperature, the Boltzmann factors involve an additional dimension to be traversed.

Since it is understood that the partition function contains all thermodynamic information about a system, the above argument implies that a zero-temperature critical point in a system of $n$ dimensions will have the same behaviour as a conventional critical point in a system of $n + 1$ dimensions. And since the mean field approximation is independent of the system's dimensionality, we conclude that the mean field approximation will give the same critical exponents for the classical and the corresponding quantum critical point.

We know that mean field approaches give behaviour that, while qualitatively true, are often not in quantitative agreement with observed critical behaviour. For each system (hamiltonian), there is a marginal spatial dimensionality $d^*$ such that when the number of dimensions is greater than $d^*$, the results of mean field theory are exact [31]. When the number of dimensions is less than $d^*$, then mean field theory is quantitatively wrong; then fluctuations are significant. When the number of dimensions is equal to $d^*$, there are only logarithmic corrections to the results of mean field theory. Thus, the designation *marginal* dimensionality.

The marginal dimensionality for a dipole-coupled Ising system, such as $LiHoF_4$ described above, is 3. Thus, we expect mean field theory to give, to first order, the correct critical behaviour for the classical critical behaviour. And it will then certainly be correct for the quantum critical behaviour. In other words, for *this* system the quantum and the classical critical behaviour should be the same: the mean field behaviour. We saw above that the mean field calculation gave the same value for the exponent $\beta$: one half.

We found that the susceptibility critical exponent $\gamma$ had the value 1 for the mean field Ising ferromagnet. So, we conclude that for $LiHoF_4$ both the classical and the quantum value of the exponent $\gamma$ should have the value 1. Figure 4.56 [30] shows susceptibility measurements in the vicinity

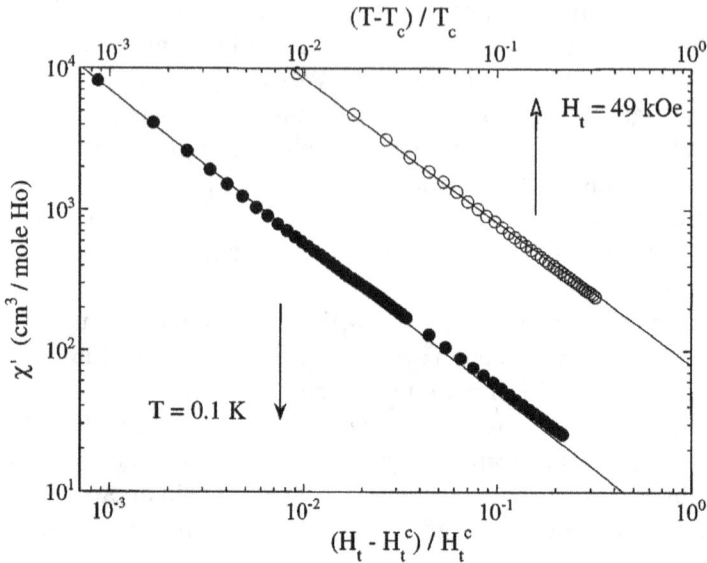

Fig. 4.56. Quantum (below) and classical (above) critical behaviour of LiHoF$_4$ [30].

of both the classical and the quantum critical points. The lines through the data both have a slope of $-1$ indicating $\gamma = 1$ in both cases.

## 4.9   Retrospective

We have now completed our survey of different types of phase transitions. You will have observed similarities between systems and you will have observed differences. In this final section, we shall look back over the various examples considered and we shall make some general observations and comments.

### 4.9.1   *The existence of order*

Ising discovered that his model exhibited no phase transition in one dimension. In other words, the one-dimensional Ising model displays long-range order only at $T = 0$. Onsager discovered that the Ising model did have a phase transition in two dimensions. In other words, the Ising model has a transition to an ordered state at finite temperatures in two (and higher) dimensions.

Somewhat surprisingly, Mermin and Wagner [32] argued that the Heisenberg model would not exhibit long-range order in two, but it would in three and higher dimensions. Thus, the Ising model has a transition in two dimensions, but the Heisenberg model does not. Why should this be? The key difference is the dimensionality of the order parameter $n$. For the Ising model $n = 1$, while for the Heisenberg model $n = 3$. We also note that for the spherical model ($n = \infty$), Kac reached the same conclusions as for the Heisenberg model.

The Mermin–Wagner theorem, sometimes attributed to Berezinskii, Hohenberg or Peierls, states that there can be no long-range order resulting from a broken continuous symmetry ($n = 2$ or above) in two or lower spatial dimensions. But there can be long-range order resulting from a broken discrete symmetry ($n = 1$) in a system of two spatial dimensions — the Ising model is an example of this. However, in one spatial dimension there can be no long-range order at all. This may be established by an energy–entropy argument similar to that in Section 4.4.3 above for the 1d Ising model. We may summarise as follows:

1. For $d \geq 3$, you can always have an ordered phase — a broken discrete or continuous symmetry.
2. For $d = 2$, you cannot break a continuous symmetry, only a discrete symmetry. So, only the Ising model has a phase transition in 2d; there is no ordered phase for the Heisenberg model or the spherical model.
3. For $d = 1$, you can't break any symmetry; no ordered phase possible in 1d above $T = 0$.

The borderline case corresponds to $d = 2$ and $n = 2$. In this case, it is possible to have *orientational* order; this is the Kosterlitz–Thouless [33] transition. Unfortunately, this is outside the scope of this book. Figure 4.57 indicates the various possibilities.

Finally, we mention the question (using the magnetic description) of the magnitude of the spin vector. For a spin $S = 1/2$, each moment has one of two projections along a given axis. For higher spin, there will be a larger number of projections. The spin magnitude is another "dimension" which characterises systems. Of particular importance, when $S \to \infty$, the moments are classical and there are no quantum effects.

### 4.9.2 *Validity of mean field theory*

In Section 4.8.7, we learned that for each system (hamiltonian) there is a marginal spatial dimensionality $d^*$ such that when the number of

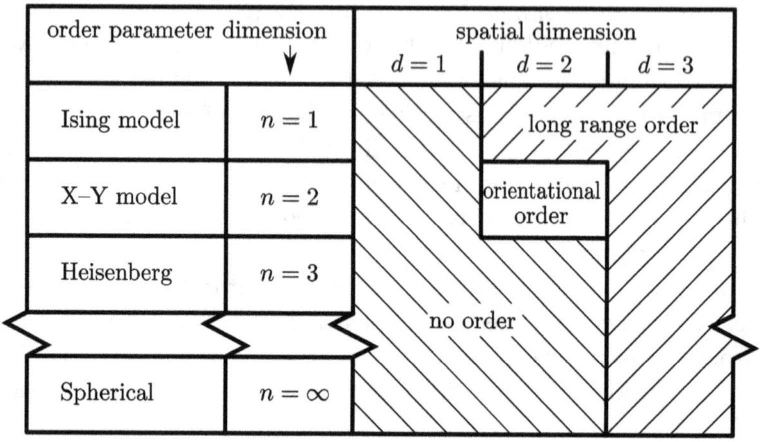

| order parameter dimension | | spatial dimension | | |
|---|---|---|---|---|
| | ↓ | $d = 1$ | $d = 2$ | $d = 3$ |
| Ising model | $n = 1$ | | long range order | |
| X–Y model | $n = 2$ | | orientational order | |
| Heisenberg | $n = 3$ | | no order | |
| Spherical | $n = \infty$ | | | |

Fig. 4.57.  Dimensionality and order.

dimensions is greater than $d^*$, the results of mean field theory are exact. We have also seen that the spatial dimension is crucial in determining whether a system will undergo a phase transition. It is therefore surprising that the spatial dimension enters nowhere in the mean field modelling of phase transitions. In fact, we saw that for a spatial dimension of four or greater, the mean field description becomes exact.

It is usually the case that for short-range interactions the marginal dimensionality $d^*$ is 4. However, when the interactions have longer range, such as dipole or Coulomb interactions, then the marginal dimensionality can be less; an example is the dipole-coupled Ising system treated in Section 4.8.

There is, however, a practical question when considering the validity of mean field theory. How close to the transition must one be to observe the (possible) breakdown of mean field theory? There are some systems, such as superconductors, where one needs to be impossibly close to the transition to observe the breakdown. For such systems, experimentally, they *appear* to follow mean field behaviour even though $d^* = 4$. The Ginzburg criterion specifies how close one must be; an extensive discussion of all these issues is given in the paper by Als-Nielsen and Birgeneau [31].

### 4.9.3   *Universality classes*

As we stated at the start of this chapter, systems undergoing phase transitions exhibit properties of *universality*: many properties of systems in the

Table 4.3. Critical exponents for some universality classes.

| | $d$ | $n$ | $\alpha$ | $\beta$ | $\gamma$ | $\delta$ | $\nu$ | $\eta$ |
|---|---|---|---|---|---|---|---|---|
| Mean field | Any | Any | 0 | ½ | 1 | 3 | ½ | 0 |
| 2d Ising | 2 | 1 | 0 | ⅛ | 7/4 | 15 | 1 | ¼ |
| 3d Ising | 3 | 1 | 0.11 | 0.33 | 1.24 | 4.79 | 0.63 | 0.04 |
| 3d XY | 3 | 2 | −0.01 | 0.35 | 1.32 | 4.78 | 0.67 | 0.04 |
| 3d Heisenberg | 3 | 3 | −0.12 | 0.37 | 1.40 | 4.8 | 0.71 | 0.04 |
| 3d spherical | 3 | ∞ | −1 | ½ | 2 | 5 | 1 | 0 |

Table 4.4. Critical exponents for Bose–Einstein condensation.

| | $d$ | $n$ | $\alpha$ | $\beta$ | $\gamma$ | $\delta$ | $\nu$ | $\eta$ |
|---|---|---|---|---|---|---|---|---|
| Mean field | Any | Any | 0 | ½ | 1 | 3 | ½ | 0 |
| 3d BEC | 3 | — | −1 | 1 | 1 | 2 | 1 | 1 |

vicinity of the critical point do not depend on microscopic details, but are shared by dissimilar systems. We have now seen various examples of this.

The key parameters are the nature of the order parameter and the spatial dimensionality. Dissimilar systems whose order parameters have the same mathematical structure, in space of a given dimensionality, will behave similarly in the vicinity of the critical point.

The order parameters of all the examples we have considered may be regarded as vectors; in these cases the nature of the order parameter is its dimensionality $n$. We have denoted the spatial dimension as $d$. So, for our examples, the *universality class* is specified by the pair $d, n$. Table 4.3, extracted from the compilation of Mattis [34], gives the critical exponents for many of the systems we have discussed.

The liquid–gas transition (Section 4.2), for which $d = 3, n = 1$, has critical exponents corresponding to those of the 3d Ising model, as expected.

The Bose–Einstein condensation is an anomaly. The critical exponents can be calculated exactly, but the results are surprising. We have $\beta = 1$ from Eq. (2.6.22) and Problem 4.4, and we have $\alpha = -1$ (for $T < T_B$) from Eq. (2.6.37) and Problem 4.3. These, and the complete set of BEC critical exponents, are calculated in reference [35]; they are given in Table 4.4.

| | first order | | second order | symmetry broken |
|---|---|---|---|---|
| **conserved order parameter** | liquid–gas (but ∃ critical point) | → | liquid–gas along critical isochore | none |
| | binary alloy (but ∃ critical point) | → | binary alloy at critical concentration | none |
| | solid–fluid: no critical point | → | – – – – – | translational invariance |
| **non-conserved order parameter** | – – – – – | → | ferromagnet | rotational invariance |
| | ferroelectric at high pressure | tricritical ← point → | ferroelectric at low pressure | inversion symmetry |

Fig. 4.58.   Phase transition examples and characteristics.

The simplest view is that the BEC transition should belong to the 3d Ising universality class, taking the order parameter to be the superfluid density ($n = 1$, a real scalar). This would give $\alpha = 0.11, \beta \approx 0.33$, rather different from the exact $\alpha = -1, \beta = 1$. So, BEC is *not* of the 3d Ising universality class. Some claim the BEC transition should belong to the XY universality class, arguing that the order parameter should be the condensate wave function ($n = 2$, a complex scalar). This would give $\alpha \approx -0.01, \beta \approx 0.35$, again rather different from the exact calculations. So also BEC is *not* of the XY universality class.

### 4.9.4   *Features of different phase transition models*

The main classification criterion we have used in discussing phase transitions has been the order of the transition: first order or second order. Recall that this is essentially a generalisation/abuse of Ehrenfest's original classification; in reality, we distinguish discontinuous and continuous transitions. The second classification related to whether the order parameter was conserved or non-conserved. In this chapter, we have seen examples of all four possible combinations of these possibilities.

The liquid–gas transition is an example of a first-order transition in a system with a conserved order parameter. This transition has a critical point, thus one observes a second-order transition when travelling along the critical isochore. The solid–fluid transition is another example of a first-order transition in a system with a conserved order parameter. However in this case, because there is a change in symmetry between the phases, there is no critical point and the transition never becomes second order.

The ferromagnet and the ferroelectric are both examples of transitions with non-conserved order parameters. The ferroelectric transition can be either first-order or second-order, depending on the conditions. And the changeover is at the tricritical point. The ferromagnetic transition is second order. However, there is no reason why such a transition may not become first-order under the right circumstances. Figure 4.58 summarises the examples treated in this chapter.

## Problems

4.1 The scaling expression for the reduced (magnetic) free energy is given in Section 4.1.9 by

$$f(t, B) = A \, |t|^{2-\alpha} \, Y \left( D \frac{B}{|t|^{\Delta}} \right).$$

(a) Show that the heat capacity is given by

$$C \sim \frac{\partial^2 f(t, B)}{\partial t^2}$$

and

(b) hence identify $\alpha$ as the heat capacity critical exponent (when $B = 0$).

4.2 Using the scaling expression for the reduced free energy in the previous question,

(a) show that the magnetisation is given by

$$M \sim \frac{\partial f(t, B)}{\partial B}$$

and hence

(b)   show that the order parameter exponent $\beta$ is given by

$$\beta = 2 - \alpha - \Delta.$$

(c)   Show that the magnetic susceptibility is given by

$$\chi \sim \frac{\partial^2 f(t, B)}{\partial B^2}$$

and hence

(d)   show that the susceptibility exponent $\gamma$ is given by

$$\gamma = -2 + \alpha + 2\Delta.$$

4.3   For temperatures below $T_B$, the heat capacity of a 3d Bose–Einstein gas is given by Eq. (2.6.37):

$$C_V = \frac{15}{4} Nk \frac{\zeta\left(\frac{5}{2}\right)}{\zeta\left(\frac{3}{2}\right)} \left(\frac{T}{T_B}\right)^{3/2}.$$

Express $C_V$ as a function of $t = (T - T_B)/T_B$ and hence show that the heat capacity critical exponent $\alpha$ (for $T < T_B$) is $-1$.

4.4   The Bose–Einstein condensation, treated in Chapter 2, is an example of a non-interacting system which, nevertheless, exhibits a phase transition. Assume the ground state fraction is the order parameter of the transition.

(a)   Obtain the order parameter critical exponent $\beta$ for both the 3d free gas.

(b)   Obtain also the order parameter critical exponent for the 3d gas trapped in a harmonic potential.

(c)   You should find these two critical exponents to be the same. Comment on this.

(d)   Also comment on the value of this critical exponent.

4.5   Plot some isotherms of the Clausius equation of state: $p(V - Nb) = NkT$. How do they differ from those of an ideal gas? Does this equation of state exhibit a critical point? Explain your reasoning.

4.6   (a)   Show that for a van der Waals fluid the critical parameters are given by $V_c = 3Nb$, $p_c = a/27b^2$, $kT_c = 8a/27b$.

(b) Show that for a van der Waals fluid the critical compressibility factor

$$z_c = p_c V_c / N k T_c \text{ has the value } 3/8 = 0.375.$$

4.7 In Problem 3.20, the van der Waals parameters $a$ and $b$ were approximated in terms of the Lennard-Jones interaction potential parameters $\sigma$ and $\varepsilon$:

$$a = \frac{16}{9} \pi \sigma^3 \varepsilon,$$

$$b = \frac{2}{3} \pi \sigma^3.$$

Use these expressions to estimate the van der Waals critical quantities in terms of the Lennard-Jones parameters.

4.8 Show that for the Dieterici fluid the critical parameters are given by $V_c = 2Nb$, $p_c = a/4b^2 e^2$, $kT_c = a/4b$, and the critical compressibility factor has the value $z_c = 2/e^2 = 0.271$.

4.9 Show that for the Berthelot fluid the critical parameters are given by $V_c = 3Nb$, $p_c = \sqrt{a/b^3}/6\sqrt{6}$, $kT_c = (2/3)^{3/2}\sqrt{a/b}$, and the critical compressibility factor has the value $z_c = 3/8 = 0.375$, just as for the van der Waals equation.

4.10 Show that for the Redlich–Kwong fluid the critical parameters are given by $V_c = 3.847Nb$, $p_c = 0.0299a^{2/3}/b^{5/3}$, $kT_c = 0.345(a/b)^{2/3}$, and the critical compressibility factor has the value $z_c = 1/3$.

4.11 The discussion around Fig. 4.19 argued that the transition temperature of a ferromagnet could be estimated from measurements at high temperatures by plotting $1/\chi$ against temperature and extrapolating the line to the axis. While this is reliable for the mean field $\gamma = 1$ case, show that for the realistic case where $\gamma > 1$, the actual transition temperature will be lower than the mean field estimate.

You should draw the Curie–Weiss line, as in Fig. 4.19, and note that it has slope of $\gamma = 1$. You should then show how low temperature deviations above and below $\gamma = 1$ alter the extrapolation to $1/\chi \to 0$.

4.12   Obtain an expression for the (Landau) Helmholtz free energy for the Weiss model in zero external magnetic field, in terms of the magnetisation. Plot $F(M)$ for $T > T_c$, $T = T_c$ and $T < T_c$.

4.13   Show that $F = \frac{1}{2}Nk\left\{(T - T_c)m^2 + \frac{1}{6}T_cm^4 + \cdots\right\}$ for the Weiss model ferromagnet in the limit of small $m$. Explain the appearance of $T_c$ in the $m^4$ term.

4.14   Show that $d^2F/d\varphi^2 > 0$ below $T_c$ at the two roots $\varphi = \pm\sqrt{-F_2/2F_4}$ in the Landau model. Show that $d^2F/d\varphi^2 < 0$ below $T_c$ and $d^2F/d\varphi^2 > 0$ above $T_c$ at the single root $\varphi = 0$. What is the physical meaning of this?

4.15   In the Landau theory of second-order transitions calculate the behaviour of the order parameter below the critical point, $\varphi(T)$, when the *sixth*-order term in the free energy expansion is not discarded. What influence does this term have on the critical exponent $\beta$? Comment on this.

4.16   Show that the Landau free energy is consistent with the scaling free energy of Problem 4.1 above, with $\alpha = 0$. Comment on the value of $\Delta$ required. Start by considering the $B = 0$ case.

4.17   A ferroelectric has a free energy of the form
$$F = \alpha(T - T_c)P^2 + bP^4 + cP^6 + DxP^2 + Ex^2$$
where $P$ is the electric polarisation and $x$ represents the strain. Minimise the system with respect to $x$. Under what circumstances is there a first-order phase transition for this system?

4.18   The first-order ferroelectric has *two* spinodals, one above and one below the equilibrium transition temperature $T_{tr}$.
   (a)   At the spinodal the first and *second* derivatives of $F(\varphi)$ vanish. Explain this.
   (b)   Show that these correspond to $\varphi^{sp} = \pm\sqrt{\frac{-F_4}{3F_6}}$, $F_2^{sp} = \frac{F_4^2}{3F_6}$ and $\varphi^{sp} = 0$, $F_2^{sp} = 0$.
   (c)   The temperatures of these spinodals correspond to $T_{sp}^u = T_c + \frac{1}{3a}\frac{F_4^2}{F_6}$ and $T_{sp}^l = T_c$, where the u and l superscripts indicate the *upper* and the *lower* temperature spinodal. Derive these

expressions and show their variation with $F_4$ by including them on a plot similar to Fig. 4.37.

(d) Now plot the order parameter as a function of temperature, as in Fig. 4.38, indicating the hysteretic jumps at $T_{sp}^l$ and $T_{sp}^u$. Why are these called *hysteretic* jumps?

4.19 Consider a one-dimensional binary alloy where the concentration of A atoms varies slowly in space: $x = x(z)$. Show that the spatial variation of $x$ results in an additional term in the free energy per bond proportional to $a^2 \varepsilon (dx/dz)^2$, where $a$ is the spacing between atoms and $\varepsilon$ is the energy parameter defined in Section 4.7.3. Determine the numerical coefficient.

4.20 Show that in the vicinity of the critical point the free energy of mixing of the binary alloy may be written as

$$F_m = F_0 + Nk \left\{ \frac{1}{2}(T - T_c)\varphi^2 + \frac{1}{12}T_c\varphi^4 + \frac{1}{30}T_c\varphi^6 + \cdots \right\}$$

where $\varphi = 2x - 1$.

Discuss the Landau truncation of this expression; in particular, explain at which term the series may/should be terminated.

# References

[1]  C. Yang and T. Lee, Statistical theory of equations of state and phase transitions. I. Theory of Condensation, *Phys. Rev.*, **87** (1952) 404–409.

[2]  T. Lee and C. Yang, Statistical theory of equations of state and phase transitions. II. Lattice Gas and Ising Model, *Phys. Rev.*, **87** (1952) 410–419.

[3]  K. Huang, *Statistical Mechanics*, 2nd ed. (John Wiley, New York, 1987).

[4]  M. E. Fisher, The nature of critical points, *University of Colorado, Summer School*, vol. VII C (1965).

[5]  G. L. Jones, Complex temperatures and phase transitions, *J. Math. Phys.*, **7** (1966) 2000–2005.

[6]  G. Jaeger, The Ehrenfest classification of phase transitions: Introduction and evolution, *Arch. Hist. Exact Sci.*, **53**, 1, (1998) 51–81.

[7]  V. L. Ginzburg and L. D. Landau, On the theory of superconductivity, *Zh. Eksp. Teor. Fiz*, **20** (1950) 1064. Translation in Collected Papers of L. D. Landau, pp. 546–568 (Pergamon Press, 1965).

[8] J. W. Gibbs, On the equilibrium of heterogeneous substances, *Trans. Conn. Acad. Arts Sci.*, **3** (1875–1878) 108. Reprinted in J. W. Gibbs, The Collected Works, Vol. 2, Longman and Green (New York, 1928).

[9] M. E. Fisher, Scaling, universality and renormalization group theory, in *Critical Phenomena: Proceedings of the Summer School Held at the University of Stellenbosch, South Africa January 18–29, 1982* (F. J. W. Hahne, ed.), pp. 1–139 (Berlin, Heidelberg: Springer Berlin Heidelberg, 1983).

[10] G. F. Mazenko, *Equilibrium Statistical Mechanics* (John Wiley, New York, 2000).

[11] H. E. Stanley, *Phase Transitions and Critical Phenomena* (OUP, Oxford, 1971).

[12] E. A. Guggenheim, The principle of corresponding states, *J. Chem. Phys.*, **13**, 7 (1945) 253.

[13] E. A. Guggenheim, *Thermodynamics* (North Holland, 1949).

[14] E. A. Guggenheim, *Applications of Statistical Mechanics* (Oxford University Press, 1966).

[15] V. L. Kulinskii, The critical compressibility factor of fluids from the global isomorphism approach, *J. Chem. Phys.*, **139**, 18 (2013) 184119.

[16] R. J. Sadus, Equations of state for fluids: The Dieterici approach revisited, *J. Chem. Phys.*, **115**, 3 (2001) 1460–1462.

[17] J. Wilks, *The Properties of Liquid and Solid Helium* (Clarendon Press, Oxford, 1967).

[18] X. Y. Li, Y. H. Huang, G. B. Chen, and von Arp, Density equation for saturated helium-3, *20th Intl. Cryo. Engr. Conf. (ICEC20)* (2004) 3–6.

[19] J. de Boer, Quantum theory of condensed permanent gases 1: The law of corresponding states, *Physica*, **14** (1948) 139–148.

[20] I. Hahn, M. Weilert, F. Zhong, and M. Barmatz, High-resolution measurements of the coexistence curve very near the $^3$He liquid-gas critical point, *J. Low Temp. Phys.*, **137**, 5–6 (2004) 579–598.

[21] W. H. Zurek, Decoherence, einselection, and the quantum origins of the classical, *Rev. Mod. Phys.*, **75**, 3 (2003) 715–775.

[22] S. Brush, History of the Lenz-Ising model, *Rev. Mod. Phys.*, **39** (1967) 883–893.

[23] C. Kittel, *Introduction to Solid State Physics*. John Wiley, New York, 7th. ed., 1996.

[24] M. Plische and B. Bergersen, *Equilibrium Statistical Physics* (World Scientific, 1994).

[25] M. E. Fisher, Renormalization group theory: Its basis and formulation in statistical physics, *Rev. Mod. Phys.*, **70** (1998) 653–681.

[26] R. J. Baxter, *Exactly Solved Models in Statistical Mechanics* (Academic Press, London, 1982).

[27] G. Burns, *Solid State Physics* (Academic Press, New York, 1986).

[28] L. D. Landau and E. M. Lifshitz, *Statistical Physics* (Pergamon Press, 1970).

[29] J. C. Slater, *Introduction to Chemical Physics* (McGraw-Hill, New York, 1939).

[30] D. Bitko, T. Rosenbaum, and G. Aeppli, Quantum critical behavior for a model magnet, *Phys. Rev. Lett.*, **77** (1996) 940–943.

[31] J. Als-Nielsen and R. J. Birgeneau, Mean field theory, the Ginzburg criterion, and marginal dimensionality of phase transitions, *Am. J. Phys.*, **45** (1977) 554.

[32] N. D. Mermin and H. Wagner, Absence of ferromagnetism or anti-ferromagnetism in one- or two-dimensional isotropic Heisenberg models, *Phys. Rev. Lett.*, **17**, 22 (1966) 1133–1136.

[33] J. M. Kosterlitz and D. J. Thouless, Ordering, metastability and phase transitions in two-dimensional systems, *J. Phys. C: Solid State Phys.*, **6**, 7 (1973) 1181.

[34] D. C. Mattis, *Statistical mechanics made simple* (World Scientific, Singapore, 2003).

[35] I. Reyes-Ayala, F. J. Poveda-Cuevas, and V. Romero-Rochín, Non-classical critical exponents at Bose–Einstein condensation, *J. Stat. Mech.: Theory Exp.*, **2019**, 11 (2019) 113102.

# CHAPTER 5

# FLUCTUATIONS AND DYNAMICS

---

The traditional subject of thermo*dynamics* is wrongly named. It deals, essentially, with equilibrium states and the relations between these states. The existence of states of equilibrium is essential to our understanding of the macroscopic world at the quantitative level. Although macroscopic systems must be described, at the microscopic level, by a very large number of variables, the *equilibrium* states of such systems are specified in terms of only a few thermodynamic parameters. The existence of equilibrium states is thus a fortunate aspect of nature; and it is upon this aspect that the discipline of thermodynamics is based.

Classical thermodynamics relates different equilibrium states of a system. One state might be transformed into another by passing "quasi-statically" through a sequence of intermediate (almost) equilibrium states. Or the process might be completely irreversible, such as the Joule–Kelvin throttling, where all one can speak of are the initial and final equilibrium states. In the former case, we can draw an "indicator diagram" showing the variation of thermodynamic variables in the process. In the latter case, all we can plot are the initial and final points. Classical thermodynamics and its microscopic underpinning of equilibrium statistical mechanics recognises the existence of initial and final states, but it can tell nothing about the *dynamics* of the transformation processes. Thus, the designation *thermodynamics* is a misnomer.

Equilibrium systems have no dynamics, almost by definition. This is true "on the average", but of course there are fluctuations. By contrast, non-equilibrium systems do have dynamics. Common experience (and indeed the Second Law of thermodynamics) tells us that a non-equilibrium system will evolve towards a state of equilibrium. The subject

313

of non-equilibrium statistical mechanics deals with this *approach* to equilibrium.

Before making direct appeal to the microscopics of particular systems it is of interest to ask what deductions can be made from specifically macroscopic considerations. As with equilibrium "thermodynamics", the macroscopic approach can relate different properties of a system, but it cannot go much further than this. Adkins's *Equilibrium Thermodynamics* [1] discusses thermoelectricity from this perspective and the limitations of the approach are clearly discussed. Further progress in the macroscopic approach can be made through the addition of some new "laws". Traditionally, Onsager's "reciprocity law" is used. More modern is the use of the law of minimum entropy production. Both of these are discussed in Kondepundi and Prigogine's *Modern Thermodynamics* [2].

The behaviour of specific systems must be treated, as in the equilibrium case, by looking at the microscopics, and applying statistical arguments. In particular, this is necessary if we want to study the time dependence of thermodynamic variables in non-equilibrium systems and their variation as the system evolves towards equilibrium. MacDonald [3] coined the phrase "time-dependent statistical mechanics" for this.

## 5.1  Fluctuations

### 5.1.1  *Probability distribution functions*

When a thermodynamic property is fixed, its conjugate variable fluctuates about its mean value. Thus, we saw, in Section 1.4.6, that when the temperature was fixed, there were fluctuations in energy. In a small region of a system there will be fluctuations in all extensive variables. Here, we are regarding the small region as a subsystem in equilibrium with a reservoir comprising the remainder of the system. Thus, the conditions for the equilibrium of the subsystem are that all intensive variables are fixed: temperature, pressure, chemical potential, etc.

To examine the fluctuations in the subsystem, we must consider the appropriate thermodynamic potential; here it is the Gibbs free energy $G(T, p, \mu)$, since $T$, $p$, and $\mu$ are fixed. We ask the question: What is the probability for the occurrence of a fluctuation of magnitude $X$?

When considering an *isolated* system, we saw that the Boltzmann relation $S = k \ln \Omega$ could be inverted, following Einstein, to give the probability

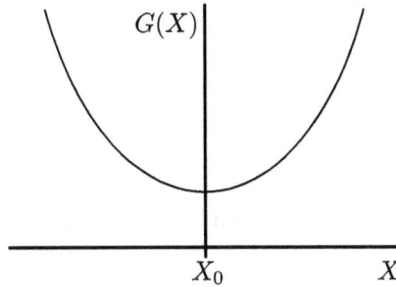

Fig. 5.1. Equilibrium value of Gibbs free energy.

of observing a fluctuation of magnitude $X$ as

$$P(X) \propto e^{S(X)/k}. \tag{5.1.1}$$

An *open system*, that is, one having fixed values of its intensive variables (our subsystem) is specified by its Gibbs free energy and the probability of a fluctuation of magnitude $X$ is

$$P(X) \propto e^{-G(X)/kT}. \tag{5.1.2}$$

(The probability would be given by $P(X) \propto e^{S(X)/k}$ where $S(X)$ is the total entropy of the system *plus* reservoir. It becomes $P(X) \propto e^{-G(X)/kT}$ where $G(X)$ is the Gibbs free energy of the system alone.)

In equilibrium, $G$ is a minimum; let us specify this to occur at $X = X_0$, Fig. 5.1. Here, $X$ could be a general extensive variable or an order parameter of a phase transition. For convenience, we shall put $X_0 = 0$ so that now the variable $X$ measures the *deviation* from the mean (equilibrium) value.

For small deviations of $X$ from its equilibrium value, we may expand

$$G(X) = G(0) + X \left.\frac{\partial G}{\partial X}\right|_0 + \frac{X^2}{2} \left.\frac{\partial^2 G}{\partial X^2}\right|_0 + \cdots. \tag{5.1.3}$$

Since the equilibrium position is a minimum, the linear term vanishes. The leading term in $X$ is thus the quadratic. To leading order in $X$, the probability is then

$$P(X) \propto \exp - \left\{ G(0)/kT + G''(0)X^2/2kT \right\} \tag{5.1.4}$$

or

$$P(X) \propto e^{-G''(0)X^2/2kT}. \tag{5.1.5}$$

This is a Gaussian probability distribution, which has the general form

$$P(X) \propto e^{-X^2/2\langle X^2 \rangle}. \tag{5.1.6}$$

And from this we can identify the mean-square measure of the fluctuations of $X$ as

$$\langle X^2 \rangle = kT \left. \frac{\partial^2 G}{\partial X^2} \right|_{X=X_0}^{-1}. \tag{5.1.7}$$

We see that the *broader* the minimum in $G$, the greater the magnitude of the fluctuations. So, in particular, at a *critical point* where we saw that $G'' \rightarrow 0$, the fluctuations become infinite. This is an important property of a critical point. We note also that the fluctuations will diverge at the *spinodal point* of a first-order transition.

### 5.1.2   *Average behaviour of fluctuations*

The mean value of a thermodynamic quantity $X$ is constant for a system in equilibrium; this is essentially the definition of equilibrium. However the *instantaneous* value of $X$ will fluctuate with time as dictated by the equations of motion of the system; a typical example is shown in Fig. 5.2. The deviations of $X$ from its mean will average to zero (indeed this is embodied in the definition of the mean), since the deviations are equally likely to be positive or negative. But the *square* of the deviations of $X$ from its mean will have a non-zero average; moreover, this can be calculated within the framework of standard equilibrium thermodynamics/ statistical mechanics. For convenience, let us re-scale $X$ by subtracting its mean value; our newly scaled $X$ has zero mean: $\langle X \rangle = 0$. But as mentioned above, $\langle X^2 \rangle \neq 0$, and in general, certainly for even $n$, we will have

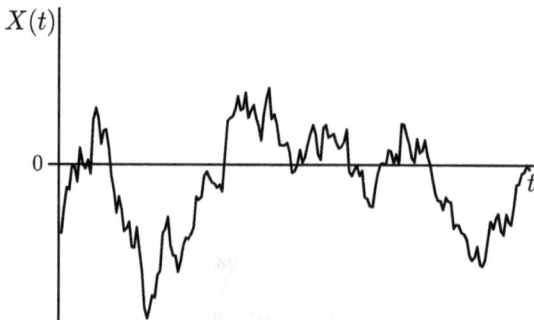

Fig. 5.2.   Fluctuations in $X$ as a function of time.

$\langle X^n \rangle \neq 0$. These "moments" can often be calculated without too much difficulty, as treated in Problem 1.9. These moments tell us about the likelihood of fluctuations of different sizes, but they give no indication about the *time-dependence* of the fluctuations.

> **Ensembles and averages:** It is not immediately obvious what is meant by taking an average in this case. If many copies of the system are imagined, each having the same values for the macroscopic observables, then one can consider the average evaluated over this collection of copies. The imaginary collection of such copies is referred to as an *ensemble*, the Gibbs ensemble of Chapter 1, and the average is called an *ensemble average*.

What can we say about the time variation of the fluctuations without completely solving the equations of motion for the system? Can we specify a quantity that describes the typical time evolution of the variations?

**Mean value:** At a particular time $t_0$, we may observe that the variable $X$ has the value $a^i$:

$$X(t_0) = a^i. \tag{5.1.8}$$

This will subsequently develop in time so that the observed mean becomes zero. We can talk of the mean time dependence of $X$ from this value $a^i$ by taking a sub-ensemble from a complete ensemble — the sub-ensemble consisting of all those elements that at time $t_0$ have the value $a^i$, that is, the elements of the sub-ensemble obey

$$X_j^i(t_0) = a^i. \tag{5.1.9}$$

The upper index of $X$ labels the sub-ensemble and the lower index indicates which element within the sub-ensemble. We can then define the mean regression from the initial value $a^i$ at time $t_0$ by the average

$$f^i(t_0 + t) = \frac{1}{n_i} \sum_{j=1}^{n_i} X_j^i(t_0 + t) \tag{5.1.10}$$

where $n_i$ is the number of elements in the $i^{\text{th}}$ sub-ensemble.

For sufficiently long times $t$, the value of $f^i(t_0 + t)$ will go to zero since there will then be no means of distinguishing that this sub-ensemble is not a representative one. Thus, we expect the function $f^i(t_0 + t)$ to behave as in Fig. 5.3. This function describes "on the average" how $X$ varies if at time $t_0$ it had the value $a^i$. However, since we are considering a system in

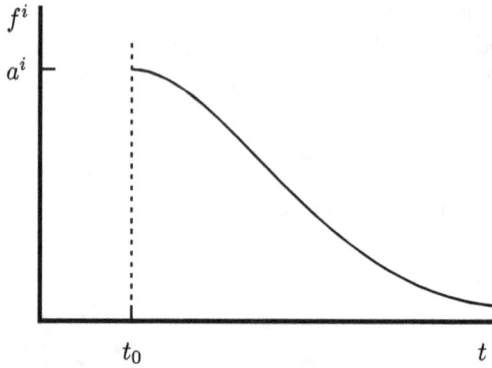

Fig. 5.3.  Mean regression of a random fluctuation from an initial value.

equilibrium, that is, it has time translation invariance — it follows that the time evolution of $f^i(t_0 + t)$ is not peculiar to the time $t_0$. Whenever $X$ is observed to take on the value $a^i$, the subsequent mean time dependence will be given by $f^i$. So, we may ignore $t_0$ and we shall put it to zero in our future discussions.

The behaviour does, however, refer specifically to the particular value $a_i$. To find the mean regression of any fluctuation, we may think of averaging over all the initial values $a_i$ with appropriate weights. But this is simply an average over the complete ensemble and this, we know, must give zero. Physically we can see that this is so since there will be many positive and negative values of $f^i$ which will average to zero, as in Fig. 5.4.

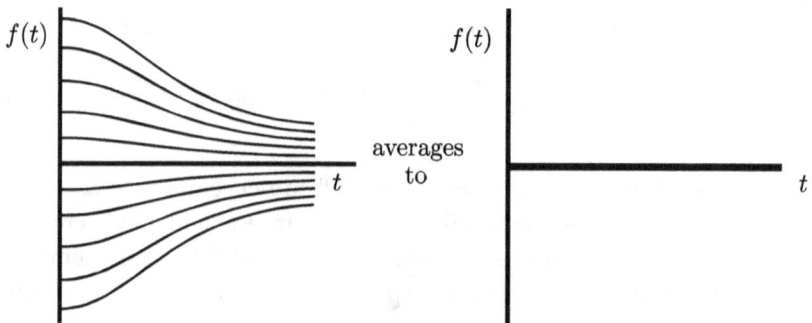

Fig. 5.4.  Average regression of many different initial states.

Mathematically, this follows since

$$\langle f(t) \rangle = \sum_i w_i f^i(t) = \sum_i \frac{w_i}{n_i} \sum_j X_j^i(t) \tag{5.1.11}$$

where $w_i$ is the weight factor for the $i$th sub-ensemble. Now, this weight factor is the proportion of the whole ensemble that the $i$th sub-ensemble represents, that is $n_i/N$ where $N$ is the total number of elements in the ensemble: $N = \sum_i n_i$. Thus,

$$\langle f(t) \rangle = \sum_i \frac{n_i}{N n_i} \sum_j X_j^i(t)$$

$$= \frac{1}{N} \sum_{i,j} X_j^i(t), \tag{5.1.12}$$

the common mean over the full ensemble which is zero.

**Mean square:** Instead of the direct mean, we could evaluate the mean of the squares of the regressions of the sub-ensembles. In this way, positive and negative variations will all contribute without cancelling each other. We then have

$$\langle f^2(t) \rangle = \sum_i w_i \{f^i(t)\}^2$$

$$= \sum_i \frac{w_i}{n_i} \left\{ \sum_j X_j^i(t) \right\}^2$$

$$= \sum_i \frac{w_i}{n_i} \sum_{j,k} X_j^i(t) X_k^i(t) \tag{5.1.13}$$

$$= \frac{1}{N} \sum_{i,j,k} X_j^i(t) X_k^i(t)$$

but this involves cross terms between members of the ensemble. The idea of correlations between different elements of an ensemble is completely un-physical, the mathematics is becoming complicated, so we shall reject this approach.

**Magnitude:** Another possibility is to take the average of the magnitude of the $f^i(t)$, defining

$$\langle f(t)\rangle_{\text{mag}} = \sum_i w_i \left|f^i(t)\right|$$

$$= \frac{1}{N} \sum_i \left|\sum_j X_j^i(t)\right|. \tag{5.1.14}$$

Unfortunately, there are mathematical difficulties in manipulating modulus functions, but this approach gives a good lead. In fact, this method of defining an average is equivalent to taking a mean with a weight function $\varepsilon_i$ which is positive for positive $a^i$ and negative for negative $a^i$:

$$\langle f(t)\rangle_{\text{mag}} = \frac{1}{N} \sum_i \varepsilon_i \sum_j X_j^i(t)$$

$$= \frac{1}{N} \sum_{i,j} \varepsilon_i X_j^i(t). \tag{5.1.15}$$

Written in this form, the expression for the average behaviour of the regression seems promising. The only difficulty is with the strange weight function

$$\varepsilon_i = +1 \text{ if } \quad a^i > 0$$
$$= -1 \text{ if } \quad a^i < 0. \tag{5.1.16}$$

**Autocorrelation:** Instead of this discontinuous function $\varepsilon$ multiplying the components, there is a much more straightforward weight function that satisfies the fundamental requirement of respecting the sign of the $a^i$: we could use the $a^i$ themselves. In other words, in evaluating the average behaviour of the natural fluctuations, we weigh each element of the ensemble by its initial value. We have then

$$\frac{1}{N} \sum_{i,j} a^i X_j^i(t) \tag{5.1.17}$$

or, since $X_j^i(0) = X^i(0) = a^i$, it can be written as

$$\frac{1}{N} \sum_{i,j} X_j^i(0) X_j^i(t) \tag{5.1.18}$$

and in this form we have achieved an important progress, in that the average is over the complete ensemble (with no cross terms) and we can write

it as

$$\langle X(0)X(t)\rangle. \tag{5.1.19}$$

This is our chosen expression for the "average" regression of a fluctuation from some initial value back to the mean.

### 5.1.3   *The autocorrelation function*

The above expression is the mean over the ensemble where each element is weighted in proportion to its initial value. This function of the randomly varying quantity $X(t)$ is known as the *autocorrelation function*. We shall denote it by the symbol $G(t)$:

$$G = \langle X(0)X(t)\rangle. \tag{5.1.20}$$

In Fig. 5.5, we show the fluctuations of Fig 5.2, $X(t)$, together with the corresponding autocorrelation function, $G(t)$.

In the sense described above, the function $G(t)$ describes the mean time variation of the fluctuations in $X(t)$. Observe the smooth behaviour of the autocorrelation function. In a sense, this function has distilled the fundamental essence of the random function $X(t)$ without its wealth of unimportant fine detail.

The zero time value of $G(t)$ has an immediate interpretation. From the definition of $G(t)$, we have

$$G(0) = \langle X^2\rangle, \tag{5.1.21}$$

the mean-square value of the fluctuating variable. For long times, as we have argued above, $G(t)$ should go to zero.

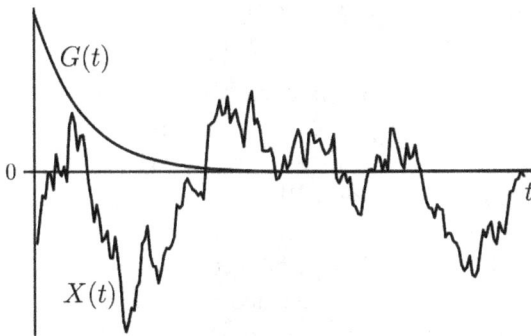

Fig. 5.5.   Fluctuations in $X$ together with the autocorrelation function.

If $X$ is not defined so that its average $\langle X \rangle$ is zero, then we must modify the definition of the autocorrelation function to

$$G(t) = \langle (X(0) - \langle X \rangle)(X(t) - \langle X \rangle) \rangle$$

$$= \langle X(0)X(t) \rangle - \langle X(0) \langle X \rangle \rangle - \langle \langle X \rangle X(t) \rangle + \langle X^2 \rangle$$

$$= \langle X(0)X(t) \rangle - \langle X^2 \rangle$$

since

$$\langle X(0) \langle X \rangle \rangle = \langle \langle X \rangle X(t) \rangle = \langle X^2 \rangle.$$

The time-translation invariance property, which we stated to be a property of equilibrium systems, now becomes the stationarity principle:

$$\langle X(\tau)X(t + \tau) \rangle = \langle X(0)X \rangle \tag{5.1.22}$$

i.e.

$$\text{equilibrium} \Rightarrow \text{stationarity}. \tag{5.1.23}$$

And stationarity implies the time-reversal behaviour. From stationarity we have:

$$\langle X(-\tau)X(0) \rangle = \langle X(0)X(\tau) \rangle \tag{5.1.24}$$

but classically the $X$ commute, so that

$$\langle X(-\tau)X(0) \rangle = \langle X(\tau)X(0) \rangle \tag{5.1.25}$$

or

$$G(t) = G(-t). \tag{5.1.26}$$

**A paradox:** The correlation function $G(t)$ has the physical significance that a "large" fluctuation will be expected on average to die out according to $G(t)$. A large fluctuation will occur infrequently and on the "large" scale $X$ may be assumed to decay to its equilibrium value.

We say that $G(t)$ traces the mean decay of a fluctuation. Now, according to this view one would expect $G(-t)$ to tell us of the past history of the fluctuation. But we have the time-reversal rule $G(t) = G(-t)$. Our zero of time always seems to coincide with the peak value of the fluctuation — but how can this be consistent with time-translation invariance? There seems to be a paradox; the time origin is important. The resolution is that the zero of time *is* unique because according to the values of $X$ at *this* time the sub-ensembles are selected. That is, the weighting factors of the averaging procedure are selected at the zero of time.

### 5.1.4 *The correlation time*

The autocorrelation function, as we have seen, starts at a finite value and it decays to zero as the time argument increases. This function indicates to us the average way that a fluctuation dies out. In other words, the correlation function indicates the time scale of the variations of the random variable. We shall quantify this time scale by introducing a *correlation time* $\tau_c$ such that for times much shorter than this $G(t)$ will hardly have changed, while for times much longer than this $G(t)$ will have gone to zero.

$$G(t) \approx G(0) \quad t \ll \tau_c$$
$$\approx 0 \quad t \gg \tau_c. \tag{5.1.27}$$

(Note: This is not a *definition* of what $\tau_c$ *is*; it is the essential condition that any suitably defined $\tau_c$ must satisfy.)

The correlation time is indicated in Fig. 5.6. It is convenient to have a precise mathematical specification and we shall see that the definition we adopt will make direct connection with future considerations.

The correlation time is a "rough measure" of the *width* of the correlation function. Now, we may regard the area of the correlation function as a rough measure of its width multiplied by a rough measure of its height. A rough measure of the height of the correlation function is its initial height $G(0)$. And its area is most conveniently expressed as the integral. Thus, we are saying

$$\int_0^\infty G(t)\,dt = \tau_c\, G(0) \tag{5.1.28}$$

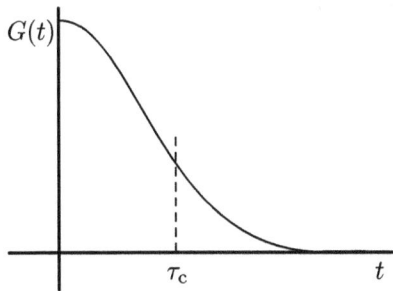

Fig. 5.6.  Correlation time of an autocorrelation function.

or

$$\tau_c = \frac{1}{G(0)} \int_0^\infty G(t)dt. \tag{5.1.29}$$

This expression is taken as the definition of the correlation time.

The autocorrelation function is an important function of a random variable. We will encounter autocorrelation functions frequently in what follows. We will see also that the area under the autocorrelation function is an important quantity and, because of this, we will make frequent use of the correlation time as defined above.

### 5.1.5 Spectral density — The Wiener–Khintchine theorem

Let us now consider the fluctuating variable in the frequency domain. We would *like* to introduce the Fourier transform of $X(t)$, which would be defined by

$$x(\omega) = \int_{-\infty}^\infty X(t)e^{i\omega t}\, dx. \tag{5.1.30}$$

If, however, $X(t)$ is a continually varying function (statistically time-translationally invariant), then this integral will diverge and the Fourier transform will not exist. In that case, it is expedient to consider a "truncated" version of $X(t)$ defined by

$$\begin{aligned} X_T(t) &= X(t) &&|t| < T/2 \\ &= 0 &&\text{otherwise.} \end{aligned} \tag{5.1.31}$$

The Fourier transform of this *does* exist:

$$\begin{aligned} x_T(\omega) &= \int_{-\infty}^\infty X_T(t)e^{i\omega t}\, dx \\ &= \int_{-T/2}^{T/2} X(t)e^{i\omega t}\, dx \end{aligned} \tag{5.1.32}$$

and the inverse transform is then

$$X_T(t) = \frac{1}{2\pi} \int_{-\infty}^\infty x_T(\omega)e^{-i\omega t}\, d\omega. \tag{5.1.33}$$

In terms of this Fourier decomposition, the autocorrelation function $G(t)$ is given by

$$G(t) = \frac{1}{T} \int_{-T/2}^{T/2} X^*(s)X(s+t)\,ds$$

$$= \frac{1}{T}\frac{1}{(2\pi)^2} \int_{-T/2}^{T/2} ds \left\{ \int_{-\infty}^{\infty} x_T^*(\omega_1)e^{i\omega_1 s}\,d\omega_1 \right\}\left\{ \int_{-\infty}^{\infty} x_T(\omega_2)e^{-i\omega_2(s+t)}\,d\omega_2 \right\}.$$

$$(5.1.34)$$

This can be simplified to

$$G(t) = \frac{1}{T}\frac{1}{(2\pi)^2} \int_{-T/2}^{T/2} ds \int_{-\infty}^{\infty} d\omega_1 \int_{-\infty}^{\infty} d\omega_2 \, x_T^*(\omega_1)x_T(\omega_2)\, e^{i(\omega_1-\omega_2)s}\,e^{-i\omega_2 t}.$$

$$(5.1.35)$$

We may perform the integral over $s$; and here we may extend the limits of the integral to infinity. Then:

$$\int_{-\infty}^{\infty} e^{i(\omega_1-\omega_2)s}\,ds = 2\pi\delta(\omega_1 - \omega_2). \tag{5.1.36}$$

We use the sifting property of the delta function, and so the expression for $G(t)$ becomes

$$G(t) = \frac{1}{T}\frac{1}{2\pi} \int_{-\infty}^{\infty} d\omega \, x_T^*(\omega)x_T(\omega)e^{-i\omega t}. \tag{5.1.37}$$

This is a Fourier transform expression; we may write it as

$$G(t) = \frac{1}{2\pi} \int_{-\infty}^{\infty} J(\omega)e^{-i\omega t}\,d\omega, \tag{5.1.38}$$

where

$$J(\omega) = \frac{1}{T}|x_T(\omega)|^2. \tag{5.1.39}$$

Just as $x_T(\omega)$ is the Fourier transform of $X_T(x)$, so $J(\omega) = |x_T(\omega)|^2\big/ T$ is the Fourier transform of $G(t)$.

The quantity $J(\omega) = |x_T(\omega)|^2\big/ T$ is the *power spectrum* or *spectral density* of the fluctuating variable $X(t)$. It should be noted that in the limit of large $T$, $|x_T(\omega)|^2\big/ T$ will tend to a constant value.

The above results, Eqs. (5.1.38) and (5.1.39), are the essence of the Wiener–Khintchine theorem: the spectral density $J(\omega)$ is the Fourier transform of the autocorrelation function $G(t)$.

## 5.2  Brownian Motion

In 1828, Robert Brown observed that tiny grains of pollen suspended in water underwent a perpetual random motion. He went on to observe this effect in a whole range of suspended powdered substances including house dust and pulverised rock. Brown's (erroneous) conclusion was that the motion was due to some "life force" in all inorganic matter. Today the importance of his work is in the *universality* of the effect. It was Einstein, in 1905 [4], who explained the origin of the motion as arising from the continual bombardment of the particles by the atoms or molecules of the fluid in which it is immersed.

The key point about Brownian motion is that it is the motion of a *macroscopic* body arising from *microscopic* impacts from atoms/molecules of the surrounding fluid.

The macroscopic body, which nevertheless might be quite small, may be a pollen grain in water, a smoke particle in air, a flake of ink pigment in water, even the mirror of a traditional galvanometer. The random motion of all these may be observed; Fig. 5.7 shows a typical example. For the present, we shall not consider examples such as the galvanometer

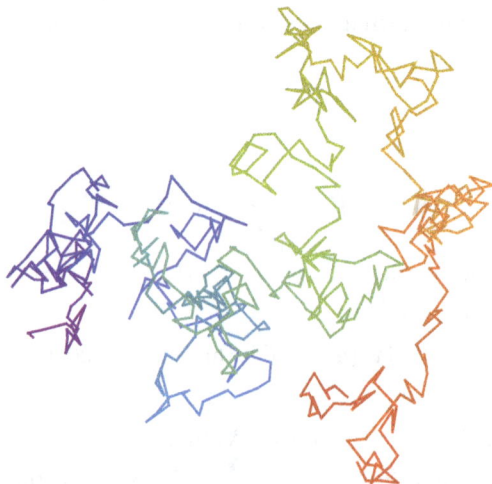

Fig. 5.7.   Typical Brownian motion of a particle in two dimensions. The colour changes with advancing time.

mirror where there is a "restoring force" acting on the Brownian particle. This will be considered in Problem 5.4.

### 5.2.1 *Kinematics of a Brownian particle*

For simplicity, we shall consider motion of a Brownian particle in one dimension; the generalisation to two and three dimensions is straightforward. The location of the particle is not bounded and as time increases, it will travel further and further from the origin. The square of the displacement for a typical trajectory is shown in Fig. 5.8. We see that the mean-square displacement increases (even though in one dimension there is always a finite probability of the particle returning to its starting point).

We shall investigate the mean-square displacement of the particle. At this stage, we consider the *kinematics* of the particle. That is, we will treat things at a descriptive (but quantitative) level without delving into the dynamics and the forces involved.

The distance travelled by the Brownian particle in a time $t$ may be found by integrating up its velocity:

$$x = \int_0^t v(\tau)\mathrm{d}\tau. \tag{5.2.1}$$

Here, $v(\tau)$ is the particle's velocity at time $\tau$.

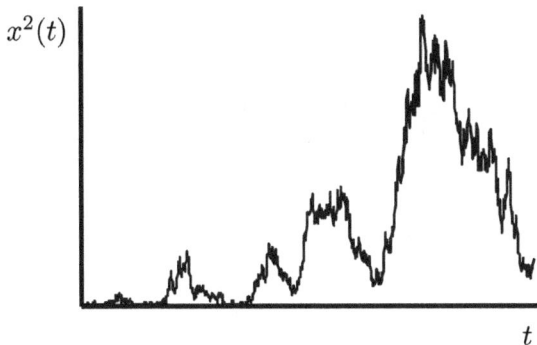

Fig. 5.8. Typical squared displacement of Brownian particle.

The square of the displacement is then

$$x^2(t) = \left\{ \int_0^t v(\tau)d\tau \right\}^2$$

$$= \int_0^t v(\tau_1)d\tau_1 \int_0^t v(\tau_2)d\tau_2$$

$$= \int_0^t d\tau_1 \int_0^t d\tau_2 v(\tau_1)v(\tau_2),$$

so the mean-square displacement is

$$\langle x^2 \rangle = \int_0^t d\tau_1 \int_0^t d\tau_2 \langle v(\tau_1)v(\tau_2) \rangle. \tag{5.2.2}$$

We see that the mean-square displacement is given in terms of the velocity autocorrelation function.

$$G_v(\tau_1 - \tau_2) = \langle v(\tau_1)v(\tau_2) \rangle. \tag{5.2.3}$$

Here, we have used the subscript $v$ to indicate that it is the autocorrelation function of the *velocity*. The stationarity of the random velocity (a consequence of thermal equilibrium) is indicated in the argument $(\tau_1 - \tau_2)$. This allows us to take a further step in the expression for the mean-square displacement. We may change variables in the double integral to

$$\begin{aligned} \tau &= \tau_1 - \tau_2 \\ T &= \tau_1 + \tau_2 \end{aligned} \tag{5.2.4}$$

whereupon we may integrate over the variable $T$. This is a non-trivial procedure, detailed in Appendix D; the result is

$$\langle x^2 \rangle = 2 \int_0^t (t - \tau)G_v(\tau)d\tau. \tag{5.2.5}$$

This is a remarkably useful expression, as we shall see. It is worthwhile to re-emphasise what has been achieved at this stage. Using only kinematics we have found an expression for the mean-square displacement of the

Brownian particle in terms of the particle's velocity autocorrelation function. This also reinforces the idea that autocorrelation functions are useful quantities when considering random processes.

Pursuing the kinematical arguments further we will see that the above expression for the mean-square displacement can be simplified in two limiting cases.

### 5.2.2 Short-time limit

The natural time scale for the process we are considering (the only time scale we have at this stage) is the velocity correlation time, which we shall denote by $\tau_v$. The specification of $\tau_v$ is such that

$$G_v(\tau) \approx G_v(0) \quad \text{when} \quad \tau \ll \tau_v; \quad (5.2.6)$$

at these short times the autocorrelation function will have changed negligibly from its initial value. So, when we consider times much shorter than $\tau_v$, we may replace $G_v(\tau)$ by $G_v(0)$ in the integral expression for $\langle x^2 \rangle$. But in this case, $G_v(0)$ comes out of the integral and we have

$$\langle x^2 \rangle = 2G_v(0) \int_0^t (t - \tau) d\tau. \quad (5.2.7)$$

Now, the integral may be evaluated simply, giving

$$\langle x^2 \rangle = 2G_v(0) \left[ t \int_0^t d\tau - \int_0^t \tau d\tau \right]$$
$$= 2G_v(0) \left[ t^2 - t^2/2 \right] \quad (5.2.8)$$
$$= G_v(0)t^2.$$

The mean-square displacement is proportional to the square of the time interval. We may write our result as

$$\langle x^2 \rangle = \langle v^2 \rangle t^2. \quad (5.2.9)$$

This indicates that the Brownian particle is moving essentially freely; at these short times there have not been sufficient atomic impacts to have any significant effect on the particle. This is referred to as the *ballistic* régime.

### 5.2.3 Long-time limit

The other statement about the correlation time is

$$G_v(\tau) \approx 0 \qquad \tau \gg \tau_v; \tag{5.2.10}$$

the autocorrelation function will have decayed to zero at long times. So, when we consider times much longer than $\tau_v$, $G_v(\tau)$ will be zero and in the expression for $\langle x^2 \rangle$ we will make no error by extending the upper limit of the integral to infinity.

$$\langle x^2 \rangle = 2 \int_0^\infty (t - \tau) G_v(\tau) d\tau. \tag{5.2.11}$$

The integral may be rearranged as

$$\langle x^2 \rangle = 2t \int_0^\infty G_v(\tau) d\tau - 2 \int_0^\infty \tau \, G_v(\tau) d\tau. \tag{5.2.12}$$

The second term, a constant, is negligible compared with the first at long times, so we conclude that in the long time limit

$$\langle x^2 \rangle = 2t \int_0^\infty G_v(\tau) d\tau. \tag{5.2.13}$$

Now, we see that the mean-square displacement of the Brownian particle is proportional to time (rather than the $t^2$ of the ballistic régime).

You should recall that a mean-square displacement proportional to time is characteristic of a *diffusive* process. And in fact in 1d the solution of the diffusion equation gives directly

$$\langle x^2 \rangle = 2Dt \tag{5.2.14}$$

where $D$ is the diffusion coefficient.

Thus, we conclude that in the long time limit the motion of the Brownian particle is diffusive, and its diffusion coefficient is given by

$$D = \int_0^\infty G_v(t) \, dt \tag{5.2.15}$$

or

$$D = \int_0^\infty \langle v(0)v(t) \rangle \, dt. \tag{5.2.16}$$

The diffusion coefficient is given by the area under the velocity autocorrelation function. The long-time limit, when $t \gg \tau_v$, is called the *diffusive* régime.

Recall that the definition of the correlation time was given in terms of the area under the autocorrelation function, Eq. (5.1.29):

$$\tau_v = \frac{1}{G_v(0)} \int_0^\infty G_v(t)\,dt. \tag{5.2.17}$$

From this it follows that we may write the diffusion coefficient as

$$D = G_v(0)\tau_v \tag{5.2.18}$$

or

$$D = \langle v^2 \rangle \tau_v. \tag{5.2.19}$$

Again we emphasise that the preceding discussion is purely kinematical. All the quantities we have considered are properties of the Brownian particle. The random atomic bombardment causes the velocity of the particle to vary randomly, but we have not, as yet, considered the dynamics of the collision processes.

### 5.2.4 *Equipartition*

Although we shall not consider the dynamics of the collision processes in this section, since the system of Brownian particle plus surrounding fluid is regarded as being in thermal equilibrium, we may apply the equipartition theorem to the Brownian particle. The objection might be raised that equipartition is a classical result which becomes invalid when issues of indistinguishability and multiple occupation of states become important. However, we will apply equipartition specifically to the Brownian particle, not to the surrounding medium. And since the Brownian particle is a macroscopic object, its behaviour may be understood purely in classical terms. As it is in thermal equilibrium with a bath at a temperature $T$, the equipartition theorem tells us that

$$\frac{1}{2}M \langle v^2 \rangle = \frac{1}{2}kT \tag{5.2.20}$$

in one dimension, where $M$ is the mass of the Brownian particle. The mean-square velocity is then

$$\langle v^2 \rangle = \frac{kT}{M}. \tag{5.2.21}$$

This is a *consequence* of the microscopic atomic bombardment from the surrounding fluid, but the expression is a purely thermodynamic result independent of the details of the interaction. It is sufficient that there is thermal equilibrium.

Equipartition allows us to write the diffusion coefficient of the Brownian particle as

$$D = \frac{kT}{M} \tau_v. \tag{5.2.22}$$

This does not mean that the diffusion coefficient is proportional to temperature since the velocity correlation time will, in general, depend on temperature. At a certain level, this is still a kinematical result about the Brownian particle since $\tau_v$ is also a property of the Brownian particle. What we really want to know is how the interactions with the atoms of the surrounding medium affect the particle's motion. For this, we need to consider the dynamics of the process.

## 5.3 Langevin's Equation

### 5.3.1 *Introduction*

The task of studying the *dynamics* of Brownian motion was initiated by Langevin in 1908. Langevin wrote down an equation of motion for the Brownian particle. Essentially this was an equation of the form $F = ma$, but Langevin's special insight was in the way he viewed the force acting on the Brownian particle. Our treatment is inspired by the papers of Uhlenbeck and Ornstein [5], Wang and Uhlenbeck [6] and both were reprinted in the collection [7] and MacDonald's book [3].

Langevin wrote the force acting on the particle as

$$F = f(t) - \frac{1}{\mu} v(t). \tag{5.3.1}$$

He regarded the force $F$ acting on the particle as being made up of two contributions: a random part $f(t)$ and a systematic or friction force proportional to and opposing the particle's velocity $v(t)$. The constant $\mu$ in the friction force is known as the *mobility*. This view of the forces acting is eminently sensible; we know that there will be random atomic bombardments and that a body moving in a fluid will experience friction.

Langevin's equation of motion is then

$$M\frac{dv(t)}{dt} = f(t) - \frac{1}{\mu}v(t),$$ (5.3.2)

solution of which will give the time variation of the velocity $v(t)$.

Considering the specific problem of Brownian motion as outlined in the previous sections, it is to be expected that from the solution of the Langevin equation, an expression for the velocity autocorrelation function of the Brownian particle may be found in terms of the random force $f(t)$ acting upon it. This will be done in what follows.

The Langevin equation is, however, capable of much more. In particular, it will show how the random force and the friction force are related. This is a result of considerable generality and importance since it relates in a fundamental way the random fluctuations in the system $f(t)$ and the dissipation characterised by the friction (or the mobility). This connection, in its general form, is known as the *fluctuation–dissipation theorem*.

### 5.3.2 *Separation of forces*

It is commonly stated that it is a *hypothesis* of Langevin's approach that the force on the Brownian particle may be decomposed as the sum of a random part and a systematic friction part proportional to velocity. We shall see that this decomposition may actually be justified and understood in terms of the different centre of mass frames of the fluid and the Brownian particle [8].

A Brownian particle at rest in the centre of mass frame of the fluid medium suffers bombardments from the atoms of the fluid. These bombardments will result in a random force. On average, there will be as many impacts in each direction so the average of the force will be zero.

Now, consider the particle moving with respect to the centre of mass frame of the fluid. Then the impacts from the front will be at a greater relative velocity and the impacts from the rear will be at a lesser relative velocity. This will result in a mean force acting on the particle in opposition to its motion. We can see this from a simple model calculation. Let us consider two impacts, one from the rear and one from the front, where the atoms are moving with velocities $+v$ and $-v$ with respect to the fluid centre of mass. This is shown in Fig. 5.9.

The impact from the atom to the left transfers momentum $m\Delta v$ to the Brownian particle. Assuming the atom mass $m$ is very much less than

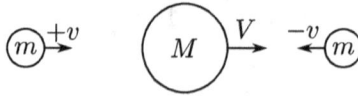

Fig. 5.9.   Bombardment of Brownian particle.

that of the Brownian particle $M$, its velocity will be reversed. Its change of velocity is then twice the relative velocity, $2(v - V)$, so the momentum transferred is then

$$\Delta p_{\text{left}} = 2m(v - V). \tag{5.3.3}$$

In the impact from the right, the change of velocity of the atom will similarly be twice the relative velocity, in this case $2(v + V)$. So, the momentum transferred from this impact is

$$\Delta p_{\text{right}} = 2m(v + V). \tag{5.3.4}$$

The net momentum transfer is the difference between these

$$\Delta p = -4mV, \tag{5.3.5}$$

independent of $v$. Then if there are $n$ impacts per unit time, the net force on the Brownian particle will be

$$-4nmV. \tag{5.3.6}$$

On average, in the centre of mass frame of the fluid, there will be equal impacts from the left and right, leading to an average force of the above form.

It is of interest to observe that we have *derived* a force of friction proportional to velocity. This is in contrast to elementary discussions which state that the friction force is *assumed* to be proportional to velocity. The hidden assumption in our treatment is that the motion of the Brownian particle does not disturb the atomic motions of the fluid. This should be true if the velocity is small or of the fluid is not too dense. This is, essentially, a "linear" approximation since if the motion does affect the fluid, then that will feed back as an additional force on the particle, changing its motion, which will feed back to the fluid and ... round and round.

We have seen that Langevin's decomposition of the forces acting on the Brownian particle may be understood in terms of the different centre of mass frames of the fluid and the particle. The random force $f(t)$, whose mean value is zero, is supplemented by a mean force proportional to and opposing the velocity of the particle. Thus, we have justified writing the force as

$$F = f(t) - \frac{1}{\mu}v(t) \tag{5.3.7}$$

and we then expect that the mobility $\mu$ should be related to $f(t)$ in a fundamental way. This relation will follow from our consideration of the solution of the Langevin equation.

### 5.3.3 The Langevin equation

We write the Langevin equation as

$$M\frac{dv(t)}{dt} + \frac{1}{\mu}v(t) = f(t), \tag{5.3.8}$$

which emphasises its structure as an inhomogeneous linear first-order ordinary differential equation with source $f(t)$. It is convenient to make a simplification by the substitutions

$$A(t) = \frac{f(t)}{M}$$

$$\gamma = \frac{1}{M\mu}. \tag{5.3.9}$$

Then the Langevin equation becomes

$$\frac{dv(t)}{dt} + \gamma v(t) = A(t); \tag{5.3.10}$$

and $\gamma$ is identified as the dissipation rate. This equation has the solution

$$v(t) = v(0)e^{-\gamma t} + \int_0^t e^{-\gamma(t-\tau)}A(\tau)\,d\tau. \tag{5.3.11}$$

The first term represents the transient part of the solution: that which depends on the initial conditions and which arises from the solution to the corresponding homogeneous equation. This is the *complementary function*. The second term represents the steady state response to the source "force" $A(t)$. This is the *particular integral* and this part persists when all memory of the initial condition has gone.

It is conventional to enunciate properties of the (scaled) random force $A(t)$ [5]. These are listed as follows:

1. $\langle A(t) \rangle = 0$. The mean of $A$ is zero. This follows, in our treatment, from the considerations of the centre of mass frame of the fluid.
2. $\langle A(t_1)A(t_2) \rangle = 0$ unless $t_1$ is close to $t_2$. In other words, the correlation time of the random force, $\tau_f$, is short.
3. $\langle A(t)^2 \rangle$ has some definite value (independent of $t$).

We may develop property 2 by approximating

$$\langle A(t_1)A(t_2) \rangle = \mathscr{A}^2 \delta(t_1 - t_2). \tag{5.3.12}$$

If we integrate this, we obtain

$$\mathscr{A}^2 = \int_{-\infty}^{\infty} \langle A(0)A(t) \rangle \, dt \tag{5.3.13}$$

so that $\mathscr{A}^2$ is the area under the (scaled) random force correlation function. Then

$$\mathscr{A} = \langle A(t)^2 \rangle \tau_f, \tag{5.3.14}$$

in terms of the force correlation time $\tau_f$.

As a simple application of the above results, we can examine the mean value of $v(t)$. We find for a given initial condition

$$\langle v \rangle = v(0)e^{-\gamma t} \tag{5.3.15}$$

since by property 1, $\langle A \rangle = 0$. This tends to zero as time proceeds and memory of the initial condition fades.

### 5.3.4 Velocity autocorrelation function

In our kinematical analysis of Brownian motion, we saw that the motion of the Brownian particle was conveniently expressed in terms of the velocity autocorrelation function. From Eq. (5.3.11), we have

$$\langle v(t)v(t + \tau) \rangle = \langle v^2(0) \rangle e^{-\gamma(2t+\tau)}$$

$$+ e^{-\gamma(2t+\tau)} \left\{ \int_0^t + \int_0^{t+\tau} \right\} du \, \langle v(0)A(u) \rangle \tag{5.3.16}$$

$$+ e^{-\gamma(2t+\tau)} \int_0^t du \int_0^{t+\tau} dw \, e^{\gamma(u+w)} \langle A(u)A(w) \rangle.$$

The first term is the transient response which dies away at long times; it is of no interest. The second term vanishes since there is no correlation between $v(t)$ and $A(t)$. The third term is of interest since it describes the equilibrium state of the particle, independent of the initial conditions.

In order to evaluate this term, we make use of property 2, that $\tau_f$ is small, and approximate the force autocorrelation function by the delta function expression, Eq. (5.3.12). This forces $w = u$ when the integral over $w$ is performed. Thus, we obtain

$$\langle v(t)v(t + \tau)\rangle = \frac{\mathscr{A}^2}{2\gamma}e^{-\gamma\tau} \tag{5.3.17}$$

or

$$G_v = \langle v^2(0)\rangle e^{-\gamma t}. \tag{5.3.18}$$

We conclude that the correlation time for the velocity autocorrelation function is simply the damping time associated with the friction force

$$\tau_v = \gamma^{-1}. \tag{5.3.19}$$

We should note that in reality the force correlation time will have a finite size and the approximation of the force autocorrelation function by a delta function, Eq. (5.3.12), must be questioned. The approximation is valid whenever the velocity correlation time $\tau_v$ is much greater than $\tau_f$. This condition:

$$\tau_v \gg \tau_f, \tag{5.3.20}$$

is known as the *separation of timescales*. It underlies many of the results connected with the Langevin method; it says that although the random force and the friction force are *related*, they are not *correlated*. Moreover, the emergence of the irreversible behaviour of macroscopic quantities from the reversible dynamics at the microscopic level is intimately connected with this separation off timescales.

### 5.3.5 *Mean-square velocity and equipartition*

The mean-square velocity is the zero time value of the velocity autocorrelation function. We set $\tau = 0$ in Eq. (5.3.17), giving

$$\langle v^2(t)\rangle = \mathscr{A}^2/2\gamma \tag{5.3.21}$$

independent of time.

The importance of this expression becomes apparent when we exploit the equipartition theorem. This tells us, as we have seen,

$$\langle v^2 \rangle = \frac{kT}{M} \qquad (5.3.22)$$

so that

$$\gamma = \frac{M}{2kT} \mathscr{A}^2 \qquad (5.3.23)$$

which provides us with a relation between the mobility (contained in $\gamma$) and the random force (contained in $A(t)$). From the definition of $\gamma$, that for $\mathscr{A}^2$ and that for $A(t)$, we can then express the mobility as

$$\frac{1}{\mu} = \frac{1}{2kT} \int_{-\infty}^{\infty} \langle f(0)f(t) \rangle \, dt. \qquad (5.3.24)$$

This expression achieves the objective of relating the two forces in the Langevin equation, the mobility or friction force and the random force of atomic bombardment. The structure of this expression is that the systematic/dissipative force is expressed in terms of the autocorrelation function of the random/fluctuation force. This is a very general result, called *the fluctuation–dissipation theorem*. The $kT$ factor that appears in the relation between the macroscopic and the microscopic force is, recall, a consequence of equipartition.

### 5.3.6 *Diffusion coefficient*

We saw that the diffusion coefficient of the Brownian particle was given in terms of $\tau_v$ by

$$D = \frac{kT}{M} \tau_v, \qquad (5.3.25)$$

which we can now re-express as

$$D = \frac{kT}{M\gamma} \qquad (5.3.26)$$

or

$$D = \mu kT. \qquad (5.3.27)$$

This connection between the diffusion coefficient and the mobility is known as the *Einstein relation*.

That is fine; it is purely kinematical and descriptive. But the real advance is that the fluctuation dissipation theorem of the previous section allows us to express this in terms of the fluctuating microscopic forces. That is the true content of the Einstein relation.

**A paradox:** Thermal equilibrium is intimately connected with time translation invariance. And on general grounds, one can argue that correlation functions such as that for the velocity must then decay exponentially (in any time interval of a given duration the function decreases by the same factor). Our calculation above for $G_v(t)$ conforms to this. This general result is known as Doob's theorem [9]. However, we have a problem. Such behaviour is incompatible with microscopic reversibility which implies that $G_v(t) = G_v(-t)$, so that assuming $G_v(t)$ is analytic at the origin, all its odd derivatives would have to vanish. The paradox is resolved, however, when we recall we used the delta function approximation for the force autocorrelation function. This causes the bad behaviour at $t = 0$. For a smoothed $G_f(t)$, we would find that $G_v(t)$ was exponential over *most* of its range, but for very short times, it would flatten off. Doob's theorem would not apply at very short times; it is a thermodynamic/hydrodynamic result so that is no problem.

### 5.3.7 *Harmonically bound particle*

To study the Brownian motion of a harmonically bound particle, we add a restoring force $Kx(t)$ to Eq. (5.3.8), giving

$$M\frac{d^2x(t)}{dt^2} + \frac{1}{\mu}\frac{dx(t)}{dt} + Kx(t) = f(t). \qquad (5.3.28)$$

Here, $K$ is the spring constant. As previously, $M$ is the mass of the Brownian particle, $\mu$ its mobility, and $f(t)$, the random force acting on the particle. And as previously we make the simplifications, through the substitutions

$$A(t) = \frac{f(t)}{M}$$

$$\gamma = \frac{1}{M\mu}. \qquad (5.3.29)$$

To these, we now add the substitution

$$\omega_z = \sqrt{\frac{K}{M}};\qquad(5.3.30)$$

here, $\omega_z$ is the oscillation (angular) frequency (that would occur for zero damping).

Now, the equation of motion is

$$\frac{d^2x(t)}{dt^2} + \gamma\frac{dx(t)}{dt} + \omega_z^2 x(t) = A(t).\qquad(5.3.31)$$

With an oscillator it is often convenient to express the damping in terms of the Q-factor

$$Q = \omega_z/\gamma.\qquad(5.3.32)$$

Equation (5.3.31) has the solution for the displacement $x(t)$

$$x(t) = \frac{\gamma x(0) + 2v(0)}{2\omega_f}e^{-\gamma t/2}\sin\omega_f t + x(0)e^{-\gamma t/2}\cos\omega_f t$$

$$+ \frac{1}{\omega_f}\int_0^t e^{-\gamma(t-u)/2}\sin\omega_f(t-u)A(u)du$$

$$(5.3.33)$$

and for the velocity $v(t) = dx(t)/dt$

$$v(t) = -\frac{2\omega_z^2 x(0) + \gamma v(0)}{2\omega_f}e^{-\gamma t/2}\sin\omega_f t + v(0)e^{-\gamma t/2}\cos\omega_f t$$

$$+ \frac{1}{\omega_f}\int_0^t e^{-\gamma(t-u)/2}\left\{-\frac{\gamma}{2}\sin\omega_f(t-u) + \omega_f\cos\omega_f(t-u)\right\}A(u)du$$

$$(5.3.34)$$

where $\omega_f$ is the free-ringing frequency, slightly lower than the zero-damping frequency $\omega_z$:

$$\omega_f^2 = \omega_z^2 - \frac{\gamma^2}{4} = \omega_z^2\left(1 - \frac{1}{4Q^2}\right).\qquad(5.3.35)$$

The autocorrelation functions for the displacement and the velocity are then found to be

$$G_x(\tau) = \langle x(t)x(t+\tau)\rangle = \frac{\mathscr{A}^2}{\pi}\int_0^\infty \frac{\cos\omega\tau}{(\omega_z^2 - \omega^2)^2 + \gamma^2\omega^2}d\omega$$

$$(5.3.36)$$

$$= \frac{\mathscr{A}^2}{2\gamma\omega_z^2}e^{-\gamma\tau/2}\left(\cos\omega_f\tau + \frac{\gamma}{2\omega_f}\sin\omega_f\tau\right)$$

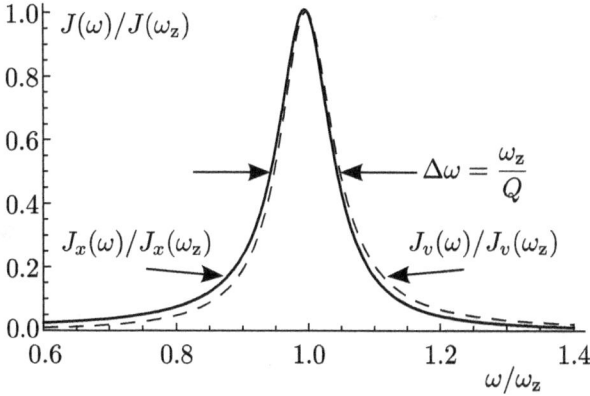

Fig. 5.10. Displacement and velocity power spectrum. Plot corresponds to an oscillator with a $Q$ factor of 10.

and

$$G_v(\tau) = \langle v(t)v(t + \tau)\rangle = \frac{\mathscr{A}^2}{\pi} \int_0^\infty \frac{\omega^2 \cos \omega \tau}{(\omega_z^2 - \omega^2)^2 + \gamma^2 \omega^2} d\omega$$

$$= \frac{\mathscr{A}^2}{2\gamma} e^{-\gamma \tau/2}\left(\cos \omega_f \tau - \frac{\gamma}{2\omega_f} \sin \omega_f \tau\right). \tag{5.3.37}$$

These have the form of a Fourier transform, so we can immediately write down the corresponding spectral density functions by inspection:

$$J_x(\omega) = \mathscr{A}^2 \frac{1}{(\omega_z^2 - \omega^2)^2 + \omega_z^2 \omega^2/Q^2}$$

$$J_v(\omega) = \mathscr{A}^2 \frac{\omega^2}{(\omega_z^2 - \omega^2)^2 + \omega_z^2 \omega^2/Q^2}. \tag{5.3.38}$$

The spectral densities have the conventional Lorentzian resonance profiles, centred at $\omega_z$ with quality factor $Q$. These are plotted in Fig. 5.10 for the case $Q = 10$. At this low $Q$ value, the small difference between the displacement and the velocity spectra can be discerned.

### 5.3.8 *Equipartition and mean-square values*

When the equipartition theorem is applied to the harmonic oscillator, we must remember that both the velocity and the displacement have their equipartition contributions since both the kinetic energy and the potential energy give quadratic terms to the hamiltonian. Thus, we have

$$\left\langle v^2 \right\rangle = \frac{kT}{M}$$

$$\left\langle x^2 \right\rangle = \frac{kT}{K} = \frac{kT}{M\omega_z^2}. \tag{5.3.39}$$

In the expression for $\left\langle x^2 \right\rangle$, we have also given the formula in terms of $M$, so that

$$\left\langle v^2 \right\rangle = \omega_z^2 \left\langle x^2 \right\rangle. \tag{5.3.40}$$

The expressions for the autocorrelation functions above are written in terms of $\mathscr{A}$, defined in Eq. (5.3.13). Now, using equipartition, and taking the zero-time values of $G_x(\tau)$ and $G_v(\tau)$, we can write the $\mathscr{A}^2$ prefactor as

$$\mathscr{A}^2 = 2\gamma \frac{kT}{M} = \frac{2\omega_z kT}{QM}. \tag{5.3.41}$$

The autocorrelation functions are then

$$G_x(\tau) = \frac{kT}{M\omega_z^2} e^{-\gamma\tau/2} \left( \cos\omega_f\tau + \frac{\gamma}{2\omega_f} \sin\omega_f\tau \right)$$

$$G_v(\tau) = \frac{kT}{M} e^{-\gamma\tau/2} \left( \cos\omega_f\tau - \frac{\gamma}{2\omega_f} \sin\omega_f\tau \right) \tag{5.3.42}$$

and the corresponding spectral density functions are

$$J_x(\omega) = \frac{2\omega_z kT}{QM} \frac{1}{(\omega_z^2 - \omega^2)^2 + \omega_z^2\omega^2/Q^2}$$

$$J_v(\omega) = \frac{2\omega_z kT}{QM} \frac{\omega^2}{(\omega_z^2 - \omega^2)^2 + \omega_z^2\omega^2/Q^2}. \tag{5.3.43}$$

From the zero-time value of Eq. (5.1.38), we have $G(0)$ in terms of the area under the spectral density function:

$$\int_0^\infty J(\omega)\, d\omega = \pi G(0), \tag{5.3.44}$$

so that the "total power" in the displacement and the velocity fluctuations are

$$\int_0^\infty J_x(\omega)\, d\omega = \pi \left\langle x^2 \right\rangle = \pi \frac{kT}{M\omega_z^2}$$

$$\int_0^\infty J_v(\omega)\, d\omega = \pi \left\langle v^2 \right\rangle = \pi \frac{kT}{M}. \tag{5.3.45}$$

The on-resonance values of $J_x(\omega)$ and $J_v(\omega)$ are

$$J_x(\omega_z) = \frac{2kT}{M} \frac{Q}{\omega_z^3}$$

$$J_v(\omega_z) = \frac{2kT}{M} \frac{Q}{\omega_z}$$

(5.3.46)

so for both

$$\int_0^\infty J_{x/v}(\omega) \, d\omega = \frac{\pi \omega_z}{2Q} J_{x/v}(\omega_z).$$

(5.3.47)

This is telling us that the power in the thermal fluctuations is spread over a bandwidth of order $\omega_z/Q$. This is indicated in Fig. 5.10.

### 5.3.9 *Electrical analogue of the Langevin equation*

The Langevin equation, Eq. (5.3.8):

$$M\frac{dv(t)}{dt} + \frac{1}{\mu}v(t) = f(t)$$

describes the velocity $v(t)$ of the Brownian particle of mass $M$ in terms of the mobility $\mu$ (inverse friction) and the random force $f(t)$. However, the real achievement of the approach was in relating the mobility to the random force, Eq. (5.3.24). In this section, we shall explore an electrical analogue of this.

Imagine an electrical circuit comprising an inductor of inductance $L$ connected in series with a resistor of resistance $R$. The current $I$ flowing in the circuit results from the motion of a very large number of electrons; in this respect, we regard $I$ as a *macroscopic* quantity. The voltage $V$ across the circuit is given by

$$L\frac{dI(t)}{dt} + RI(t) = V(t).$$

(5.3.48)

The random motion of the electrons will result in a randomly fluctuating voltage $V(t)$. Then the analogy between the two above equations is quite clear. We can follow the analogy through the arguments of the previous sections. The equipartition expression, Eq. (5.3.22), now becomes

$$\langle I^2 \rangle = \frac{kT}{L} \tag{5.3.49}$$

and the analogue of the fluctuation-dissipation result, Eq. (5.3.24), gives

$$R = \frac{1}{2kT} \int_{-\infty}^{\infty} \langle V(0)V(t) \rangle \, dt. \tag{5.3.50}$$

This shows how the resistance (dissipation) is related to the fluctuations of voltage. This will be explored further in Section 5.5.2 where we shall consider Nyquist's theorem.

In Problem 5.6, you will examine a different electrical analogue of the Langevin equation which results in relating resistance to *current* fluctuations:

$$\frac{1}{R} = \frac{1}{2kT} \int_{-\infty}^{\infty} \langle I(0)I(t) \rangle \, dt. \tag{5.3.51}$$

## 5.4 Linear Response I — Phenomenology

### 5.4.1 *Definitions and assumptions*

In this section, we turn to the question of the response of a system to an externally applied disturbance. This is fundamentally a non-equilibrium problem and so it is outside the area of applicability of *equilibrium* statistical mechanics. A special case of this question is the way a system relaxes to its equilibrium state from an initial non-equilibrium configuration. The general case involves the introduction of the concept of the dynamic susceptibility; for magnetic systems this is a generalisation of the magnetic susceptibility discussed in Chapter 2.

We consider an arbitrary system, to which we apply a generalised force $B$. And we observe the response $M$ to this force. This is shown schematically in Fig. 5.11. In the magnetic case, $B$ could be an applied magnetic field and $M$ the magnetisation response. But the discussion will remain more general than this.

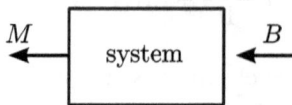

Fig. 5.11.   Excitation $B$ and its response $M$.

The response depends on the force: $M$ is a function of $B$. Thus, we write

$$M = f(B). \tag{5.4.1}$$

We shall now assume that our system obeys the following three rules:

**(a) Linearity:** If $M_1$ is the response to the force $B_1$ and $M_2$ is the response to the force $B_2$, then the response to the force $\alpha B_1 + \beta B_2$ is given by $\alpha M_1 + \beta M_2$. In other words, the response function $f(B)$ obeys the rule

$$f(\alpha B_1 + \beta B_2) = \alpha f(B_1) + \beta f(B_2). \tag{5.4.2}$$

Physically, we would expect the response to be linear for sufficiently small excitation $B$.

If we now consider time variation, then (very generally) we expect that $M$ at a given time will depend on $B$ at other times. The condition of linearity means that $M$ at a given time will depend *linearly* on $B$ at other times. Thus, $M(t)$ will be a linear *functional* of $B(t)$:

$$M(t) = \int X(t, \tau) B(\tau) \, d\tau. \tag{5.4.3}$$

The function $X()$ is known as the time-domain dynamical susceptibility or simply the response function.

**(b) Stationarity:** If the time variation of $B(t)$ is shifted by an amount $t'$, then the response $M(t)$ will be shifted by the same amount $t'$. This is the physical requirement of time translation invariance. If we apply this time shift to Eq. (5.4.3), we find that time translation invariance requires that

$$X(t + t', \tau + t') = X(t, \tau) \tag{5.4.4}$$

for any time $t'$. If, in particular, we set $t' = -\tau$, then we find

$$X(t - \tau, 0) = X(t, \tau). \tag{5.4.5}$$

This indicates that the response function $X$ depends only on the time difference $t - \tau$. And then Eq. (5.4.3) can be written as

$$M(t) = \int X(t - \tau) B(\tau) \, d\tau \tag{5.4.6}$$

or, by change of variable, as

$$M(t) = \int X(\tau) B(t - \tau) \, d\tau. \tag{5.4.7}$$

In many respects the response function $X(\tau)$ describes the *memory* of the system. $M(t)$ will depend strongly on excitations $B(t)$ when $t$ is close to $\tau$, and less so as $t$ becomes more distant from $\tau$, as indicated in Fig. 5.12.

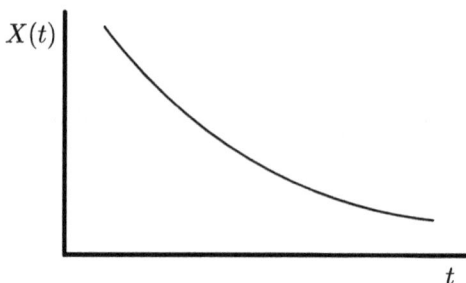

Fig. 5.12.   General form of the response function.

**(c) Causality:**   It is a fundamental principle of Physics that effect cannot precede cause. Thus, in Eqs. (5.4.6) and (5.4.7), we see that this requires that $X(t)$ be zero for negative $t$:

$$X(t) = 0 \quad t < 0. \tag{5.4.8}$$

This requirement also means that the domain of integration in Eq. (5.4.6) may be taken to be $-\infty < \tau < t$:

$$M(t) = \int_{-\infty}^{t} X(t - \tau)B(\tau)\,d\tau, \tag{5.4.9}$$

and in Eq. (5.4.7) to be $0 < \tau < \infty$:

$$M(t) = \int_{0}^{\infty} X(\tau)B(t - \tau)\,d\tau. \tag{5.4.10}$$

These equations give the general expression for the linear response of $M$ to the excitation $B$.

### 5.4.2   *Response to a harmonic excitation*

Many experimental techniques are based upon observing the response of a system to a harmonic excitation. This is the method of conventional *spectroscopy*.

We shall consider our system to be excited by a cosine of frequency $\omega$:

$$B(t) = b\cos(\omega t). \tag{5.4.11}$$

Then the response, from Eq. (5.4.9), will be

$$M(t) = b \int_{0}^{\infty} X(\tau)\cos\omega(t - \tau)\,d\tau. \tag{5.4.12}$$

Upon expanding the cosine, we have

$$M(t) = b\cos(\omega t) \int_0^\infty X(\tau)\cos\omega(\tau)\,d\tau + b\sin(\omega t)\int_0^\infty X(\tau)\sin\omega(\tau)\,d\tau.$$

$$(5.4.13)$$

This is telling us that the response to the monochromatic excitation $b\cos(\omega t)$ will be monochromatic at the same frequency $\omega$. However, in general the phase of the response will *lag behind* that of the excitation; particularly at higher frequencies the response will not be able to "keep up" with the excitation. We shall define:

$$\chi'(\omega) = \int_0^\infty X(t)\cos(\omega t)\,dt \qquad (5.4.14)$$

and

$$\chi''(\omega) = \int_0^\infty X(t)\sin(\omega t)\,dt. \qquad (5.4.15)$$

Then the response is expressed as a component $b\,\chi'(\omega)$ in phase with the excitation and a component $b\,\chi''(\omega)$ in quadrature (90° out of phase):

$$M(t) = b\chi'(\omega)\cos(\omega t) + b\chi''(\omega)\sin(\omega t). \qquad (5.4.16)$$

We see that $\chi'(\omega)$ and $\chi''(\omega)$ correspond to the Fourier cosine and sine transforms of $X(t)$.

## Complex representation

Since the system is linear, it will prove mathematically convenient to consider the response to the un-physical complex exponential excitation[1]

$$B_c(t) = be^{-i\omega t} = b\cos(\omega t) - ib\sin(\omega t). \qquad (5.4.17)$$

The *mathematical* response to this complex excitation will be a complex response $M_c(t)$. We identify the *physical* excitation as the real part of the complex $B_c(t)$. And then the linearity condition, Eq. (5.4.2), means that the *physical* response, Eq. (5.4.16), will be the real part of the complex $M_c(t)$.

---

[1]We write the elementary complex harmonic oscillation as $e^{-i\omega t}$ (with a $-$ sign), consistent with the usual expression for a propagating wave: $e^{i(kx-\omega t)}$. This conforms with the physicist's usual Fourier transform convention, Eq. (5.4.27), but it contrasts with engineering practice. Engineers prefer to express the elementary oscillation with a $+$ sign as $e^{j\omega t}$.

The complex representation proves particularly convenient since it obviates the need to keep track of the sines and cosines. Note, however, that this simplification cannot be used in the case of a *nonlinear* response.

The response to the complex sinusoidal excitation will be, from Eq. (5.4.7),

$$M_c(t) = b \int_{-\infty}^{\infty} X(\tau) e^{-i\omega(t-\tau)} d\tau$$

$$= b e^{-i\omega t} \int_{-\infty}^{\infty} X(\tau) e^{i\omega\tau} d\tau. \tag{5.4.18}$$

Here, we have written the integration range $-\infty < \tau < \infty$, so the integral is the complex Fourier transform; we must therefore account for causality in the response function $X(\tau)$, as in Eq. (5.4.8), separately. So, now, we see that the response to the (hypothetical) complex monochromatic excitation $be^{-i\omega t}$ is a complex response proportional to the same complex sinusoid $e^{-i\omega t}$. And the proportionality factor is the Fourier transform of $X(t)$, which we denote by $\chi(\omega)$:

$$\chi(\omega) = \int_{-\infty}^{\infty} X(t) e^{i\omega t} dt. \tag{5.4.19}$$

Thus,

$$M_c(t) = b e^{-i\omega t} \chi(\omega). \tag{5.4.20}$$

The real (physical) response is then found by taking the real part of this equation

$$M(t) = \Re M_c(t)$$

$$= b \cos(\omega t) \int_{-\infty}^{\infty} X(\tau) \cos(\omega\tau) d\tau + b \sin(\omega t) \int_{-\infty}^{\infty} X(\tau) \sin(\omega\tau) d\tau \tag{5.4.21}$$

since $X(t)$ is real. However, since causality requires the domain of integration to be restricted to $0 < \tau < \infty$, the integrals here correspond to the $\chi'(\omega)$ and $\chi''(\omega)$ of Eqs. (5.4.14) and (5.4.15). Thus, we have precisely the result obtained in Eq. (5.4.13), where now $\chi'(\omega)$ and $\chi''(\omega)$ are seen to be the real and imaginary parts of the complex susceptibility, the (complex) Fourier transform of $X(t)$:

$$\chi(\omega) = \chi'(\omega) + i\chi''(\omega).^2 \qquad (5.4.22)$$

### 5.4.3 Fourier representation

Any excitation $B(t)$ can be made up from a superposition of sinusoids:

$$B(t) = \frac{1}{2\pi} \int_{-\infty}^{\infty} b(\omega) e^{-i\omega t} d\omega. \qquad (5.4.23)$$

Now, the response to $b(\omega)e^{-i\omega t}$ is given by Eq. (5.4.20). Let us call this $m(\omega)e^{-i\omega t}$. Then clearly

$$m(\omega) = \chi(\omega)b(\omega) \qquad (5.4.24)$$

so that on multiplying by $e^{i\omega t}$ and integrating over $\omega$, we obtain

$$\int_{-\infty}^{\infty} m(\omega)e^{i\omega t} d\omega = \int_{-\infty}^{\infty} \chi(\omega)b(\omega)e^{i\omega t} d\omega \qquad (5.4.25)$$

where the left-hand side of this equation is the Fourier transform of $M(t)$, that is,

$$M(t) = \frac{1}{2\pi} \int_{-\infty}^{\infty} m(\omega)e^{-i\omega t} d\omega. \qquad (5.4.26)$$

Thus, Eq. (5.4.24) expresses the relation between the Fourier transform of the excitation and the response.

$$m(\omega) = \int_{-\infty}^{\infty} M(t)e^{i\omega t} dt$$

$$M(t) = \frac{1}{2\pi} \int_{-\infty}^{\infty} m(\omega)e^{-i\omega t} d\omega. \qquad (5.4.27)$$

The *same* Fourier transform relations hold also for $\{b(\omega), B(t)\}$ and $\{\chi(\omega), X(t)\}$.

---

[2]We note that the engineer's convention for the complex sinusoid: $e^{+j\omega t}$ (footnote 1 on p. 347), requires the complex conjugate form $\chi(\omega) = \chi'(\omega) - j\chi''(\omega)$.

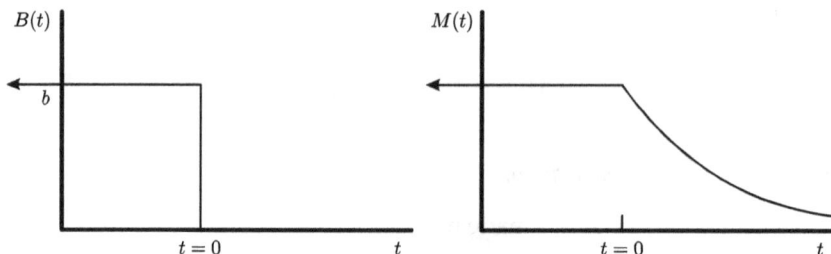

Fig. 5.13.   Step excitation and its response.

### 5.4.4  *Response to a step excitation*

Let us consider the response to a step excitation, Fig. 5.13. We imagine a constant excitation to have been applied back into the past and we remove this at time $t = 0$. Clearly, for negative times the response $M(t)$ will be a constant. And the interest then focuses on how $M(t)$ relaxes to its new equilibrium value when $B$ drops to zero.

Here, the excitation is specified to be

$$B(t) = b \quad t < 0$$
$$= 0 \quad t > 0 \tag{5.4.28}$$

or

$$B(t) = b\,\theta_-(t) \tag{5.4.29}$$

where $\theta_-(t)$ is the unit step (off) function. Then the response, according to Eq. (5.4.6), is given by

$$M_{\text{step}}(t) = b \int_{-\infty}^{0} X(t - \tau)\,d\tau \tag{5.4.30}$$

or, upon change of variables

$$M_{\text{step}}(t) = b \int_{t}^{\infty} X(\tau)\,d\tau. \tag{5.4.31}$$

Of course, this holds only for times $t > 0$.

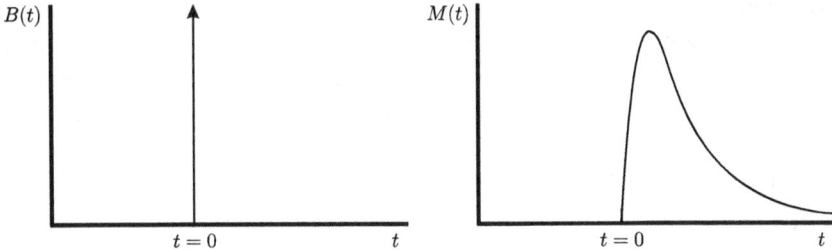

Fig. 5.14.  Delta function excitation and its response.

### 5.4.5  *Response to a "shock" or delta function excitation*

Let us now consider the response to a delta function, or spike, excitation, Fig. 5.14. Now, $B(t)$ is specified as

$$B(t) = b\,\delta(t) \tag{5.4.32}$$

where $\delta(t)$ is the Dirac delta function.

Then the response, according to Eq. (5.4.6), is given by

$$M_\delta(t) = b \int_{-\infty}^{\infty} X(t - \tau)\,\delta(\tau)\,d\tau \tag{5.4.33}$$

and, upon using the properties of the delta function, this becomes

$$M_\delta(t) = bX(t). \tag{5.4.34}$$

This is just the response function $X(t)$ introduced above. In other words, $X(t)$ may be regarded as the response to a delta function of unit strength.

The delta function is minus the derivative of the unit step function. The linearity condition relates the responses to these excitations. And indeed directly from Eqs. (5.4.31) and (5.4.34), we see that

$$M_\delta = -\frac{d}{dt} M_{\text{step}}. \tag{5.4.35}$$

It will prove useful to define the function $\Phi(t)$ as the response to a unit step excitation, just as $X(t)$ is the response to a delta function excitation. Then

$$X(t) = -\frac{d}{dt}\Phi(t). \tag{5.4.36}$$

Conventional spectroscopy, as discussed in Section 5.4.2, observes the response of a system to a succession of pure sinusoids. Making such measurements can be time-consuming. A delta-function excitation, however, contains spectral components at *all* frequencies. So, if we were to excite our system with such a shock, its response would contain the responses at all frequencies. As an example, one may measure the harmonic characteristics of a concert hall by carefully recording the response to sinusoidal excitations of different frequencies. Alternatively, one may simply fire off a gun: the Fourier transform of the response to this excitation gives the frequency response over the entire spectrum (almost instantaneously)!

### 5.4.6   *Response to a noise excitation**

There is an alternative to a delta-function excitation in order to excite at all frequencies simultaneously. White noise, also, contains spectral components at *all* frequencies. So, if we were to excite our system with such noise, its response would contain the responses at all frequencies. In this case, we need to know how to "disentangle" all these responses.

We shall express the system's response to an excitation as in Eq. (5.4.9), writing the excitation as $n$ to indicate that it is noise:

$$M_n(t) = \int_0^\infty X(\tau)n(t-\tau)\,d\tau. \qquad (5.4.37)$$

The details of the response $M_n(t)$ depend on the details of the noise excitation. In other words, we might say that the $M_n(t)$ response is *correlated* with the $n(t)$ excitation. We would study this correlation through the $M_n - n$ cross-correlation function $\langle M_n(t)n(T)\rangle$:

$$\langle M_n(t)n(T)\rangle = \int_0^\infty X(\tau)\,\langle n(T)n(t-\tau)\rangle\,d\tau. \qquad (5.4.38)$$

Inside the integral, we observe the noise *auto*correlation function. If the noise is regarded as "white", then its correlation is a delta function, as in Eq. (5.3.12),

$$\langle n(T)n(t-\tau)\rangle = a^2\delta(t-\tau-T). \qquad (5.4.39)$$

Here $a$, a measure of the strength of the noise, is to be determined. The effect of the delta function is to select out the $\tau = t - T$ value from the

---

*This section may be omitted on a first reading.

integral so that

$$\langle M_n(t)n(T)\rangle = a^2 X(t-T) \tag{5.4.40}$$

then invoking stationarity (time-translation invariance), we obtain

$$\langle M_n(t)n(0)\rangle = a^2 X(t). \tag{5.4.41}$$

Thus, the $M_n - n$ cross-correlation function gives the response function directly.

In order to determine $a$, we write the area under the noise autocorrelation function as

$$\int_{-\infty}^{\infty} \langle n(t)n(0)\rangle \, dt = \langle n^2 \rangle \tau_n, \tag{5.4.42}$$

where $\langle n^2 \rangle$ is the mean-square strength of the noise excitation and $\tau_n$ is its correlation time. Then by comparing Eq. (5.4.39) with Eq. (5.4.42), we identify

$$a^2 = \langle n^2 \rangle \tau_n \tag{5.4.43}$$

so that

$$\langle M_n(t)n(0)\rangle = \langle n^2 \rangle \tau_n X(t). \tag{5.4.44}$$

This is telling us that the dynamical response function $X(t)$ may be determined by exciting the system with noise and cross-correlating the resultant response with the driving noise

$$X(t) = \frac{\langle M_n(t)n(0)\rangle}{\langle n^2 \rangle \tau_n}. \tag{5.4.45}$$

There appears to be a contradiction. Although we regarded the noise to be *white*, we then referred to its correlation time $\tau_n$. But, by definition, the correlation time of white noise is zero. In reality, there is no such thing as truly white noise; its spectrum cannot extend to infinite frequencies! The resolution of the contradiction is in the *separation of timescales*. The timescale for the noise fluctuations, $\tau_n$, must be much less than the timescale for the response function, that is

$$\frac{X(t+\tau_n) - X(t)}{X(t)} \ll 1. \tag{5.4.46}$$

When this condition is satisfied, the noise is *effectively* white and the arguments above are valid.

The shock excitation and the white noise excitation, both applying all frequencies simultaneously, provide the same information. The relative advantages of these methods involve signal-to-noise considerations; unfortunately these are beyond the scope of this book.

### 5.4.7 Consequence of the reality of X(t)

In the following few sections, we shall revisit the use of complex quantities in linear response questions. The response to a real time-dependent excitation $B(t)$ is a real quantity $M(t)$. And these are related by the integral expression, Eq. (5.4.3). Thus, we conclude that the kernel $X(t)$ must be a real function. However, its Fourier transform, $\chi(\omega)$, the dynamical susceptibility, can be complex. We shall see that the reality of the time function has important consequences for the frequency function.

In Eq. (5.4.19):

$$\chi(\omega) = \int_{-\infty}^{\infty} X e^{i\omega t}\, dt \tag{5.4.47}$$

let us take the complex conjugate. This gives

$$\chi^*(\omega) = \int_{-\infty}^{\infty} X e^{-i\omega t}\, dt \tag{5.4.48}$$

since the real $X(t)$ remains unchanged. And if we now replace $\omega$ by $-\omega$, we have

$$\chi^*(-\omega) = \int_{-\infty}^{\infty} X e^{i\omega t}\, dt = \chi(\omega) \tag{5.4.49}$$

so we conclude that

$$\chi^*(-\omega) = \chi(\omega). \tag{5.4.50}$$

This is the consequence of the reality of $X(t)$.

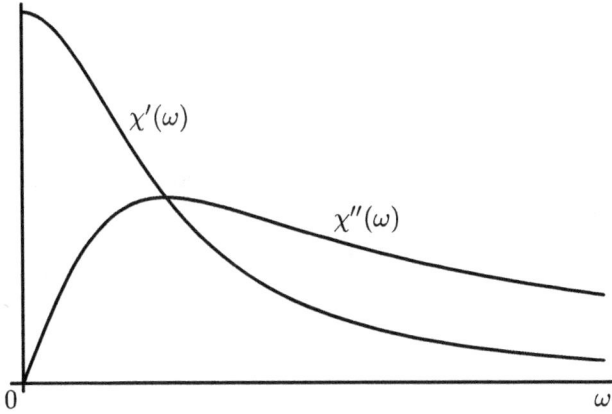

Fig. 5.15.   Real and imaginary part of the dynamical susceptibility.

In terms of the real and imaginary parts of the susceptibility, this gives the pair of relations

$$\chi'(\omega) = \chi'(-\omega)$$
$$\chi''(\omega) = -\chi''(-\omega).$$

(5.4.51)

Thus, $\chi'(\omega)$ is an even function and $\chi''(\omega)$ 0-is an odd function. This means that $\chi''(0)$ must be zero; obviously there cannot be a quadrature response at zero frequency, the static case. These are shown in Fig. 5.15.

### 5.4.8   Consequence of causality

The requirement of causality was expressed in Eq. (5.4.8), that $X(t)$ be zero for negative times. We can express this condition by

$$\theta_+(t)X(t) = X(t)$$

(5.4.52)

where $\theta_+(t)$ is the unit step (on) function. We also have the result

$$\theta_-(t)X(t) = 0.$$

(5.4.53)

These equations put an important restriction on the dynamic susceptibility $\chi(\omega)$. This is most easily seen from the convolution theorem for Fourier transforms:

$$\int_{-\infty}^{\infty} F(t)\, G(t) e^{i\omega t}\, dt = \frac{1}{2\pi} \int_{-\infty}^{\infty} f(\omega')\, g(\omega - \omega')\, d\omega' \qquad (5.4.54)$$

where $\{F(t),\, f(\omega)\}$ and $\{G(t),\, g(\omega)\}$ are Fourier transform pairs, as defined by Eq. (5.4.27).

Now, the Fourier transform of the two step functions is

$$\left.\begin{aligned} \mathscr{F}\{\theta_+\} &= \frac{i}{\omega + i\varepsilon} \\[2mm] \mathscr{F}\{\theta_-\} &= \frac{-i}{\omega - i\varepsilon} \end{aligned}\right\} \quad \varepsilon \to 0_+. \qquad (5.4.55)$$

So, the Fourier transforms of Eqs. (5.4.52) and (5.4.53), using the convolution expression Eq. (5.4.54), are

$$\frac{i}{2\pi} \int_{-\infty}^{\infty} \frac{\chi(\omega')\, d\omega'}{\omega - \omega' + i\varepsilon} = \chi(\omega) \qquad (5.4.56)$$

and

$$\frac{i}{2\pi} \int_{-\infty}^{\infty} \frac{\chi(\omega')\, d\omega'}{\omega - \omega' - i\varepsilon} = 0. \qquad (5.4.57)$$

We define the *principal part* of the integral along the real line by

$$\mathscr{P} \int_{-\infty}^{\infty} \frac{f(\omega')\, d\omega'}{\omega - \omega'} = \lim_{\delta \to 0} \left\{ \int_{-\infty}^{\omega-\delta} \frac{f(\omega')\, d\omega'}{\omega - \omega'} + \int_{\omega+\delta}^{\infty} \frac{f(\omega')\, d\omega'}{\omega - \omega'} \right\} \qquad (5.4.58)$$

and we shall accept this as equal to

$$\frac{1}{2} \left\{ \int_{-\infty}^{\infty} \frac{f(\omega')\, d\omega'}{\omega - \omega' + i\varepsilon} + \int_{-\infty}^{\infty} \frac{f(\omega')\, d\omega'}{\omega - \omega' - i\varepsilon} \right\}. \qquad (5.4.59)$$

Then from Eqs. (5.4.56) and (5.4.57), we obtain

$$\chi(\omega) = \frac{i}{\pi} \mathscr{P} \int_{-\infty}^{\infty} \frac{\chi(\omega')\, d\omega'}{\omega - \omega'}. \qquad (5.4.60)$$

The appearance of the $i$ in this equation means that if we separate the real and imaginary parts, $\chi'$ is given in terms of $\chi''$ and *vice versa*.

$$\chi'(\omega) = -\frac{1}{\pi}\mathscr{P}\int_{-\infty}^{\infty}\frac{\chi''(\omega')\,d\omega'}{\omega - \omega'}$$

$$\chi''(\omega) = \frac{1}{\pi}\mathscr{P}\int_{-\infty}^{\infty}\frac{\chi'(\omega')\,d\omega'}{\omega - \omega'}.$$

(5.4.61)

These are known as the Kramers–Kronig relations; they are purely a consequence of causality. They tell us that the real and imaginary parts of the susceptibility are not independent; the real part may be found from knowledge of the imaginary part *at all frequencies* and *vice versa*.

Problem 5.8 explores a demonstration of the Kramers–Kronig relations that does not require an excursion into the complex plane.

### 5.4.9 *Energy considerations*

If we regard the excitation $B(t)$ as a force and the response $M(t)$ as a displacement, then the element of work done *on* the system is given by

$$dW = -B\,dM.$$

(5.4.62)

And the power dissipated, the rate of doing work *by* the system is then

$$B\frac{dM}{dt}.$$

(5.4.63)

For a periodic excitation, we can find the mean power dissipated by averaging this expression over a cycle,

$$P = \left\langle B\frac{dM}{dt}\right\rangle.$$

(5.4.64)

From Eq. (5.4.16), we see that the response to an excitation $B = b\cos\omega t$ is

$$M = b(\chi'(\omega)\cos\omega t + \chi''(\omega)\sin\omega t).$$

(5.4.65)

The mean power dissipated is then

$$P = \omega b^2\left(-\chi'(\omega)\langle\cos\omega t\ \sin\omega t\rangle + \chi''(\omega)\langle\cos^2\omega t\rangle\right).$$

(5.4.66)

The $\cos\times\sin$ average is zero and the $\cos^2$ average is $1/2$. Thus, we find

$$P = \frac{1}{2}b^2\omega\chi''(\omega);$$

(5.4.67)

the power dissipated depends solely on the imaginary part of the dynamical susceptibility. In other words, dissipation occurs in the response that is 90° out of phase with the excitation. Since the energy dissipation must be positive, a requirement of the Second Law of thermodynamics, this puts a further general restriction on the dynamical response: $\omega \chi''(\omega)$ must be positive at all frequencies (in conformity with Eq. (5.4.51)).

There is, of course, a certain arbitrariness in what we choose to call a generalised force or excitation and in what we call a generalised displacement or response. However, we shall use Eq. (5.4.62) as a paradigmatic relation, in that, in general, the excitation $B$ and the response $M$ are related by

$$B = \frac{\partial \, \text{energy}}{\partial M}.$$  (5.4.68)

In other words, they comprise the regular extensive–intensive conjugate pairs of thermodynamics.

We can use this result to examine the dimensions of the generalised susceptibility. The defining equation for the dynamical response function gives us the result

$$[M] = [\chi][B]$$  (5.4.69)

and the energy expression above tells us that

$$[B] = [\text{energy}]/[M].$$  (5.4.70)

By eliminating $[B]$, we then find that

$$[\chi] = [M^2]/[\text{energy}];$$  (5.4.71)

the dynamical susceptibility thus has the dimensions of the square of the response divided by energy. In a similar way, we see that the dimensions of the dynamical response function are

$$[X] = [M^2]/[\text{energy}][\text{time}]$$  (5.4.72)

and the dimensions of the step response function are

$$[\Phi] = [M^2]/[\text{energy}].$$  (5.4.73)

### 5.4.10 *Static susceptibility*

The static susceptibility is the response to a constant excitation. This may be regarded as the zero-frequency limit of the frequency-dependent susceptibility $\chi(\omega)$. We denote the static susceptibility by $\chi_0$. Thus,

$$\chi_0 = \chi(\omega \to 0). \tag{5.4.74}$$

We can relate the static susceptibility to the time-domain response function through the Fourier transform relation, Eq. (5.4.19), taking the zero-frequency limit. Then we find

$$\chi_0 = \int_0^\infty X(t)\,dt; \tag{5.4.75}$$

the static susceptibility is the area under the time response function. We also note that the static susceptibility is equal to the zero-time value of the step response function, defined in Section 5.4.5.

$$\Phi(0) = \chi_0. \tag{5.4.76}$$

This may be seen from Eq. (5.4.36).

We know that the imaginary part of $\chi$ vanishes at zero frequency. So, the static susceptibility may be expressed as the zero-time value of the real part of the susceptibility:

$$\chi_0 = \chi'(\omega \to 0). \tag{5.4.77}$$

Now, the zero-frequency value of $\chi'$ may be found from the Kramers–Kronig relation:

$$\chi_0 = \frac{1}{\pi} \int_{-\infty}^\infty \frac{\chi''(\omega)}{\omega}\,d\omega. \tag{5.4.78}$$

This is the basis of resonance methods used to measure very small static susceptibilities in terms of power absorption. Observe that there is no need to take the principal part of the integral as $\chi''$ vanishes (sufficiently fast) at $\omega = 0$, so that there is no pole in the integral.

We conclude this section by making a connection between the generalised susceptibility discussed here and the magnetic susceptibility treated in Chapter 2. The magnetic susceptibility is defined, in the SI system of units, as the ratio of the magnetisation per unit volume to the applied magnetic $H$ field. It is not our place here to discuss the rights and wrongs of this definition. But we note two points that follow. Firstly, since it is magnetisation *per unit volume*, this is an intensive quantity. The magnetic field is an intensive quantity and this means that the magnetic susceptibility is *intensive*. Secondly, since the magnetisation per unit volume and the $H$ field have the same dimensions, this means that the magnetic susceptibility is dimensionless. These are the consequences of the SI definition of electromagnetic units. Now, let us look at the generalised susceptibility of this chapter, in the magnetic case. The response here is the total magnetisation — an extensive quantity. The magnetic field is intensive, so this means that the generalised susceptibility is *extensive*. And here the appropriate magnetic field is the $B$ field, to ensure the product of the excitation and the response has the dimensions of energy. We then conclude that the magnetic susceptibility and the generalised susceptibility are related by

$$\chi_{\text{gen}} = \frac{V}{\mu_0}\chi_{\text{m}}. \tag{5.4.79}$$

In Chapter 2, we found the Curie law static magnetic susceptibility of an assembly of $N$ magnetic moments $\mu$. We can now recast the results expressed in Eq. (2.10.12) as the generalised static susceptibility of this chapter, using Eq. (5.4.79). The result is

$$\chi_0 = \frac{N}{kT}\mu^2 = \frac{1}{kT}\left\langle M^2 \right\rangle. \tag{5.4.80}$$

### 5.4.11 *Relaxation time approximation*

The memory of a system to a disturbance is most directly specified in the step response function, $\Phi(t)$, schematically displayed in Fig. 5.13. This is a decaying function and the more rapidly it decays, the less memory the system has of distantly occurring excitations. Indeed, the characteristic time of this relaxation may be regarded as the "memory time" of the system. Subsequent discussion will relate the step response function to the autocorrelation function of the response variable. Thus, the memory time is also the correlation time, introduced in Section 5.1.4.

It is often observed that the relaxation of the step response function follows (at least approximately) a decaying exponential

$$\Phi(t) = \chi_0\,e^{-t/\tau}; \tag{5.4.81}$$

we write the $\chi_0$ prefactor since we know that the zero-time value of the step response function is the static susceptibility. In this section, we shall explore the consequences of this exponential form.

The dynamical response function is minus the derivative of $\Phi(t)$.

$$X = \chi_0 \frac{1}{\tau} e^{-t/\tau} \tag{5.4.82}$$

and the frequency domain susceptibility is given by the Fourier transform. Thus,

$$\chi(\omega) = \frac{\chi_0}{1 - i\omega\tau}, \tag{5.4.83}$$

its real and imaginary parts being given by

$$\chi'(\omega) = \chi_0 \frac{1}{1 + \omega^2\tau^2}$$

$$\chi''(\omega) = \chi_0 \frac{\omega\tau}{1 + \omega^2\tau^2}. \tag{5.4.84}$$

It is these functions that were plotted in Fig. 5.15.

This is often called the Debye form for the dynamical susceptibility. The electric susceptibility of many dielectrics follows this form to a *reasonable* extent.

## 5.5 Linear Response II — Microscopics

### 5.5.1 *Onsager's hypothesis*

We have discussed many formal aspects of the dynamical response function $X(t)$ and its Fourier transform, the dynamical susceptibility $\chi(\omega)$. These provide a description for the evolution of a system from a non-equilibrium state, following a disturbance and they also give the energy dissipation under the application of such a disturbance. The formalism is quite elegant, but it does not tell us how to calculate any of these things from microscopic first principles. In this respect, it is more properly of the nature of thermodynamics rather than of statistical mechanics.

It is possible to calculate such response functions from microscopic first principles, certainly in the linear régime, using statistical-mechanical perturbation theory. However, such techniques are fairly complicated; they require methods beyond the scope of this book. The interested reader is referred particularly to the works of Ryogo Kubo [10, 11]. We shall give a flavour of these methods in Sections 5.5.3 and 5.5.4.

Fortunately, there is another way, perhaps a more intuitive way, of tackling the question; this was pioneered by Lars Onsager. Imagine looking at the equilibrium fluctuations in a system, perhaps observing through a microscope the fluctuations in a fluid. The observed quantity will vary in some random way about its mean value. Sometimes there will be small excursions and sometimes, large excursions from the mean. These excursions return, on the average, to the mean value. Now, consider that someone has applied a disturbance to this system, driving it from equilibrium, and then the disturbance is removed. The system will then return to its equilibrium state. Through our microscope we would see the observed quantity fluctuating while it returns, on average, to its equilibrium value. Onsager made the remarkable hypothesis [12, 13] that one would not be able to distinguish between the two situations. In other words, he assumed that the relaxation of a system following a disturbance is the same as the average regression of a fluctuation in an equilibrium system.

This is indeed remarkable. It says that the behaviour of a *non-equilibrium* system may be understood by studying the properties of the corresponding *equilibrium* system. Alternatively, one may say that at the *microscopic* level there is no distinction between equilibrium and non-equilibrium. But then, of course, equilibrium is a thoroughly *macroscopic* concept.

The point, for us, is that we already know the average way fluctuations in an equilibrium system regress towards the mean. This is given by the autocorrelation function of the fluctuating quantity, $\langle M(0) M(t) \rangle$, discussed and explained in Sections 5.1.2 and 5.1.3. The relaxation following the removal of a disturbance is described by the step response function $\Phi(t)$. Thus, Onsager's hypothesis may be expressed as

$$\Phi(t) = \beta \langle M(0)M(t) \rangle \tag{5.5.1}$$

where $\beta$ is a constant to be determined. From the considerations of Section 5.4.9, we know that $\beta$ has the dimensions of inverse energy.

A full microscopic calculation is needed to determine the value of $\beta$. For inspiration, we shall look at the zero-time value of Eq. (5.5.1). Since $\Phi(0)$ is equal to the static susceptibility $\chi_0$, we have

$$\chi_0 = \beta \langle M^2 \rangle. \tag{5.5.2}$$

Compare this with the magnetic case, expressed in Eq. (5.4.80). We are thus led to identify $\beta$ with $1/kT$. It is *plausible* that this result is more generally applicable; we assert that it is.

The equation for the step response function

$$\Phi(t) = \frac{1}{kT} \langle M(0)M(t) \rangle, \tag{5.5.3}$$

or, equivalently for the dynamical response function

$$X(t) = -\frac{1}{kT} \frac{d}{dt} \langle M(0)M(t) \rangle \tag{5.5.4}$$

is indeed the correct *classical* result, as could be derived using statistical-mechanical perturbation theory. A quantum-mechanical calculation gives a more complicated temperature dependence, which reduces to the above expression in the high temperature limit. The $kT$ factor is essentially a reflection of equipartition.

The dynamical susceptibility is then found from the Fourier transform of $X(t)$:

$$\chi(\omega) = -\frac{1}{kT} \int_0^\infty \langle M(0)\dot{M}(t) \rangle e^{i\omega t} dt \tag{5.5.5}$$

or, upon integration by parts

$$\chi(\omega) = \chi_0 + \frac{i\omega}{kT} \int_0^\infty \langle M(0)M(t) \rangle e^{i\omega t} dt. \tag{5.5.6}$$

This shows how the dynamical susceptibility depends on the autocorrelation function of the equilibrium fluctuations in the response variable.

### 5.5.2 Nyquist's theorem

As early as 1906, Einstein [14, 15] predicted that the random motion of charge carriers in a conductor would lead to fluctuations of voltage or current in any system in thermal equilibrium. This was first observed by Johnson [16–18] and subsequently explained theoretically by Nyquist [19], who calculated the power spectrum of the fluctuations.

In order to frame this treatment within the discussion of the previous sections, the generalised force will now be taken as the applied voltage $V$ and the generalised response will be the electric charge $Q$. Then the linear response relation between these two will be

$$Q(t) = \int X(t-\tau)V(\tau)d\tau \tag{5.5.7}$$

or, in the frequency domain

$$q(\omega) = \chi(\omega)v(\omega). \tag{5.5.8}$$

Thus in electrical terms we identify the generalised susceptibility in this case as a frequency-dependent capacitance. Then the imaginary part of this will be related to the conductance and thus to the dissipation. Recall we saw in Section 5.4.9 that the imaginary part of the generalised susceptibility was connected with dissipation.

The admittance $Y(\omega)$ is given by $i\omega$ multiplied by the capacitance, so that from Eq. (5.5.5), this is

$$Y(\omega) = -\frac{i\omega}{kT} \int_0^\infty \langle Q(0)\dot{Q}(t)\rangle e^{i\omega t} dt. \tag{5.5.9}$$

Note, here, that $\dot{Q}(t)$ is the electric current $I(t)$. Let us integrate the above equation by parts in such a way as to differentiate the other $Q$. This will give

$$Y(\omega) = \frac{1}{kT} \int_0^\infty \langle I(0)I(t)\rangle e^{i\omega t} dt. \tag{5.5.10}$$

This result tells us that the electrical admittance of an object is related to the equilibrium current fluctuations in it. Indeed we can decompose the admittance into its real and imaginary parts: the conductance $G$ and the susceptance $S$

$$Y(\omega) = G(\omega) + iS(\omega) \tag{5.5.11}$$

so that

$$G(\omega) = \frac{1}{kT} \int_0^\infty \langle I(0)I(t)\rangle \cos\omega t \, dt$$
$$S(\omega) = \frac{1}{kT} \int_0^\infty \langle I(0)I(t)\rangle \sin\omega t \, dt. \tag{5.5.12}$$

The expression for $G(\omega)$ is a generalisation of Eq. (5.3.51) derived through the Langevin equation discussion.

Nyquist's result follows from inverting the expression for the conductance $G(\omega)$.

$$\langle I(0)I(t)\rangle = kT\frac{2}{\pi}\int_0^\infty G(\omega)\cos\omega t \; d\omega. \qquad (5.5.13)$$

If we now let $t$ tend to zero, then we obtain the mean-square value for the current fluctuations

$$\langle I^2\rangle = \frac{2kT}{\pi}\int_{-\infty}^\infty G(\omega)d\omega. \qquad (5.5.14)$$

Thus in a frequency band $\Delta f = \Delta\omega/2\pi$, there will be a contribution to the mean-square current of

$$\langle I^2\rangle_{\Delta f} = 4kTG\Delta f. \qquad (5.5.15)$$

Alternatively, in a resistance of $R = 1/G$, there will be mean-square voltage fluctuations of

$$\langle V^2\rangle_{\Delta f} = 4kTR\Delta f. \qquad (5.5.16)$$

This so-called "Johnson noise" is commonly heard as the hiss from the speakers of a hi-fi amplifier when there is no signal output.

We first encountered Johnson noise as the thermal equilibrium black body radiation in the (one-dimensional) terminated transmission line, in Section 2.9.5. That treatment obtained the full quantum-mechanical result, of which Eqs. (5.5.15) and (5.5.16) are the high temperature/equipartition limit. We then saw it from the perspective of the electrical analogue of the Langevin equation, in Section 5.3.9.

### 5.5.3 *Calculation of the step response function*

We introduced the step response function in Section 5.4.4. We imagined a constant excitation $B = b$ to have been applied back into the past, and we remove this at time $t = 0$. For negative times the response $M(t)$ will be a constant. And interest focuses on how $M(t)$ relaxes to its new equilibrium value when $B$ drops to zero, Fig. 5.16.

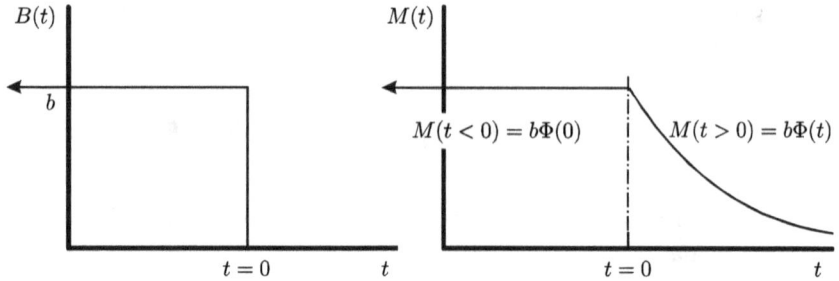

Fig. 5.16.   Step response.

The step response function $\Phi(t)$ is then defined, for $t > 0$, as

$$\Phi(t) = \frac{1}{b}M(t). \tag{5.5.17}$$

And the question is then to (try to) calculate this from first principles. In this section, we shall show how $\Phi(t)$ is related to the autocorrelation function of $M(t)$, obtaining the Onsager expression. By no means can the derivation be regarded as rigorous, but it is hoped that the flavour of the procedure will be appreciated. The question of how the autocorrelation function may be calculated will be considered in the following section.

The mean value of the response $M(t)$ is found by evaluating the average, weighted by the appropriate Boltzmann factor:

$$\langle M \rangle = \frac{1}{Z} \sum_i M_i e^{-E_i/kT}. \tag{5.5.18}$$

Now, the system has been *prepared* by applying the "field" $b$ for negative times. This has the effect of altering the energy levels by adding the term $bM_i$. Then in calculating the mean values we must use the appropriately-modified Boltzmann factor

$$\langle M \rangle = \frac{1}{Z} \sum_i M_i e^{-(E_i+bM_i)/kT}. \tag{5.5.19}$$

The point about *linear* response is that we take the excitation $b$ to be very small. In this case, we can make the approximation

$$e^{-(E_i+bM_i)/kT} \sim e^{-E_i/kT}(1 + bM_i/kT) \tag{5.5.20}$$

and then

$$\langle M \rangle = \frac{1}{Z} \sum_i M_i e^{-E_i/kT} + \frac{b}{kT}\frac{1}{Z} \sum_i M_i(0)M_i(t)e^{-E_i/kT}. \tag{5.5.21}$$

The first term is the equilibrium value of $M$ in the *absence* of the applied field; this will be zero. The second term may be written as an average over the equilibrium ensemble, and then

$$\langle M(t) \rangle = \frac{b}{kT} \langle M(0)M(t) \rangle. \tag{5.5.22}$$

So, the step response function is

$$\Phi = \frac{1}{kT} \langle M(0)M(t) \rangle, \tag{5.5.23}$$

as we proposed in Section 5.5.1.

The important point about this result — we have mentioned it before, but it is worth repeating it — is that the relaxation from a non-equilibrium state is expressed in terms of the fluctuations in the equilibrium state.

### 5.5.4 *Calculation of the autocorrelation function*

The calculation of the dynamical susceptibility from microscopic first principles is clearly a very difficult problem. As with our introductory discussion of the foundations of statistical mechanics, some sort of statistical treatment will be needed to solve any real problem. Moreover, almost all calculational procedures will involve the use of approximations.

There are some general methods of approach, often using a Boltzmann equation, to calculate the evolution of a system in phase space from specified initial conditions. Unfortunately this is beyond the scope of this book. However, we note that the discussions of Section 5.4. impose various constraints on any autocorrelation function so-calculated. In terms of the dynamical susceptibility $X(t)$ and its Fourier transform, the frequency-dependent susceptibility $\chi(\omega)$, we know that:

1. Causality places restrictions on the relationship between the real part $\chi'(\omega)$ and the imaginary part $\chi''(\omega)$.
2. Energy considerations restrict $\chi''(\omega)$ to be positive for all (positive) frequencies.
3. The reality of $X(t)$ requires that $\chi''(0)$ be zero.
4. Causality and microscopic reversibility require that the first derivative $X'(0)$ be zero; the reason for this is explored in Problem 5.7.

So, any approximately calculated susceptibility must satisfy these requirements.

Hydrodynamic considerations will often give further information; the long-time behaviour of $X(t)$ and the low-frequency behaviour of $\chi(\omega)$ will thereby have further constraints imposed upon them. The way in which such hydrodynamic constraints may be incorporated into the description of correlation functions has been extensively explored by Kadanoff and Martin [20].

In a very few cases a full quantum-mechanical description of the dynamics might be appropriate. Spin diffusion in a paramagnet is a classical example of this [21]. It is then possible to calculate the short-time behaviour of $X(t)$ as a power series in time. Higher-order derivatives become increasingly difficult to evaluate, but knowledge of the first few derivatives may be combined with long-time hydrodynamic information through the use of an appropriate interpolation procedure. The reader is referred to the author's book on Nuclear Magnetic Resonance [22], particularly the section dealing with solid $^3$He, for further details. Section 6.3.7 of that book shows also how short-time expansions actually place constraints on the behaviour of the correlation functions at long times.

## Problems

5.1 A random quantity has an exponential autocorrelation function $G(t) = G(0)e^{-\gamma t}$. What are the dimensions of $\gamma$? Calculate the correlation time of $G(t)$ using the usual definition.

Calculate the correlation time for a gaussian autocorrelation function $G(t) = G(0)e^{-\gamma^2 t^2/2}$.

5.2 Show that the autocorrelation function of a periodically varying quantity $m = m\cos\omega t$ is given by

$$G = \frac{m^2}{2}\cos\omega t.$$

Show that the autocorrelation function is independent of the *phase* of $m(t)$. In other words, show that if $m = m\cos(\omega t + \varphi)$, then $G(t)$ is independent of $\varphi$.

5.3 The mean-square displacement of a Brownian particle at long times was given by Eq. (5.2.12)

$$\langle x^2 \rangle = 2t \int_0^\infty G_v(\tau)d\tau - 2 \int_0^\infty \tau G_v(\tau)d\tau.$$

It was stated in the text that the second term was negligible, compared with the first, and so could be ignored.

(a) In that case, show that the mean-square displacement may be expressed

$$\langle x^2 \rangle = 2G_v(0)\tau_v\, t,$$

where $\tau_v$ is the correlation time associated with $G_v(\tau)$.

(b) Show that the second integral above may be expressed, approximately, as

$$\int_0^\infty \tau G_v(\tau)d\tau \approx G_v(0)\tau_v^2.$$

There is a choice of ways for demonstrating this. You might approximate $G_v(\tau)$ by a decaying exponential $G_v(0)e^{-\tau/\tau_v}$ before doing the integral.

(c) Using this approximate expression, show that without neglecting the second term, the expression for $\langle x^2 \rangle$ becomes

$$\langle x^2 \rangle = 2G_v(0)\tau_v\,(t - \tau_v).$$

Hence, justify the neglect of the second term.

5.4 A small mirror is suspended from a quartz fibre whose torsion constant is $\kappa$. When the mirror is rotated an angle $\theta$ the torque exerted by the fibre is $\Gamma = -\kappa\theta$. The moment of inertia of the mirror about the suspension axis is $I$. The mirror reflects a beam of light so that the angular fluctuations caused by the impact of surrounding molecules can be read on a suitable scale. The position of the equilibrium is $\langle \theta \rangle = 0$. The average value $\langle \theta^2 \rangle$ is observed. From this the goal is to find the Avogadro constant (or, what is the same thing since the gas constant $R$ is known, to determine the Boltzmann constant).

The following are the data: At $T = 287\,$K, for a fibre with $\kappa = 9.43 \times 10^{-16}\,$N m it was found that $\langle \theta^2 \rangle = 4.20 \times 10^{-6}$.

Calculate the Avogadro constant.

5.5 The assembly of the previous question is placed in a chamber from which the air may be evacuated. Can the amplitude of these fluctuations be reduced by reducing air density? Describe the change in the behaviour of the fluctuations as the air pressure is lowered. In particular, discuss the behaviour as the pressure goes to zero.

5.6 In Section 5.3.9, we considered an electrical analogue of the Langevin Equation based on a circuit comprising an inductor and a resistor in series. In this problem, we shall examine a different analogue: a circuit of a capacitor and a resistor in parallel. Show that the equation analogous to the Langevin equation, in this case, is

$$C\frac{dV}{dt} + \frac{1}{R}V = I.$$

Hence, show that the fluctuation-dissipation result relates the resistance to the current fluctuations through

$$\frac{1}{R} = \frac{1}{2kT} \int_{-\infty}^{\infty} \langle I(0)I(t) \rangle \, dt.$$

5.7 The dynamical response function $X(t)$ must vanish at zero times, as shown in Fig. 5.13. What is the physical explanation of this? What is the consequence for the step response function $\Phi(t)$? Is this compatible with an exponentially decaying $\Phi(t)$?

5.8 We saw in Section 5.4.2 that $\chi'(\omega)$ and $\chi''(\omega)$ may be regarded as the cosine and sine transforms of the dynamical susceptibility $X(t)$. Now, these real Fourier transforms may be inverted; $X(t)$ may be found equivalently from either $\chi'(\omega)$ or $\chi''(\omega)$. The point about this is that $\chi'(\omega)$ and $\chi''(\omega)$ are not independent; they both come (invertibly) from $X(t)$. So, in particular, $\chi'(\omega)$ may be found from $\chi''(\omega)$ or *vice versa* (the Kramers–Kronig relations).

(a) In this way, derive the following expressions:

$$\chi'(\omega) = -\frac{2}{\pi} \int_0^{\infty} \frac{\omega' \chi''(\omega')}{\omega'^2 - \omega^2} \, d\omega'$$

$$\chi''(\omega) = \frac{2}{\pi} \int_0^{\infty} \frac{\omega \chi'(\omega')}{\omega'^2 - \omega^2} \, d\omega'.$$

(b) These are not *quite* the same as those in Eq. (5.4.61). Why is this?

(c) Where, exactly, does the requirement of causality enter into this derivation?

5.9 Show that for the Debye susceptibility, the relation

$$\chi_0 = \frac{1}{\pi} \int_{-\infty}^{\infty} \frac{\chi''(\omega)}{\omega} d\omega$$

holds. Demonstrate that $\chi''$ vanishes sufficiently fast at $\omega = 0$, so there is no pole in the integral and there is thus no need to take the principal part of the integral in the Kramers–Kronig relations.

5.10 In Section 5.4.11, we examined the form of the dynamical suscepti- bility $\chi(\omega)$ that followed from the assumption that the step response function $\Phi(t)$ decayed exponentially. In this question, consider a step response function that decays with a gaussian profile: $\Phi = \chi_0 e^{-t^2/2\tau^2}$. Evaluate the real and imaginary parts of the dynamical susceptibil- ity and plot them as a function of frequency. The real part of the susceptibility is difficult to evaluate without a symbolic mathemat- ics system such as *Mathematica*. Compare and discuss the differences and similarities between this susceptibility and that deduced from the exponential step response function (Debye susceptibility).

5.11 The Debye form for the dynamical susceptibility, Eq. (5.4.84), is

$$\chi'(\omega) = \chi_0 \frac{1}{1 + \omega^2 \tau^2}$$

$$\chi''(\omega) = \chi_0 \frac{\omega\tau}{1 + \omega^2 \tau^2}.$$

Plot the imaginary part against the real part and show that the figure corresponds to a semicircle. This pictorial representation is known as a Cole-Cole plot.

5.12 Plot the Cole-Cole plot (Problem 5.11) for the dynamical susceptibil- ity considered in Problem 5.10. How does it differ from that of the Debye susceptibility?

5.13    The full quantum-mechanical calculation of the Johnson noise of a
resistor gives

$$\langle v^2 \rangle_{\Delta f} = 4R \frac{hf}{e^{hf/kT} - 1} \Delta f.$$

Show that this reduces to the classical Nyquist expression at low frequencies. At what frequency will there start to be serious deviations
from the Nyquist value? Estimate the value of this frequency.

# References

[1]    C. Adkins, *Equilibrium Thermodynamics* (Cambridge University
Press, 1983).

[2]    D. Kondepundi and I. Prigogine, *Modern Thermodynamics* (John
Wiley, 1998).

[3]    D. K. C. MacDonald, *Noise and Fluctuations*. John Wiley, 1962. (Dover
reprint, 2009).

[4]    A. Einstein, *Investigations on the Theory of the Brownian Movement*.
Dover, 1956. (English translations of five papers written by Einstein
between 1905 and 1908).

[5]    G. Uhlenbeck and L. Ornstein, On the theory of the Brownian
motion, *Phys. Rev.*, **36** (1930) 823–841. (Reprinted in Ref. [7]).

[6]    M. Wang and G. Uhlenbeck, On the theory of the Brownian motion
II, *Rev. Mod. Phys.*, **17** (1945) 323–342. (Reprinted in Ref. [7]).

[7]    N. Wax, *Selected Papers on Noise and Stochastic Processes* (Dover, New
York, 1954).

[8]    B. Cowan, *Classical Mechanics* (Springer, 1984).

[9]    J. L. Doob, The Brownian movement and stochastic equations, *Ann.
Math.*, **43** (1942) 351–369. (Reprinted in Ref. [7]).

[10]   R. Kubo, Some aspectsof the statistical-mechanical theory of irreversible processes, *University of Colorado, Summer School*, **1** (1959)
120–203.

[11]   R. Kubo, The fluctuation-dissipation theorem, *Rep. Prog. Phys.*, **29**
(1966) 255–284.

[12]   L. Onsager, Reciprocal relations in irreversible processes I, *Phys.
Rev.*, **38** (1931) 2265–2279.

[13]   L. Onsager, Reciprocal relations in irreversible processes. II, *Phys.
Rev.*, **38** (1931) 2265.

[14]   A. Einstein, A new determination of molecular dimensions, *Ann.
Phys.*, **19** (1906) 289–306. Paper III of Ref. [4].

[15]   A. Einstein, On the theory of the Brownian movement, *Ann. Phys.*, **19** (1906) 371–381. Paper II of Ref. [4].

[16]   J. B. Johnson, Thermal agitation of electricity in conductors, *Phys. Rev.*, **29** (1927) 367–368.

[17]   J. B. Johnson, Thermal agitation of electricity in conductors, *Nature*, **119**, 2984 (1927) 50–51.

[18]   J. B. Johnson, Thermal agitation of electricity in conductors, *Phys. Rev.*, **32**, 1 (1928) 97–109.

[19]   H. Nyquist, Thermal agitation of electric charge in conductors, *Phys. Rev.*, **32** (1928) 110–113.

[20]   L. P. Kadanoff and P. C. Martin, Hydrodynamic equations and correlation functions, *Ann. Phys.*, **24** (1963) 419–469.

[21]   B. Cowan and M. Fardis, Direct measurement of spin diffusion from spin relaxation times in solid $^3$He, *Phys. Rev. B*, **44** (1991) 4304–4313.

[22]   B. Cowan, Cumulant approach to spin echoes, *J. Phys. C: Solid State Phys.*, **10** (1977) 3383–3389.

# THE GIBBS–DUHEM RELATION

## A.1 Homogeneity of the Fundamental Relation

The Gibbs–Duhem relation follows from the fact that entropy is an extensive quantity and that it is a function of the other extensive variables of the system. An important mathematical result can be derived from *just* this extensivity. We saw that (for a $pV$ system) entropy is a function of the energy, the volume, and the number of particles in the system: $S = S(E, V, N)$ and this can, formally, be rearranged to write the energy as a function of $S$, $V$, and $N$

$$E = E(S, V, N). \tag{A.1.1}$$

This is referred to as the "fundamental relation" for the system.

Now, since $E$, $S$, $V$, and $N$ are all extensive variables, the fundamental relation function $E = E(S, V, N)$ is *homogeneous*. In other words, if the size of the system is multiplied by a constant $\lambda$, then $E$, $S$, $V$, and $N$ are all multiplied by the same $\lambda$. Thus,

$$E(\lambda S, \lambda V, \lambda N) = \lambda E(S, V, N). \tag{A.1.2}$$

## A.2 The Euler Relation

In order to investigate the consequences of this homogeneity, let us differentiate this expression with respect to $\lambda$; we find

$$\frac{\partial E(\lambda S, \lambda V, \lambda N)}{\partial(\lambda S)} \frac{\partial(\lambda S)}{\partial \lambda} + \frac{\partial E(\lambda S, \lambda V, \lambda N)}{\partial(\lambda V)} \frac{\partial(\lambda V)}{\partial \lambda}$$
$$+ \frac{\partial E(\lambda S, \lambda V, \lambda N)}{\partial(\lambda N)} \frac{\partial(\lambda N)}{\partial \lambda} = E(S, V, N) \tag{A.2.1}$$

or

$$\frac{\partial E(\lambda S, \lambda V, \lambda N)}{\partial(\lambda S)}S + \frac{\partial E(\lambda S, \lambda V, \lambda N)}{\partial(\lambda V)}V + \frac{\partial E(\lambda S, \lambda V, \lambda N)}{\partial(\lambda N)}N = E(S, V, N).$$

$$(A.2.2)$$

If we now set the value of $\lambda$ to unity, then the derivatives are simply $T$, $-p$, and $\mu$ so that

$$E = TS - pV + \mu N. \tag{A.2.3}$$

Following Callen [1], we shall call this the "Euler relation". This is appropriate because of Euler's interest in the mathematical properties of homogeneous functions.

## A.3   Two Caveats

- We remark, parenthetically, that the student should not be misled by this equation. In learning about thermodynamics one encounters the concept of a function of state, and the fact that heat and work are not functions of state; they depend on path. The Euler relation tells us that there are three contributions to the internal energy: $TS$, $-pV$, and $\mu N$. But we *cannot* identify $TS$ as the heat content, $-pV$ as the work content, and $\mu N$ as the chemical content. In other words, the path-dependent $\int T\mathrm{d}S$ is different from the function of state $TS$, etc.
- If we had taken $N$ to be fixed, and thereby ignored $\mu$, the derivation would not have worked since scaling the system by $\lambda$ requires that $N$ be so-scaled. In other words, $E = TS - pV$ does *not* hold.

## A.4   The Gibbs–Duhem Relation

The Gibbs–Duhem relation [2] follows from combining the Euler relation with the differential form for d$E$ (the First Law expression), namely

$$\mathrm{d}E = T\mathrm{d}S - p\mathrm{d}V + \mu\mathrm{d}N \tag{A.4.1}$$

to give

$$S\mathrm{d}T - V\mathrm{d}p + N\mathrm{d}\mu = 0. \tag{A.4.2}$$

This tells us that the intensive variables $T$, $p$, and $\mu$ are *not independent*, which may be surprising. We understand temperature and pressure as (independent) variables over which we have fairly direct control. And by varying these we can effect heat and volume "flow". Similarly, we learned that particles can be induced to flow by varying the chemical potential. Certainly, our "feel" for chemical potential is not so intuitive as that for temperature and pressure, but we are helped by the mathematical analogies with $T$ and $p$. However, now we see that if we regard $T$ and $p$ as independent variables, then $\mu$ depends on $T$ and $p$; it is not independent.

This is explored in Problem 2.5 for the case of an ideal gas.

# THERMODYNAMIC POTENTIALS

---

The discussions in this appendix draw unashamedly on the treatment of Callen [1]. That discussion is so clear and intuitive; it cannot be improved upon. The reader is most certainly advised to read Callen's book at some stage of his/her studies.

## B.1 Equilibrium States

We have seen that the equilibrium state of an isolated system — characterised by $E$, $V$, $N$, — is determined by maximising the entropy $S$. On the other hand, we know that a purely mechanical system settles down to the equilibrium state that minimises the energy: the state where the forces $F_i = -\partial E/\partial X_i$ vanish. In this section, we shall see how to relate these two ideas. And then in the following sections, we shall see how to extend them.

By a purely mechanical system, we mean one with *no* thermal degrees of freedom. This means no changing of the *populations* of the different quantum states — i.e. at constant entropy. But this should also apply for a thermodynamic system at constant entropy. So, we should be able to find the equilibrium state in two ways:

- maximise the entropy at constant energy;
- minimise the energy at constant entropy.

That these approaches are equivalent may be seen by considering the form of the $E - S - X$ surface for a system (Fig. B.1). Here $X$ is some extensive quantity that will vary when the system approaches equilibrium like the energy in or the number of particles in one half of the system.

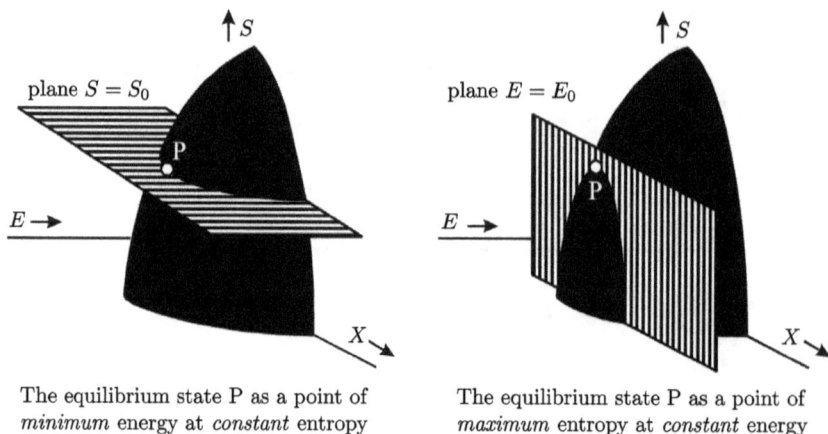

The equilibrium state P as a point of *minimum* energy at *constant* entropy

The equilibrium state P as a point of *maximum* entropy at *constant* energy

Fig. B.1. Alternative specification of equilibrium states (After Callen [1]).

At constant energy, the plane $E = E_0$ intersects the $E - S - X$ surface along a line of possible states of the system. We have seen that the equilibrium state — the state of maximum probability — will be the point of maximum entropy: the point P on the curve.

But now consider the same system at constant entropy. The plane $S = S_0$ intersects the $E - S - X$ surface along a different line of possible states. Comparing the two pictures, we see that the equilibrium point P is now the point of minimum energy.

Equivalence of the entropy maximum and the energy minimum principles depends on the shape of the $E - S - X$ surface. In particular, it relies on

1. $S$ having a maximum with respect to $X$ — guaranteed by the Second Law;
2. $S$ being an increasing function of $E$: $\partial S / \partial E = 1/T > 0$ means positive temperatures;
3. $E$ being a single-valued continuous function of $S$.

To demonstrate the equivalence of the entropy maximum and the energy minimum principles, we shall show that the converse would lead to a violation of the second law.

Assume that the energy $E$ is *not* the minimum value consistent with a given entropy $S$. Then let us take some energy out of the system in the form of work and return it in the form of heat. The energy is then the same but the entropy will have increased. So, the original state could not have been an equilibrium state.

The equivalence of the energy minimum and the entropy maximum principle is rather like describing the circle as having the maximum area at fixed circumference or the minimum circumference for a given area.

We shall now look at the specification of equilibrium states when, instead of energy or entropy, other variables are held constant.

## B.2 Constant Temperature and Volume: The Helmholtz Potential

To maintain a system of fixed volume at constant temperature we shall put it in contact with a heat reservoir. The equilibrium state can be determined by maximising the total entropy while keeping the total energy constant.

The total entropy is the sum of the system entropy and that of the reservoir. The entropy maximum condition is then

$$\left. \begin{array}{l} dS_t = dS + dS_{res} = 0 \\ d^2 S_t < 0. \end{array} \right\} \tag{B.2.1}$$

The entropy differential for the reservoir is

$$\begin{aligned} dS_{res} &= \frac{dE_{res}}{T_{res}} \\ &= -\frac{dE}{T} \end{aligned} \tag{B.2.2}$$

since the total energy is constant. The total entropy maximum condition is then

$$\left. \begin{array}{l} dS - \dfrac{dE}{T} = 0 \\ d^2 S - \dfrac{d^2 E}{T} < 0. \end{array} \right\} \tag{B.2.3}$$

Or, since $T$ is constant,

$$\left. \begin{array}{l} d(E - TS) = 0 \\ d^2(E - TS) < 0, \end{array} \right\} \tag{B.2.4}$$

which is the condition for a minimum in $E - TS$. But we have encountered $E - TS$ before, in the consideration of the link between the partition function and thermodynamic properties. This function is the Helmholtz free energy $F$. So, we conclude that

**at constant $N$, $V$, and $T$, $F = E - TS$ is a minimum**

— The Helmholtz minimum principle.

We can understand this as a competition between two opposing effects. At high temperatures, the entropy tends to a maximum, while at low temperatures, the energy tends to a minimum. And the balance between these competing processes is given, at general temperatures, by minimising the combination $F = E - TS$.

## B.3  Constant Pressure and Energy: The Enthalpy Function

To maintain a system at constant pressure, we shall put it in mechanical contact with a "volume" reservoir. That is, it will be connected by a movable, thermally isolated piston, to a very large volume. As before, we can determine the equilibrium state by maximising the total entropy while keeping the total energy constant. Alternatively and equivalently, we can keep the total entropy constant while minimising the total energy.

The energy minimum condition is

$$\left.\begin{aligned} dE_t = dE + dE_{res} = 0 \\ d^2 E_t > 0. \end{aligned}\right\} \tag{B.3.1}$$

In this case, the reservoir may do mechanical work on the system:

$$dE_{res} = -p_{res}dV_{res} = pdV \tag{B.3.2}$$

since the total volume is fixed. We then write the energy minimum condition as

$$\left.\begin{aligned} dE + pdV = 0 \\ d^2 E + pd^2 V > 0, \end{aligned}\right\} \tag{B.3.3}$$

or, since $p$ is constant,

$$\left.\begin{aligned} d(E + PV) = 0 \\ d^2(E + pV) < 0. \end{aligned}\right\} \tag{B.3.4}$$

This is the condition for a minimum in $E + pV$. This function is called Enthalpy and it is given the symbol $H$. So, we conclude that

**at constant $N$, $p$, and $E$, $H = E + pV$ is a minimum**
— The Enthalpy minimum principle.

## B.4 Constant Pressure and Temperature: The Gibbs Free Energy

In this case, our system can exchange both thermal and mechanical energy with a reservoir; both heat energy and "volume" may be exchanged. Working in terms of the minimum energy at constant entropy condition for the combined system + reservoir:

$$\left.\begin{aligned} dE_t = dE + dE_{res} = 0 \\ d^2E_t > 0. \end{aligned}\right\} \tag{B.4.1}$$

In this case, the reservoir exchanges heat energy and/or it may do mechanical work on our system:

$$dE_{res} = T_{res}dS_{res} - p_{res}dV_{res} = -TdS + pdV \tag{B.4.2}$$

since the total energy is fixed. We then write the energy minimum condition as

$$\left.\begin{aligned} dE - TdS + pdV = 0 \\ d^2E - Td^2S + pd^2V > 0, \end{aligned}\right\} \tag{B.4.3}$$

or, since $T$ and $p$ are constant,

$$\left.\begin{aligned} d(E - TS + PV) = 0 \\ d^2(E - TS + pV) < 0. \end{aligned}\right\} \tag{B.4.4}$$

This is the condition for a minimum in $E - TS + pV$. This function is called the Gibbs free energy and it is given the symbol $G$. So, we conclude that

**at constant $N$, $p$, and $T$, $G = E - TS + pV$ is a minimum**
— The Gibbs free energy minimum principle.

## B.5　Differential Expressions for the Potentials

The internal energy, Helmholtz free energy, enthalpy, and Gibbs free energy are called *thermodynamic potentials*. Clearly, they are all functions of state. From the definitions

$$F = E - TS \qquad \text{Helmholtz function}$$
$$H = E + pV \qquad \text{Enthalph function} \qquad (\text{B.5.1})$$
$$G = E - TS + pV \quad \text{Gibbs function}$$

and the differential expression for the internal energy

$$dE = TdS - pdV \qquad (\text{B.5.2})$$

we obtain the differential expressions for the potentials:

$$dF = -SdT - p\,dV$$
$$dH = TdS + Vdp \qquad (\text{B.5.3})$$
$$dG = -SdT + Vdp.$$

## B.6　Natural Variables and the Maxwell Relations

Each of the thermodynamic potentials has its own *natural variables*. For instance, taking $E$, the differential expression for the First Law is

$$dE = TdS - pdV. \qquad (\text{B.6.1})$$

Thus, if $E$ is known as a function of $S$ and $V$, then everything else — such as $T$ and $p$ — can be obtained by differentiation, since $T = \partial E/\partial S|_V$ and $p = -\partial E/\partial V|_S$. If, instead, $E$ were known as a function of, say, $T$ and $V$, then we would need more information, such as an equation of state, to completely determine the remaining thermodynamic functions.

All the potentials have their natural variables in terms of which the dependent variables may be found by differentiation:

$$E(S,V), \quad dE = TdS - pdV, \quad T = \left.\frac{\partial E}{\partial S}\right|_V, \quad p = -\left.\frac{\partial E}{\partial V}\right|_S,$$

$$F(T,V), \quad dF = -SdT - pdV, \quad S = -\left.\frac{\partial F}{\partial T}\right|_V, \quad p = -\left.\frac{\partial F}{\partial V}\right|_T,$$

$$\text{(B.6.2)}$$

$$H(S,p), \quad dH = TdS + Vdp, \quad T = \left.\frac{\partial H}{\partial S}\right|_p, \quad V = \left.\frac{\partial H}{\partial p}\right|_S,$$

$$G(T,P), \quad dG = -SdT + Vdp, \quad S = -\left.\frac{\partial G}{\partial T}\right|_p, \quad V = \left.\frac{\partial G}{\partial p}\right|_T.$$

If we differentiate one of these results with respect to a further variable, then the order of differentiation is immaterial; differentiation is commutative. Thus, for instance, using the energy natural variables, we see that

$$\left.\frac{\partial}{\partial V}\right|_S \left.\frac{\partial}{\partial S}\right|_V = \left.\frac{\partial}{\partial S}\right|_V \left.\frac{\partial}{\partial V}\right|_S \qquad \text{(B.6.3)}$$

and operating on $E$ with this, we obtain

$$\left.\frac{\partial}{\partial V}\right|_S \left.\frac{\partial E}{\partial S}\right|_V = \left.\frac{\partial}{\partial S}\right|_V \left.\frac{\partial E}{\partial V}\right|_S$$
$$\|\qquad\qquad\quad\|$$
$$T \qquad\qquad -p \qquad\qquad \text{(B.6.4)}$$

so that, we obtain the result

$$\left.\frac{\partial T}{\partial V}\right|_S = -\left.\frac{\partial p}{\partial S}\right|_V.$$

Such expressions are known as Maxwell relations.

Similarly, we get one relation for each potential by differentiating it with respect to its two natural variables

$$E: \quad \left.\frac{\partial T}{\partial V}\right|_S = -\left.\frac{\partial p}{\partial S}\right|_V$$

$$F: \quad \left.\frac{\partial S}{\partial V}\right|_T = \left.\frac{\partial p}{\partial T}\right|_V$$

$$H: \quad \left.\frac{\partial T}{\partial p}\right|_S = \left.\frac{\partial V}{\partial S}\right|_p \qquad\qquad (B.6.5)$$

$$G: \quad \left.\frac{\partial S}{\partial p}\right|_T = -\left.\frac{\partial V}{\partial T}\right|_p .$$

The Maxwell relations give equations between seemingly different quantities. In particular, they often connect easily measured, but uninteresting quantities to difficult-to-measure, but very interesting ones.

Another application of the Maxwell relations is in the study of glassy systems and other quasi-equilibrium states of matter. The validity of the Maxwell relations relies on the assumption of a state of thermal equilibrium. An experimental demonstration that a Maxwell relation does *not* hold is then an indication that the system under investigation is not in a state of thermal equilibrium.

# *MATHEMATICA* NOTEBOOKS

This appendix contains various *Mathematica* Notebooks for the calculation of properties of gases in the low- and high-temperature limits. *Mathematica* input lines are indicated by bold text, while Mathematica output lines are indicated by roman (normal) text.

Further details are given in the author's paper [3]. That gives formulae together with their *Mathematica* routines for the chemical potential of maxwellons, fermions, and bosons in 1, 2, and 3 dimensions, together with plots, series, etc.

## C.1 Chemical Potential of a Fermi Gas at Low Temperatures — Sommerfeld Expansion

### C.1.1 *Setting up the problem*

This *Mathematica* Notebook evaluates the low-temperature chemical potential of a Fermi gas as a power series in temperature using the Sommerfeld method discussed in Section 2.5.3.

The starting point is the Sommerfeld integral given in Eq. (2.5.50):

$$\varepsilon_F^{3/2} = \int_{-\infty}^{\infty} \frac{e^x}{(e^x+1)^2} [\mu + kTx]^{3/2} dx.$$

We pull out a factor $\mu^{3/2}$ from the square bracket and divide both sides by $(kT)^{3/2}$ giving

$$\left(\frac{kT}{\varepsilon_F}\right)^{-3/2} = \left(\frac{kT}{\mu}\right)^{-3/2} \int_{-\infty}^{\infty} \frac{e^x}{(e^x+1)^2} \left[1 + \frac{kT}{\mu}x\right]^{3/2} dx.$$

Observe this is an expression for $kT/\varepsilon_F$ in terms of $kT/\mu$. The task is to invert this to obtain $kT/\mu$ in terms of $kT/\varepsilon_F$. But first, we shall transform to reduced (dimensionless) variables:

$$t = \frac{kT}{\varepsilon_F}, \quad m = \frac{\mu}{\varepsilon_F}$$

so that

$$t^{-3/2} = \left(\frac{t}{m}\right)^{-3/2} \int_{-\infty}^{\infty} \frac{e^x}{(e^x+1)^2} \left[1 + \frac{t}{m}x\right]^{3/2} dx.$$

This is an expression for $t$ in terms of $t/m$. We need to reverse this, to obtain $t/m$ in terms of $t$, and then manipulating to get $m$ as a function of $t$.

### C.1.2   *Mathematica manipulations*

**1. Series expansion of the square bracket in powers of $tx/m$**

First, specify the order for the power series expansions $n$. We shall use $n = 6$ here, but you can go higher if required.

**n = 6**

6

Now, specify the square bracket term **brac**. Note **tom** represents $t/m$.

**brac = (1 + tom x)^(3/2)**

$(1 + \text{tom } x)^{3/2}$

and then we evaluate its power series expansion. We expand this to about $x = 0$, which is equivalent to expanding to about $\varepsilon = \mu$.

**sbrac = Series[brac, {x, 0, n}] // FullSimplify**

$1 + \frac{3\text{tom}x}{2} + \frac{3\text{tom}^2x^2}{8} - \frac{\text{tom}^3x^3}{16} + \frac{3\text{tom}^4x^4}{128} - \frac{3\text{tom}^5x^5}{256} + \frac{7\text{tom}^6x^6}{1024} + O[x]^7.$

This is a series object. *Mathematica* cannot integrate with respect to $x$ because $x$ is the expansion parameter. We must convert it to a normal polynomial first. So, make it Normal and then integrate it.

## 2. Term-by-term integration

**in = Integrate[Normal[sbrac]E^x/(E^x + 1)^2, {x, −Infinity, Infinity}] //
Apart**

$$1 + \frac{\pi^2 \text{tom}^2}{8} + \frac{7\pi^4 \text{tom}^4}{640} + \frac{31\pi^6 \text{tom}^6}{3072}$$

and then we convert it back to a series

**sin = Series[in, {tom, 0, n}]**

$$1 + \frac{\pi^2 \text{tom}^2}{8} + \frac{7\pi^4 \text{tom}^4}{640} + \frac{31\pi^6 \text{tom}^6}{3072} + O[\text{tom}]^7.$$

## 3. Raise to the −2/3 power

We need to raise this series to the −2/3 power

**iss = sin ^(−2/3) // FullSimplify**

$$1 - \frac{\pi^2 \text{tom}^2}{12} + \frac{\pi^4 \text{tom}^4}{720} - \frac{\pi^6 \text{tom}^6}{162} + O[\text{tom}]^7$$

and multiply this by **tom** $(t/m)$ to get **ttt** $(t)$ as a series in tom $(t/m)$

**ttt = tom iss**

$$\text{tom} - \frac{\pi^2 \text{tom}^3}{12} + \frac{\pi^4 \text{tom}^5}{720} - \frac{\pi^6 \text{tom}^7}{162} + O[\text{tom}]^8.$$

## 4. Reverse the series to give $t/m$ as a series in $t$

Now, we invert this series to give **tom** — now called **ttomm** — $(t/m)$ as a series in **tt** $(t)$

**ttomm = InverseSeries[ttt, tt] //FullSimplify**

$$\text{tt} + \frac{\pi^2 \text{tt}^3}{12} + \frac{7\pi^4 \text{tt}^5}{360} + \frac{79\pi^6 \text{tt}^7}{6480} + O[\text{tt}]^8.$$

## 5. Reduced chemical potential

At this stage, we have $t/m$ as a series in $t$. We require $m$ as a series in $t$. We must divide $t$ by the $t/m$ series. *Mathematica* can do this immediately since **ttomm** is a series object and *Mathematica* is happy manipulating series objects. So, **mm** $(m)$ is then

**mm = tt/ttomm//FullSimplify**

$$1 - \frac{\pi^2 tt^2}{12} - \frac{\pi^4 tt^4}{80} - \frac{247\pi^6 tt^6}{25920} + O[tt]^7.$$

### 6. Transformation back to proper variables

This is the end of the *Mathematica* manipulations. It now remains to convert from reduced units back to proper ones.

The chemical potential is

$$\mu(T) = \varepsilon_F \left\{ 1 - \frac{\pi^2}{12}\left(\frac{kT}{\varepsilon_F}\right)^2 - \frac{\pi^4}{80}\left(\frac{kT}{\varepsilon_F}\right)^4 - \frac{247\pi^6}{25920}\left(\frac{kT}{\varepsilon_F}\right)^6 + \cdots \right\}.$$

## C.2    Internal Energy of a Fermi Gas at Low Temperatures — Sommerfeld Expansion

### C.2.1   *Setting up the problem*

This *Mathematica* Notebook evaluates the low-temperature internal energy of a Fermi gas as a power series in temperature using the Sommerfeld method discussed in Section 2.5.3.

The starting point is the Sommerfeld integral given in Eq. (2.5.56):

$$E = \frac{3}{5}\frac{N}{\varepsilon_F^{3/2}} \int_{-\infty}^{\infty} \frac{e^x}{(e^x + 1)^2} [\mu + kTx]^{5/2} dx.$$

We pull out a factor $\mu^{5/2}$ from the square bracket and divide both sides by $N\varepsilon_F$ giving

$$\frac{E}{N\varepsilon_F} = \frac{3}{5}\left(\frac{\mu}{\varepsilon_F}\right)^{5/2} \int_{-\infty}^{\infty} \frac{e^x}{(e^x + 1)^2}\left[1 + \frac{kT}{\mu}x\right]^{5/2} dx.$$

Observe this is an expression for $E/N\varepsilon_F$ in terms of $\mu$ and $T$. We must perform an expansion in powers of $T$, however, we must incorporate the temperature dependence of $\mu$, as determined in the previous section.

But first, we shall transform to reduced (dimensionless) variables:

$$e = \frac{E}{N\varepsilon_F} \quad t = \frac{kT}{\varepsilon_F}, \quad m = \frac{\mu}{\varepsilon_F}$$

so that

$$e = \frac{3}{5}m^{5/2} \int_{-\infty}^{\infty} \frac{e^x}{(e^x + 1)^2} \left[1 + \frac{t}{m}x\right]^{5/2} dx.$$

### C.2.2 *Mathematica manipulations*

**1. Series expansion of the square bracket in powers of $tx/m$**

Let's first specify the order of the power series expansion $n$. We shall use $n = 6$ here, but you can go higher if required.

**$n = 6$**

6

Now, specify the square bracket term **brac**. As before **tom** represents $t/m$.

**brac = (1 + tom x)^(5/2)**

$(1 + \text{tom } x)^{5/2}$

and then evaluate its power series expansion. We expand this about $x = 0$, which is equivalent to expanding about $\varepsilon = \mu$.

**sbrac = Series[brac, {x, 0, n}]**

$1 + \frac{5\text{tom}x}{2} + \frac{15\text{tom}^2x^2}{8} + \frac{5\text{tom}^3x^3}{16} - \frac{5\text{tom}^4x^4}{128} + \frac{3\text{tom}^5x^5}{256} - \frac{5\text{tom}^6x^6}{1024} + O[x]^7.$

This is a series object. *Mathematica* cannot integrate with respect to $x$ because $x$ is the expansion parameter. We must convert it to a normal polynomial first. So, make it Normal and then integrate it.

**2. Term-by-term integration**

**in = Integrate[Exp[x]Normal[sbrac]/(Exp[x] + 1)^2, {x, −Infinity, Infinity}]
//Apart**

$1 + \frac{5\pi^2\text{tom}^2}{8} - \frac{7\pi^4\text{tom}^4}{384} - \frac{155\pi^6\text{tom}^6}{21504}$

and from this we obtain $e$:

$e = (3/5)m^\wedge(5/2)$in /. tom $\rightarrow t/m$

$$\tfrac{3}{5}m^{5/2}\left(1 + \tfrac{5\pi^2 t^2}{8m^2} - \tfrac{7\pi^4 t^4}{384m^4} - \tfrac{155\pi^6 t^6}{21504m^6}\right).$$

### 3. Incorporation of the chemical potential

Now, we substitute for the chemical potential, from the previous section. For safety, I have included a further term in the expansion of $m(t)$ calculation. This expression is "pasted in" from a previous Notebook.

$$m = 1 - \tfrac{\pi^2 t^2}{12} - \tfrac{\pi^4 t^4}{80} - \tfrac{247\pi^6 t^6}{25920} - \tfrac{16291\pi^8 t^8}{777600}$$

$$1 - \tfrac{\pi^2 t^2}{12} - \tfrac{\pi^4 t^4}{80} - \tfrac{247\pi^6 t^6}{25920} - \tfrac{16291\pi^8 t^8}{777600}.$$

Now, we will have a complicated expression for $e$, but we are interested in the low-temperature power series. So, expand $e$ once again

**Series[$e$, {$t$, 0, $n$}]**

$$\tfrac{3}{5} + \tfrac{\pi^2 t^2}{4} - \tfrac{3\pi^4 t^4}{80} - \tfrac{247\pi^6 t^6}{12096} + O[t]^7.$$

This is the expansion of $e$ in powers of $t$.

### 4. Transformation back to proper variables

This is the end of the Mathematica manipulations. It now remains to convert from reduced units back to proper ones.

The internal energy is:

$$E(T) = N\varepsilon_F\left\{\frac{3}{5} + \frac{\pi^2}{4}\left(\frac{kT}{\varepsilon_F}\right)^2 - \frac{3\pi^4}{80}\left(\frac{kT}{\varepsilon_F}\right)^4 - \frac{247\pi^6}{12096}\left(\frac{kT}{\varepsilon_F}\right)^6 + \cdots\right\}.$$

This is the required low-temperature series expression for the internal energy.

## C.3    Fugacity and Chemical Potential of the Fermi, Bose, and Maxwell Gas at High Temperatures

### C.3.1   *Setting up the problem*

This *Mathematica* Notebook evaluates the high-temperature chemical potential of an ideal gas. It will be expedient to obtain the fugacity first, before finding $\mu$, and we will be able to consider the Fermi, Bose, and Maxwell cases together. We follow the procedure discussed in Section 2.7.3. The starting point is the expression for the "quantum energy" $\varepsilon_q$ in Eq. (2.7.9):

$$\varepsilon_q^{3/2} = \frac{3}{2} \int_0^\infty n(\varepsilon) \varepsilon^{1/2} \mathrm{d}\varepsilon.$$

We shall write the occupation function as

$$n(\varepsilon) = \frac{1}{e^{\varepsilon/kT} z^{-1} + a}$$

where $z$ is the fugacity, $z = e^{\mu/kT}$, and $a$ determines the statistics of the particles:

$$a = \begin{cases} +1 & \text{for fermions} \\ 0 & \text{for maxwellons} \\ -1 & \text{for bosons.} \end{cases}$$

In this way, we treat all statistics together.

Then by a change of variables: $x = \varepsilon/kT$, we have

$$\left(\frac{\varepsilon_q}{kT}\right)^{3/2} = \frac{3}{2} \int_0^\infty \frac{x^{1/2}}{e^x z^{-1} + a} \mathrm{d}x$$

or, in terms of reduced temperature $t = kT/\varepsilon_F$:

$$t^{-3/2} = \frac{3}{2} \int_0^\infty \frac{x^{1/2}}{e^x z^{-1} + a} \mathrm{d}x.$$

This is the starting point for an expansion of $t^{-3/2}$ in powers of $z$, which will then be reversed to give a series for $z$ in (inverse) powers of $t$.

### C.3.2 *Mathematica manipulations*

**1. Series expansion in powers of $z$**

First, specify the order for the power series expansions $n$. We shall use $n = 4$ here, but you can go higher if required.

**$n = 4$**

4

The distribution function **ne** $(n(\varepsilon))$ is

ne = 1/(E^x/z + a)

$$\frac{1}{a+\frac{e^x}{z}}$$

where $a$ determines the statistics: $a = 1$ for fermions, $a = -1$ for bosons, and $a = 0$ for maxwellons. Here, $z$ is the fugacity $z = e^{\mu/kT}$. In the high-temperature limit, $ne \to 0$ so that $z$ is small. We expand **ne** in powers of $z$:

sn = Normal[Series[ne, {z, 0, n}]]

$$e^{-x}z - ae^{-2x}z^2 + a^2e^{-3x}z^3 - a^3e^{-4x}z^4$$

For convenience, we have made this Normal before integrating.

## 2. Term-by-term integration

We then perform the integral to get $\tau^{-3/2}$ called **tm3o2**

tm3o2 = (3/2)Integrate[snx^(1/2), {x, 0, Infinity}]//Apart

$$\frac{3\sqrt{\pi}z}{4} - \frac{3}{8}a\sqrt{\frac{\pi}{2}}z^2 + \frac{1}{4}a^2\sqrt{\frac{\pi}{3}}z^3 - \frac{3}{32}a^3\sqrt{\pi}z^4.$$

## 3. Series for $\tau^{-3/2}$ in powers of $z$

And now, we express this as a series again (in preparation for reversal):

stm3o2 = Series[tm3o2, {z, 0, n}]

$$\frac{3\sqrt{\pi}z}{4} - \frac{3}{8}\left(a\sqrt{\frac{\pi}{2}}\right)z^2 + \frac{1}{4}a^2\sqrt{\frac{\pi}{3}}z^3 - \frac{3}{32}\left(a^3\sqrt{\pi}\right)z^4 + O[z]^5.$$

## 4. Inversion of series — for fugacity series

We have a series for $\tau^{-3/2}$ in powers of $z$. We now invert this, to give a series for **(zz)** $z$ in powers of $\tau^{-3/2}$ (**ttm3o2**)

zz = InverseSeries[stm3o2, ttm3o2]//Simplify

$$\frac{4\text{ttm3o2}}{3\sqrt{\pi}} + \frac{4\sqrt{2}a\text{ttm3o2}^2}{9\pi} + \frac{16(9-4\sqrt{3})a^2\text{ttm3o2}^3}{243\pi^{3/2}} + \frac{8(36+45\sqrt{2}-40\sqrt{6})a^3\text{ttm3o2}^4}{729\pi^2} +$$
$$O[\text{ttm3o2}]^5.$$

Now, express this explicitly as a series in $\tau^{-3/2}$. Again first, we must make this Normal (as **ttm3o2** is the expansion parameter)

**nzz = Normal[zz] /. ttm3o2 → $t^\wedge(-3/2)$//Simplify//Apart**

$$-\frac{8(-36a^3-45\sqrt{2}a^3+40\sqrt{6}a^3)}{729\pi^2 t^6} - \frac{16(-9+4\sqrt{3})a^2}{243\pi^{3/2}t^{9/2}} + \frac{4\sqrt{2}a}{9\pi t^3} + \frac{4}{3\sqrt{\pi}t^{3/2}}.$$

And then we can re-express this as a series in inverse powers of $t$. It is expedient, for *Mathematica*, to do this by expanding in powers of $a$. This is the expression for the fugacity for fermions, bosons, and maxwellons:

**snzz = Series[nzz, {$a$,0,$n$}] // FullSimplify**

$$\frac{4}{3\sqrt{\pi}t^{3/2}} + \frac{4\sqrt{2}a}{9\pi t^3} - \frac{16(-9+4\sqrt{3})a^2}{243(\pi^{3/2}t^{9/2})} + \frac{8(36+45\sqrt{2}-40\sqrt{6})a^3}{729\pi^2 t^6} + O[a]^5.$$

## 5. Chemical potential

The (reduced) chemical potential $m$ is given by $t\ln z$. (*Mathematica* uses **Log[z]** for the natural logarithm).

**m = t Log[snzz] // Simplify**

$$t \text{Log}\left[\frac{4}{3\sqrt{\pi}t^{3/2}}\right] + \frac{\sqrt{\frac{2}{\pi}}a}{3\sqrt{t}} + \frac{(27-16\sqrt{3})a^2}{81\pi t^2} + \frac{(72+60\sqrt{2}-64\sqrt{6})a^3}{243\pi^{3/2}t^{7/2}} - \frac{(695+216\sqrt{2}-576\sqrt{3})a^4}{2187(\pi^2 t^5)} +$$
$$O[a]^5.$$

## 6. Transformation back to proper variables

This is the end of the *Mathematica* manipulations. It now remains to convert from reduced units back to proper ones.

The fugacity is:

$$z = \frac{4}{3\sqrt{\pi}}\left(\frac{\varepsilon_F}{kT}\right)^{3/2} + a\frac{4\sqrt{2}}{9\pi}\left(\frac{\varepsilon_F}{kT}\right)^3 - a^2\frac{16\left(-9+4\sqrt{3}\right)}{243\pi^{3/2}}\left(\frac{\varepsilon_F}{kT}\right)^{9/2}$$
$$+ a^3\frac{8\left(36+45\sqrt{2}-40\sqrt{6}\right)}{729\pi^2 t^6}\left(\frac{\varepsilon_F}{kT}\right)^6 + \cdots$$

and the chemical potential is:

$$\mu = kT\left\{\ln\left[\frac{4}{3\sqrt{\pi}}\left(\frac{\varepsilon_F}{kT}\right)^{3/2}\right] + a\frac{1}{3}\sqrt{\frac{2}{\pi}}\left(\frac{\varepsilon_F}{kT}\right)^{1/2} + a^2\frac{\left(27 - 16\sqrt{3}\right)}{81\pi}\left(\frac{\varepsilon_F}{kT}\right)^2\right.$$

$$+ \frac{\left(72 + 60\sqrt{2} - 64\sqrt{6}\right)a^3}{243\pi^{3/2}}\left(\frac{\varepsilon_F}{kT}\right)^{7/2} - \frac{\left(695 + 216\sqrt{2} - 576\sqrt{3}\right)a^4}{2187\pi^2}\left(\frac{\varepsilon_F}{kT}\right)^5$$

$$\left. + \cdots\right\}.$$

These expressions give the fugacity and chemical potential of fermions ($a = 1$), bosons ($a = -1$), and maxwellons ($a = 0$).

## C.4   Internal Energy of the Fermi, Bose, and Maxwell Gas at High Temperatures

### C.4.1   *Setting up the problem*

This *Mathematica* Notebook evaluates the high temperature internal energy of an ideal gas. It will be expedient to obtain the fugacity first, before finding $\mu$, and we will be able to consider the Fermi, Bose, and Maxwell cases together. We follow the procedure discussed in Section 2.7.4.

The starting point is the expression for the internal energy in Eq. (2.7.21), which we shall write as follows:

$$E = \frac{3N}{2\varepsilon_q^{3/2}}\int_0^\infty \frac{\varepsilon^{3/2}}{e^{\varepsilon/kT}z^{-1} + a}\,d\varepsilon,$$

where $z$ is the fugacity, $z = e^{\mu/kT}$, and $a$ determines the statistics of the particles so we can treat all statistics together.

Then by a change of variable: $x = \varepsilon/kT$, we have

$$E = \frac{3N}{2\varepsilon_q^{3/2}}(kT)^{5/2}\int_0^\infty \frac{x^{1/2}}{e^x z^{-1} + a}\,dx$$

or, in terms of reduced variables

$$e = \frac{E}{N\varepsilon_F} \qquad t = \frac{kT}{\varepsilon_F}$$

the reduced internal energy is

$$e = \frac{3}{2}t^{5/2}\int_0^\infty \frac{x^{3/2}}{e^x z^{-1} + a}\, dx.$$

This is the starting point for an expansion of $e$ in powers of $z$, and then we must put in the temperature dependence of $z$ from the previous section.

### C.4.2   *Mathematica manipulations*

### 1. Series expansion in powers of $z$

First, specify the order for the power series expansions $n$. We shall use $n = 4$ here, but you can go higher if required.

**$n = 4$**

4

The distribution function **ne** ($n(\varepsilon)$) is

**ne = 1/(E^x/z + a)**

$$\frac{1}{a + \frac{e^x}{z}}$$

where $a$ determines the statistics: $a = 1$ for fermions, $a = -1$ for bosons, and $a = 0$ for maxwellons. Here, $z$ is the fugacity $z = e^{\mu/kT}$. In the high-temperature limit, **ne**$\to 0$ so that $z$ is small. We expand **ne** in powers of $z$:

**sn = Normal[Series[ne, {z,0,n}]]**

$$e^{-x}z - ae^{-2x}z^2 + a^2 e^{-3x}z^3 - a^3 e^{-4x}z^4.$$

For convenience, we have made this Normal before integrating.

### 2. Term-by-term integration

We then perform the integral to get $e$, called **ee**

**ee = (3/2)t^(5/2)Integrate[snx^(3/2), {x, 0, Infinity}] // Apart**

$$\frac{9}{8}\sqrt{\pi}t^{5/2}z - \frac{9}{32}a\sqrt{\frac{\pi}{2}}t^{5/2}z^2 + \frac{1}{8}a^2\sqrt{\frac{\pi}{3}}t^{5/2}z^3 - \frac{9}{256}a^3\sqrt{\pi}t^{5/2}z^4.$$

### 3. Temperature dependence of $z$

Now, we substitute for the fugacity, from the previous section. For safety I have included a further term in the expansion of $z(t)$ calculation. This expression is "pasted in" from a previous Notebook.

$$z = \frac{4}{3\sqrt{\pi}t^{3/2}} + \frac{4\sqrt{2}a}{9\pi t^3} - \frac{16(-9+4\sqrt{3})a^2}{243(\pi^{3/2}t^{9/2})} + \frac{8(36+45\sqrt{2}-40\sqrt{6})a^3}{729\pi^2 t^6} +$$

$$\frac{(-9216\sqrt{5}+800(95+54\sqrt{2}-84\sqrt{3}))a^4}{54675\pi^{5/2}t^{15/2}} - \frac{32(6048\sqrt{10}+25(-1512-1463\sqrt{2}+336\sqrt{3}+912\sqrt{6}))a^5}{492075(\pi^3 t^9)}$$

$$-\frac{32(6048\sqrt{10}+25(-1512-1463\sqrt{2}+336\sqrt{3}+912\sqrt{6}))a^5}{492075\pi^3 t^9} + \frac{(-9216\sqrt{5}+800(95+54\sqrt{2}-84\sqrt{3}))a^4}{54675\pi^{5/2}t^{15/2}}$$

$$+ \frac{8(36+45\sqrt{2}-40\sqrt{6})a^3}{729\pi^2 t^6} - \frac{16(-9+4\sqrt{3})a^2}{243\pi^{3/2}t^{9/2}} + \frac{4\sqrt{2}a}{9\pi t^3} + \frac{4}{3\sqrt{\pi}t^{3/2}}$$

and now, we express $e$ as a series again.

### 4. Internal energy

So, the reduced internal energy $e$ is now

**se = Series[ee, {$t$, Infinity, $n$}]//Simplify**

$$\frac{3t}{2} + \frac{a\sqrt{\frac{1}{t}}}{2\sqrt{2\pi}} - \frac{(-27+16\sqrt{3})a^2}{81\pi t^2} + \frac{(18+15\sqrt{2}-16\sqrt{6})a^3(\frac{1}{t})^{7/2}}{54\pi^{3/2}}$$

$$+ \frac{2(7925+5400\sqrt{2}-7200\sqrt{3}-1728\sqrt{5})a^4}{18225\pi^2 t^5} + O\left[\frac{1}{t}\right]^{13/2}.$$

This is the expression for the reduced internal energy. The leading term is $\frac{3t}{2}$. It is convenient to pull out this as a prefactor; the remaining term is then

**se $/\left(\frac{3t}{2}\right)$ // FullSimplify**

$$1 + \frac{a(\frac{1}{t})^{3/2}}{3\sqrt{2\pi}} + \frac{(54-32\sqrt{3})a^2}{243\pi t^3} + \frac{(18+15\sqrt{2}-16\sqrt{6})a^3(\frac{1}{t})^{9/2}}{81\pi^{3/2}} +$$

$$\frac{4(7925+5400\sqrt{2}-7200\sqrt{3}-1728\sqrt{5})a^4}{54675\pi^2 t^6} + O\left[\frac{1}{t}\right]^{15/2}.$$

### 5. Transformation back to proper variables

This is the end of the *Mathematica* manipulations. It now remains to convert from reduced units back to proper ones.

The internal energy is:

$$E = \frac{3}{2}NkT\left\{1 + a\frac{1}{3\sqrt{2\pi}}\left(\frac{\varepsilon_F}{kT}\right)^{3/2} + a^2\frac{\left(54 - 32\sqrt{3}\right)}{243\pi}\left(\frac{\varepsilon_F}{kT}\right)^3\right.$$

$$+ a^3\frac{\left(18 + 15\sqrt{2} - 16\sqrt{6}\right)}{81\pi^{3/2}}\left(\frac{\varepsilon_F}{kT}\right)^{9/2}$$

$$\left. + a^4\frac{4\left(7925 + 5400\sqrt{2} - 7200\sqrt{3} - 1728\sqrt{5}\right)a^4}{54675\pi^2}\left(\frac{\varepsilon_F}{kT}\right)^6 + \cdots\right\}.$$

These expressions give the internal energy of fermions ($a = 1$), bosons ($a = -1$), and maxwellons ($a = 0$).

## C.5 Chemical Potential of the Bose Gas at Low Temperatures

### C.5.1 *Setting up the problem*

This *Mathematica* Notebook evaluates the low-temperature chemical potential of the Bose. Here, as in the high-temperature case, it will be expedient to obtain the fugacity first, before finding $\mu$. We know that $\mu = 0$ when $T < T_B$, so the following calculation, although for low temperatures, is for temperatures above $T_B$.

The starting point is the expression for the "quantum energy" in Eq. (2.7.9):

$$\varepsilon_q^{3/2} = \frac{3}{2}\int_0^\infty n(\varepsilon)\varepsilon^{1/2}\mathrm{d}\varepsilon.$$

We shall write the occupation function as

$$n(\varepsilon) = \frac{1}{e^{\varepsilon/kT}z^{-1} - 1}$$

where $z$ is the fugacity, $z = e^{\mu/kT}$. Then by a change of variables: $x = \varepsilon/kT$, we have

$$\left(\frac{\varepsilon_q}{kT}\right)^{3/2} = \frac{3}{2}\int_0^\infty \frac{x^{1/2}}{e^x z^{-1} - 1}\mathrm{d}x$$

or, in terms of reduced temperature $t = kT/\varepsilon_F$:

$$t^{-3/2} = \frac{3}{2} \int_0^\infty \frac{x^{1/2}}{e^x z^{-1} - 1} \, dx.$$

This gives (essentially) $t$ as a function of $z$ and we must invert this to get $z$ as a function of $t$.

There are a number of complications we must beware of related to Bose–Einstein condensation. For temperatures $T < T_B$, we have $\mu = 0$, $z = 1$, and then at $T = T_B$, $z$ drops from unity and $\mu$ grows negatively. So, we have to be careful with power series expansions.

Our initial small quantity (expansion parameter) will be

$$y = 1 - z;$$

thus, $y$ grows from zero as $T$ increases from $T_B$. So, our starting point is the integral

$$t^{-3/2} = \frac{3}{2} \int_0^\infty \frac{x^{1/2}}{e^x (1 - y)^{-1} - 1} \, dx.$$

### C.5.2 *Mathematica manipulations*

**1. Expression for $t$**

This is the expression for $t^{-3/2}$, called **tm3o2**. In this case, we ask *Mathematica* to evaluate the integral immediately, before doing any power series expansions.

**tm3o2 = 3/2 Integrate[Sqrt[$x$]/(Exp[$x$]/(1 − $y$) − 1), {$x$, 0, Infinity}]**

$\frac{3}{4} \sqrt{\pi} \text{PolyLog} \left[ \frac{3}{2}, 1 - y \right].$

The integral is given in terms of a PolyLog function[1]. This is the expression for $t^{-3/2}$. So, in order to obtain $t$, we must find the $-2/3$ power. Thus, $t$ is given by

**$t$ = tm3o2^(−2/3)**

$$\frac{2\left(\frac{2}{\pi}\right)^{1/3}}{3^{2/3} \text{PolyLog}\left[\frac{3}{2}, 1-y\right]^{2/3}}$$

---

[1] The polylogarithm function, written $\text{Li}_n(z)$, is defined as the infinite sum $\sum_{k=1}^\infty \frac{z^k}{k^n}$. *Mathematica* uses the notation **PolyLog[n, z]**.

## 2. Power series expansion

First, specify the order for the power series expansions $n$. We shall use $n = 2$; this seems small, but there are powers of $\sqrt{y}$, so there will be plenty of terms. Indeed, even then I will not include all terms of the *Mathematica* output.

$n = 2$

$n = 2.$

Then the series expansion for $t$ (**st**) is

st = **Series[t, {y, 0, 2}]//Simplify**

$$\frac{2\left(\frac{2}{\pi}\right)^{1/3}}{\left(3\text{Zeta}\left[\frac{3}{2}\right]\right)^{2/3}} + \frac{8\,2^{1/3}\pi^{1/6}\sqrt{y}}{3\,3^{2/3}\text{Zeta}\left[\frac{3}{2}\right]^{5/3}} + \frac{4\left(\frac{2}{\pi}\right)^{1/3}\left(10\pi + 3\text{Zeta}\left[\frac{1}{2}\right]\text{Zeta}\left[\frac{3}{2}\right]\right)y}{9\,3^{2/3}\text{Zeta}\left[\frac{3}{2}\right]^{8/3}}$$
$$+ \frac{2\,2^{1/3}\pi^{1/6}\left(320\pi + 9\text{Zeta}\left[\frac{3}{2}\right]\left(20\text{Zeta}\left[\frac{1}{2}\right] + 3\text{Zeta}\left[\frac{3}{2}\right]\right)\right)y^{3/2}}{81\,3^{2/3}\text{Zeta}\left[\frac{3}{2}\right]^{11/3}} + O[y]^{5/2}.$$

The condensation temperature is that for which $z \to 1$ or $y \to 0$. This is the $y = 0$ value of the $t$ expansion. We call this reduced temperature **tb**. Then

tb = **Normal[st] /. $y \to 0$**

$$\frac{2\left(\frac{2}{\pi}\right)^{1/3}}{\left(3\text{Zeta}\left[\frac{3}{2}\right]\right)^{2/3}}.$$

We will want to find expansions of $z$ and $\mu$ in powers of **t − tb**. So, the series expression for $t - tb$ in powers of $y$ is

sdt = **st − tb**

$$\frac{8\,2^{1/3}\pi^{1/6}\sqrt{y}}{3\,3^{2/3}\text{Zeta}\left[\frac{3}{2}\right]^{5/3}} + \frac{4\left(\frac{2}{\pi}\right)^{1/3}\left(10\pi + 3\text{Zeta}\left[\frac{1}{2}\right]\text{Zeta}\left[\frac{3}{2}\right]\right)y}{9\,3^{2/3}\text{Zeta}\left[\frac{3}{2}\right]^{8/3}}$$
$$+ \frac{2\,2^{1/3}\pi^{1/6}\left(320\pi + 9\text{Zeta}\left[\frac{3}{2}\right]\left(20\text{Zeta}\left[\frac{1}{2}\right] + 3\text{Zeta}\left[\frac{3}{2}\right]\right)\right)y^{3/2}}{81\,3^{2/3}\text{Zeta}\left[\frac{3}{2}\right]^{11/3}} + \cdots + O[y]^{5/2}.$$

## 3. Reversion of the series expansion

We have called $t - tb$ by the name **dt**. So, now, we reverse the series to obtain $y$ as a series in powers of **dt**.

**ssy = InverseSeries[sdt, dt]//Simplify**

$$\frac{27\left(\frac{3}{\pi}\right)^{1/3}\text{Zeta}\left[\frac{3}{2}\right]^{10/3}\text{dt}^2}{64\,2^{2/3}} - \frac{81\left(\text{Zeta}\left[\frac{3}{2}\right]^4\left(10\pi+3\text{Zeta}\left[\frac{1}{2}\right]\text{Zeta}\left[\frac{3}{2}\right]\right)\right)\text{dt}^3}{1024\pi}$$

$$+\,\frac{27\,3^{2/3}\text{Zeta}\left[\frac{3}{2}\right]^{14/3}\left(860\pi^2+54\pi\left(10\text{Zeta}\left[\frac{1}{2}\right]-\text{Zeta}\left[\frac{3}{2}\right]\right)\text{Zeta}\left[\frac{3}{2}\right]+135\text{Zeta}\left[\frac{1}{2}\right]^2\text{Zeta}\left[\frac{3}{2}\right]^2\right)\text{dt}^4}{32768\,2^{1/3}\pi^{5/3}}$$

$$+\,O[\text{dt}]^5.$$

This is the series for $y$ in powers of $dt = t - tb$. We require the fugacity $z = 1 - y$, and let us express this in powers of $(t - tb)/tb$, which we shall call **dd**. We must make the series Normal before substituting. We obtain:

**nsz = 1 − Normal[ssy]/.dt → dd tb.**

$$1 - \frac{9\text{dd}^2\text{Zeta}\left[\frac{3}{2}\right]^2}{16\pi} + \frac{9\text{dd}^3\text{Zeta}\left[\frac{3}{2}\right]^2\left(10\pi+3\text{Zeta}\left[\frac{1}{2}\right]\text{Zeta}\left[\frac{3}{2}\right]\right)}{64\pi^2}$$

$$-\,\frac{3\text{dd}^4\text{Zeta}\left[\frac{3}{2}\right]^2\left(860\pi^2+54\pi\left(10\text{Zeta}\left[\frac{1}{2}\right]-\text{Zeta}\left[\frac{3}{2}\right]\right)\text{Zeta}\left[\frac{3}{2}\right]+135\text{Zeta}\left[\frac{1}{2}\right]^2\text{Zeta}\left[\frac{3}{2}\right]^2\right)}{1024\pi^3}.$$

Or, as a series:

**sz = Series[nsz, {dd, 0, 3}]**

$$1 - \frac{9\text{Zeta}\left[\frac{3}{2}\right]^2\text{dd}^2}{16\pi} + \frac{9\text{Zeta}\left[\frac{3}{2}\right]^2\left(10\pi+3\text{Zeta}\left[\frac{1}{2}\right]\text{Zeta}\left[\frac{3}{2}\right]\right)\text{dd}^3}{64\pi^2} + O[\text{dd}]^4.$$

Now, we calculate $\mu$. Since **tb** has been evaluated, we use **ttb** as the symbol for the transition temperature. Then the series for the reduced $\mu$ is

**sm = ttb(dd + 1)Log[sz]//FullSimplify**

$$-\frac{9\left(\text{ttbZeta}\left[\frac{3}{2}\right]^2\right)\text{dd}^2}{16\pi} + \frac{27\text{ttbZeta}\left[\frac{3}{2}\right]^2\left(2\pi+\text{Zeta}\left[\frac{1}{2}\right]\text{Zeta}\left[\frac{3}{2}\right]\right)\text{dd}^3}{64\pi^2} + O[\text{dd}]^4.$$

## 4. Transformation back to proper variables

This is the end of the *Mathematica* manipulations. It now remains to convert from reduced units back to proper ones.

The low temperature fugacity (above $T_B$) is

$$z = 1 - \frac{9\zeta\left(\frac{3}{2}\right)^2}{16\pi}\left(\frac{T-T_B}{T_B}\right)^2 + \frac{9\zeta\left(\frac{3}{2}\right)^2\left(3\zeta\left(\frac{1}{2}\right)\zeta\left(\frac{3}{2}\right)+10\pi\right)}{64\pi^2}\left(\frac{T-T_B}{T_B}\right)^3 + \cdots$$

and the chemical potential is

$$\mu = kT_B \left\{ -\frac{9\zeta\left(\frac{3}{2}\right)^2}{16\pi} \left(\frac{T - T_B}{T_B}\right)^2 + \frac{27\zeta\left(\frac{3}{2}\right)^2 \left(\zeta\left(\frac{1}{2}\right)\zeta\left(\frac{3}{2}\right) + 2\pi\right)}{64\pi^2} \right.$$
$$\left. \times \left(\frac{T - T_B}{T_B}\right)^3 + \cdots \right\}.$$

## C.6 Internal Energy (and Heat Capacity) of the Bose Gas at Low Temperatures

### C.6.1 *Setting up the problem*

This *Mathematica* Notebook evaluates the low-temperature internal energy of the Bose gas. The internal energy is given by

$$E = \alpha \int_0^\infty \varepsilon n(\varepsilon) g(\varepsilon) \, d\varepsilon$$

and we express the density of states in terms of the quantum energy $\varepsilon_q$:

$$g(\varepsilon) = \frac{3}{2} \frac{N}{\alpha} \frac{\varepsilon^{1/2}}{\varepsilon_q^{3/2}}$$

thus,

$$E = \frac{3}{2} \frac{N}{\varepsilon_q^{3/2}} \int_0^\infty \frac{\varepsilon^{3/2}}{e^{(\varepsilon - \mu)/kT} - 1} \, d\varepsilon.$$

Upon change of variables, $x = \varepsilon/kT$, we write

$$\frac{E}{N\varepsilon_q} = \frac{3}{2} \left(\frac{kT}{\varepsilon_q}\right)^{5/2} \int_0^\infty \frac{x^{3/2}}{e^x/z - 1} \, dx \qquad \text{(C.6.1)}$$

where $z$ is the fugacity.

**Case 1: Below the condensation temperature**

When $T < T_B$, we have $\mu = 0$ and so $z = 1$. In this case, the expression for $E$ was given in Eq. (2.6.36), which we shall write as follows:

$$E = NkT_B \left(\frac{T}{T_B}\right)^{5/2} \frac{3}{2} \frac{\zeta\left(\frac{5}{2}\right)}{\zeta\left(\frac{3}{2}\right)},$$

for comparison with the $T > T_B$ case below.

**Case 2: Above the condensation temperature**

When $T > T_B$, then $\mu$ is non-zero and $z$ is different from unity. Then we must integrate up the full Eq. (C.6.1). The integral is a polylogarithm function, giving

$$\frac{E}{N\varepsilon_q} = \frac{9}{8}\sqrt{\pi} \left(\frac{kT}{\varepsilon_q}\right)^{5/2} \mathrm{Li}_{5/2}(z). \tag{C.6.2}$$

We shall eliminate $\varepsilon_q$ in favour of $T_B$ using

$$\left(\frac{\varepsilon_q}{kT_B}\right)^{3/2} = \frac{3\sqrt{\pi}}{4}\zeta\left(\tfrac{3}{2}\right),$$

so that

$$E = NkT_B \left(\frac{T}{T_B}\right)^{5/2} \frac{3}{2} \frac{1}{\zeta\left(\frac{3}{2}\right)} \mathrm{Li}_{5/2}(z).$$

Since $\mathrm{Li}_{5/2}(1) = \zeta\left(\tfrac{5}{2}\right)$, this expression for $E$ is in agreement with the $T < T_c$ case above, Eq. (C.6.1).

### C.6.2 *Mathematica manipulations*

In this notebook, we shall adopt reduced variables

$$e = \frac{E}{N\varepsilon_q} \quad \text{and} \quad t = \frac{kT}{\varepsilon_q}.$$

Then Eq. (C.6.1) becomes

$$e = \frac{3}{2} t^{5/2} \int_0^\infty \frac{x^3/2}{e^x/z - 1}\, dx \tag{C.6.3}$$

and this is the starting point for the calculations.

**1. Check the integral — to give the polylogarithm**

But first, let us check the integral

**integral = Integrate[$x$^(3/2)/($E$^$x$/$z$ − 1), {$x$, 0, Infinity}]**

$\frac{3}{4}\sqrt{\pi}\text{PolyLog}\left[\frac{5}{2}, z\right]$

and here we look at the power series expansion about $y = 1 - z$

**Series $\left[\text{PolyLog}\left[\frac{5}{2}, 1 - y\right], \{y, 0, 2\}\right]$**

$\text{Zeta}\left[\frac{5}{2}\right] - \text{Zeta}\left[\frac{3}{2}\right]y + \frac{4}{3}\sqrt{\pi}y^{3/2} + \frac{1}{2}\left(\text{Zeta}\left[\frac{1}{2}\right] - \text{Zeta}\left[\frac{3}{2}\right]\right)y^2 + O[y]^{5/2}.$

**2. Evaluation of the internal energy**

Now, we proceed with the calculations proper. Equation (C.6.3) gives the internal energy (in reduced units):

$e = 3/2t^\wedge(5/2)\textbf{Integrate}[x^\wedge(3/2)/(E^\wedge x/z - 1), \{x, 0, \textbf{Infinity}\}]$

$\frac{9}{8}\sqrt{\pi}t^{5/2}\text{PolyLog}\left[\frac{5}{2}, z\right].$

This is the polylogarithm expression for the internal energy.

Next, we import the series expression for $z$ from the previous notebook (**dd** is $(T - T_B)/T_B$):

$\textbf{sz} = 1 - \frac{9\text{Zeta}\left[\frac{3}{2}\right]^2\text{dd}^2}{16\pi} + \frac{9\text{Zeta}\left[\frac{3}{2}\right]^2\left(10\pi + 3\text{Zeta}\left[\frac{1}{2}\right]\text{Zeta}\left[\frac{3}{2}\right]\right)\text{dd}^3}{64\pi^2}$

`+ more terms not printed`

$1 - \frac{9\text{Zeta}\left[\frac{3}{2}\right]^2\text{dd}^2}{16\pi} + \frac{9\text{Zeta}\left[\frac{3}{2}\right]^2\left(10\pi + 3\text{Zeta}\left[\frac{1}{2}\right]\text{Zeta}\left[\frac{3}{2}\right]\right)\text{dd}^3}{64\pi^2}$

`+ more terms not printed`

Substitute this into the expression for **e**. We need to make the **z** series normal before substituting, and then we shall evaluate the series expansion for **e** and call this **se**

**se = FullSimplify[Series[$e$/.$z$ → Normal[sz], {dd, 0, 3}], Assumptions → {dd ∈ Reals&&dd > 0}]**

$\frac{9}{8}\sqrt{\pi}t^{5/2}\text{Zeta}\left[\frac{5}{2}\right] - \frac{81\left(t^{5/2}\text{Zeta}\left[\frac{3}{2}\right]^3\right)\text{dd}^2}{128\sqrt{\pi}} + \frac{81t^{5/2}\text{Zeta}\left[\frac{3}{2}\right]^3\left(14\pi + 3\text{Zeta}\left[\frac{1}{2}\right]\text{Zeta}\left[\frac{3}{2}\right]\right)\text{dd}^3}{512\pi^{3/2}} +$
$O[\text{dd}]^4$

**dd** is $(T - T_B)/T_B$ — that's fine. But **t** here is in "reduced" units of $\varepsilon_q$. We must change this to units of $T_B$. Copy this from the previous notebook:

$$\textbf{tb} = \frac{2\left(\frac{2}{\pi}\right)^{1/3}}{\left(3\text{Zeta}\left[\frac{3}{2}\right]\right)^{2/3}}$$

$$\frac{2\left(\frac{2}{\pi}\right)^{1/3}}{\left(3\text{Zeta}\left[\frac{3}{2}\right]\right)^{2/3}}$$

so now express $E/NkT_B$ with leading $t$ as $T/T_B$:

**see = se tb^(3/2)**

$$\frac{3t^{5/2}\text{Zeta}\left[\frac{5}{2}\right]}{2\text{Zeta}\left[\frac{3}{2}\right]} - \frac{27\left(t^{5/2}\text{Zeta}\left[\frac{3}{2}\right]^2\right)\text{dd}^2}{32\pi} + \frac{27t^{5/2}\text{Zeta}\left[\frac{3}{2}\right]^2\left(14\pi+3\text{Zeta}\left[\frac{1}{2}\right]\text{Zeta}\left[\frac{3}{2}\right]\right)\text{dd}^3}{128\pi^2} + O[\text{dd}]^4$$

or, pulling out the first term as a prefactor

$$\textbf{see} \Big/ \left(\frac{3t^{5/2}\textbf{Zeta}\left[\frac{5}{2}\right]}{2\textbf{Zeta}\left[\frac{3}{2}\right]}\right)$$

$$1 - \frac{9\text{Zeta}\left[\frac{3}{2}\right]^3\text{dd}^2}{16\left(\pi\text{Zeta}\left[\frac{5}{2}\right]\right)} + \frac{9\text{Zeta}\left[\frac{3}{2}\right]^3\left(14\pi+3\text{Zeta}\left[\frac{1}{2}\right]\text{Zeta}\left[\frac{3}{2}\right]\right)\text{dd}^3}{64\pi^2\text{Zeta}\left[\frac{5}{2}\right]} + O[\text{dd}]^4.$$

These give Eq. (C.6.4) series for $E$ in the following.

### 3. Evaluation of heat capacity

We need to go to real units (temperature tt and ree = E/N) and make it normal before the substitutions

**ree = tB Normal[see]/.{$t \rightarrow$ tt/tB, dd $\rightarrow$ (tt − tB)/tB}**

$$\text{tB}\left(-\frac{27\left(\frac{tt}{tB}\right)^{5/2}(-\text{tB}+\text{tt})^2\text{Zeta}\left[\frac{3}{2}\right]^2}{32\pi\text{tB}^2} + \frac{27\left(\frac{tt}{tB}\right)^{5/2}(-\text{tB}+\text{tt})^3\text{Zeta}\left[\frac{3}{2}\right]^2\left(14\pi+3\text{Zeta}\left[\frac{1}{2}\right]\text{Zeta}\left[\frac{3}{2}\right]\right)}{128\pi^2\text{tB}^3}\right.$$

$$\left.+\frac{3\left(\frac{tt}{tB}\right)^{5/2}\text{Zeta}\left[\frac{5}{2}\right]}{2\text{Zeta}\left[\frac{3}{2}\right]}\right)$$

then differentiate for the heat capacity

**cc = D[ree, tt]**

$$\text{tB}\left(-\frac{27\left(\frac{tt}{tB}\right)^{5/2}(-\text{tB}+\text{tt})\text{Zeta}\left[\frac{3}{2}\right]^2}{16\pi\text{tB}^2} - \frac{135\left(\frac{tt}{tB}\right)^{3/2}(-\text{tB}+\text{tt})^2\text{Zeta}\left[\frac{3}{2}\right]^2}{64\pi\text{tB}^3} + \text{more terms}\right.$$

$$\left.\text{not printed}\right)$$

and then turn it back into a series about $T_B$

**ccc = Series[cc, {tt, tB, 2}]//Simplify**

$$\frac{15\text{Zeta}\left[\frac{5}{2}\right]}{4\text{Zeta}\left[\frac{3}{2}\right]} + \frac{9\left(-3\text{Zeta}\left[\frac{3}{2}\right]^3 + 10\pi\text{Zeta}\left[\frac{5}{2}\right]\right)(\text{tt}-\text{tB})}{16\pi\text{tBZeta}\left[\frac{3}{2}\right]}$$

$$+ \frac{9\left(36\pi\text{Zeta}\left[\frac{3}{2}\right]^3 + 27\text{Zeta}\left[\frac{1}{2}\right]\text{Zeta}\left[\frac{3}{2}\right]^4 + 20\pi^2\text{Zeta}\left[\frac{5}{2}\right]\right)(\text{tt}-\text{tB})^2}{128\pi^2\text{tB}^2\text{Zeta}\left[\frac{3}{2}\right]} +$$

and finally pull out the leading term as a prefactor

**ccc $\Big/ \left(\dfrac{15\text{Zeta}\left[\frac{5}{2}\right]}{4\text{Zeta}\left[\frac{3}{2}\right]}\right)$ //FullSimplify**

$$1 + \frac{3\left(10 - \frac{3\text{Zeta}\left[\frac{3}{2}\right]^3}{\pi\text{Zeta}\left[\frac{5}{2}\right]}\right)(\text{tt}-\text{tB})}{20\text{tB}} + \frac{3\left(20 + \frac{9\text{Zeta}\left[\frac{3}{2}\right]^3\left(4\pi + 3\text{Zeta}\left[\frac{1}{2}\right]\text{Zeta}\left[\frac{3}{2}\right]\right)}{\pi^2\text{Zeta}\left[\frac{5}{2}\right]}\right)(\text{tt}-\text{tB})^2}{160\text{tB}^2} + O[\text{tt}-\text{tB}]^3.$$

These give Eq. (C.6.5) series for $C_V$ in what follows.

**4. Transformation back to proper variables**

This is the end of the *Mathematica* manipulations. It now remains to convert from reduced units back to proper ones.

The internal energy $E$ is

$$E = NkT_\text{B}\left(\frac{T}{T_\text{B}}\right)^{5/2}\frac{3}{2}\frac{\zeta\left(\frac{5}{2}\right)}{\zeta\left(\frac{3}{2}\right)}\left\{1 - \frac{9\zeta\left(\frac{3}{2}\right)^3}{16\pi\zeta\left(\frac{5}{2}\right)}\left(\frac{T-T_\text{B}}{T_\text{B}}\right)^2 + \cdots\right\}. \qquad (C.6.4)$$

The first term here is the internal energy for all temperatures up to $T_\text{B}$ and the second term gives the leading order deviation above $T_\text{B}$.

The heat capacity $C_V$ is obtained by differentiating $E$. In the vicinity of $T_\text{B}$, the heat capacity is

$$C_V = \frac{15\zeta\left(\frac{5}{2}\right)}{4\zeta\left(\frac{3}{2}\right)}Nk\left\{1 - \left(\frac{3\zeta\left(\frac{3}{2}\right)^3}{20\pi\zeta\left(\frac{5}{2}\right)} - \frac{1}{2}\right)\left(\frac{T-T_\text{B}}{T_\text{B}}\right) + \cdots\right\}. \qquad (C.6.5)$$

The linear deviation indicates that the heat capacity critical exponent $\alpha$ is $-1$ above $T_\text{B}$. (In Problem 4.3 $\alpha$ was shown to be $-1$ below $T_\text{B}$.)

# EVALUATION OF THE CORRELATION FUNCTION INTEGRAL

---

## D.1  Initial Domain of Integration

In Section 5.2.1, we made use of the integral identity

$$\int_0^t d\tau_1 \int_0^t d\tau_2\, G(\tau_1 - \tau_2) = 2 \int_0^t (t - \tau)\, G(\tau) d\tau, \tag{D.1.1}$$

which, we stated, was obtained by an appropriate change of variables. In this appendix, we shall show how this result is evaluated.

The integral

$$I = \int_0^t d\tau_1 \int_0^t d\tau_2\, G(\tau_1 - \tau_2) \tag{D.1.2}$$

is a two-dimensional integral whose domain of integration is

$$0 < \tau_1 < t, \quad 0 < \tau_2 < t. \tag{D.1.3}$$

In the $\tau_1 - \tau_2$ plane, this is the square with vertices A, B, C, D, shown in Fig. D.1.

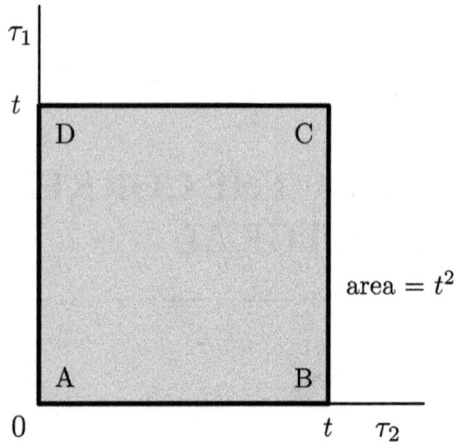

Fig. D.1.   Original domain of integration.

## D.2   Transformation of Variables

The argument of the integral is $\tau_1 - \tau_2$; this is constant along the lines $\tau_1 = \tau_2 + \text{const}$. It thus makes sense to change variables to

$$\tau = \tau_1 - \tau_2$$
$$T = \tau_1 + \tau_2 \tag{D.2.1}$$

and the domain of integration in terms of the new variables may be found from the location of the vertices:

| | $\tau_1$ | $\tau_2$ | $\tau$ | $T$ |
|---|---|---|---|---|
| A | 0 | 0 | 0 | 0 |
| B | $t$ | 0 | $t$ | $t$ |
| C | $t$ | $t$ | 0 | $2t$ |
| D | 0 | $t$ | $-t$ | 0 |

Thus, the domain of integration in the $T - \tau$ plane is as shown in Fig. D.2.

In terms of the transformed variables, the argument of the integral is $G(\tau)$ and this is independent of the $T$ variable. So, $G(\tau)$ is constant throughout each slice shown in the figure. Now, the slice at height $\tau$ has width $2(t - \tau)$ and so its area is $2(t - \tau)d\tau$.

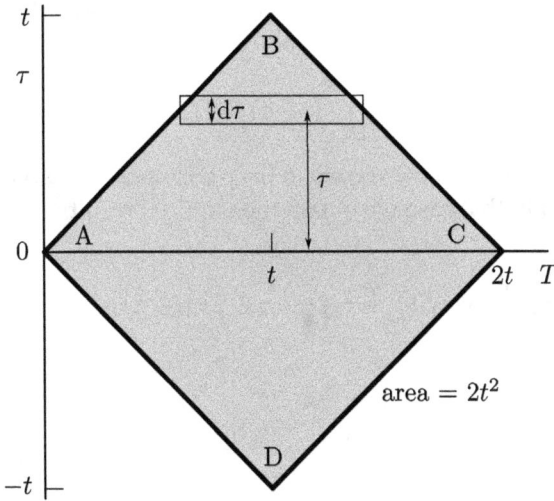

Fig. D.2.   Transformed domain of integration.

## D.3   Jacobian of the Transformation

The contribution to the integral from the slice at height $\tau$ is then $2(t - \tau)G(\tau)d\tau$. However, we need to multiply this by the Jacobian $J$ of the transformation of variables. Thus, the integral is

$$I = 2J \int_{-t}^{t} (t - \tau)G(\tau)\,d\tau. \tag{D.3.1}$$

The Jacobian is needed since the area of the transformed domain is double that of the original, whereas the value of the integral must be independent of the transformation. Consider the case where $G(\tau) = 1$. Then

$$I = \int_{0}^{t} d\tau_1 \int_{0}^{t} d\tau_2\, G(\tau_1 - \tau_2) = t^2, \tag{D.3.2}$$

the original domain area. The Jacobian is given by

$$J = \begin{vmatrix} \dfrac{\partial \tau_1}{\partial \tau} & \dfrac{\partial \tau_2}{\partial \tau} \\[2mm] \dfrac{\partial \tau_1}{\partial T} & \dfrac{\partial \tau_2}{\partial T} \end{vmatrix} = \begin{vmatrix} \dfrac{1}{2} & -\dfrac{1}{2} \\[2mm] \dfrac{1}{2} & \dfrac{1}{2} \end{vmatrix} = \dfrac{1}{2}, \tag{D.3.3}$$

precisely the required value; this ensures the transformed integral in the case where $G(\tau) = 1$ has the same value, $t^2$. Thus, we conclude that

$$I = \int_{-t}^{t} (t - \tau) G(\tau) \, d\tau. \tag{D.3.4}$$

Usually, as indeed we have argued in the previous sections, $G(\tau)$ is an even function. And in that case we obtain the required result

$$\int_{0}^{t} d\tau_1 \int_{0}^{t} d\tau_2 \, G(\tau_1 - \tau_2) = I = 2 \int_{0}^{t} (t - \tau) G(\tau) \, d\tau. \tag{D.3.5}$$

# BOSE–EINSTEIN AND FERMI–DIRAC DISTRIBUTION FUNCTIONS

---

## E.1 Simple Derivation

The derivation of the quantum distribution functions in Chapter 1 was based on the grand canonical approach. Conventional derivations use either the grand canonical or the micro canonical distributions. In many ways, the canonical distribution is the most intuitive for many applications, but it is not best-suited to treat the quantum distribution functions. In this section, I follow a method based on a discussion in Feynman's book, *Statistical Mechanics* [4]. He used it to obtain the Fermi–Dirac distribution and I have extended it to give the Bose–Einstein distribution also.

We shall find the distribution functions for particles obeying Fermi–Dirac statistics and those obeying Bose–Einstein statistics. Thus, we want to know the mean number of particles which may be found in a given quantum state. We start by considering an idealised model, of a subsystem comprising a single quantum state of energy $\varepsilon$, in thermal equilibrium with a reservoir of many particles. The mean energy of a particle in the reservoir is denoted by $\mu$ (we will tighten up on the precise definition of this energy later).

A particle may be in the reservoir or may be in the subsystem, Fig. E.1. The probability that it is in the subsystem is proportional to the Boltzmann factor $e^{-\varepsilon/kT}$, while the probability that it is in the reservoir is proportional to $e^{-\mu/kT}$. If $P(1)$ is the probability that there is one particle in the subsystem and $P(0)$ is the probability of no particles in the subsystem, then we

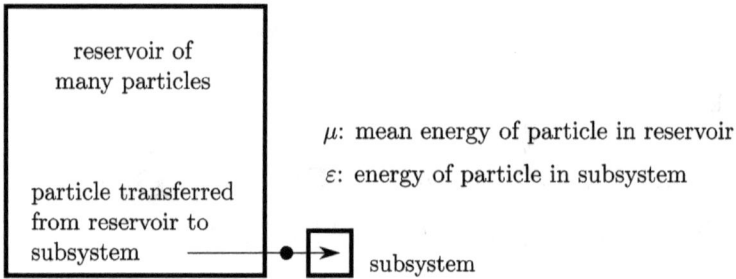

Fig. E.1.   Transfer of a particle from the reservoir to the subsystem.

may write

$$\frac{P(1)}{P(0)} = e^{-(\varepsilon-\mu)/kT} \quad \text{or} \quad P(1) = P(0)e^{-(\varepsilon-\mu)/kT}.$$

If the statistics allow (for Bosons, but not for Fermions), then we may transfer more particles from the reservoir to the subsystem. Each particle transferred will lose an energy $\mu$ and gain an energy $\varepsilon$. Associated with the transfer of $n$ particles, there will therefore be a Boltzmann factor of $e^{-n(\varepsilon-\mu)/kT}$. And so the probability of having $n$ particles in the subsystem is

$$P(n) = P(0)e^{-n(\varepsilon-\mu)/kT}. \tag{E.1.1}$$

Let us put

$$x = e^{-(\varepsilon-\mu)/kT}. \tag{E.1.2}$$

Then

$$P(n) = P(0)x^n. \tag{E.1.3}$$

Normalisation requires that all possible probabilities sum to unity. For Fermions, we know that $n$ can take on only the values 0 and 1, while for Bosons, $n$ can be any integer. Thus, we have

$$P(0) + P(1) = 1 \quad \text{for Fermions}$$

$$\sum_{n=0}^{\infty} P(n) = 1 \quad \text{for Bosons}$$

which can be written, quite generally, as

$$\sum_{n=0}^{\nu} P(n) = 1 \tag{E.1.4}$$

where $\nu = 1$ for Fermions and $\nu = \infty$ for Bosons.

Since $P(n)$ is given by Eq. (E.1.3), the normalisation requirement may be expressed as

$$P(0) \sum_{n=0}^{\nu} x^n = 1,$$

(E.1.5)

which gives us $P(0)$:

$$P(0) = \left( \sum_{n=0}^{\nu} x^n \right)^{-1}.$$

(E.1.6)

We will be encountering the above sum of powers of $x$ quite frequently, so let's denote it by the symbol $\Sigma$:

$$\Sigma = \sum_{n=0}^{\nu} x^n.$$

(E.1.7)

In terms of this

$$P(0) = \Sigma^{-1}$$

(E.1.8)

and then from Eq. (E.1.3),

$$P(n) = x^n / \Sigma.$$

(E.1.9)

What we want to know is the *mean* number of particles in the subsystem. That is, we want to calculate

$$\bar{n} = \sum_{n=0}^{\nu} n P(n)$$

(E.1.10)

which is given by

$$\bar{n} = \frac{1}{\Sigma} \sum_{n=0}^{\nu} n x^n.$$

(E.1.11)

The sum of $n x^n$ may be found by using the beta trick of Section 1.4.5. The sum differs from the previous sum $\Sigma$ which we used, because of the extra factor of $n$. Now, we can bring down an $n$ from $x^n$ by differentiation.

Thus, we write

$$nx^n = x\frac{d}{dx}x^n,$$ 

(E.1.12)

so that

$$\sum_{n=0}^{v} nx^n = x\frac{d}{dx}\sum_{n=0}^{v}x^n.$$ 

(E.1.13)

Observe that the sum on the right-hand side here is our original sum $\Sigma$. This means that $\bar{n}$ can be expressed as

$$\bar{n} = x\frac{1}{\Sigma}\frac{d\Sigma}{dx}$$ 

(E.1.14)

or

$$\bar{n} = x\frac{d\ln\Sigma}{dx}.$$ 

(E.1.15)

## E.2 Parallel Evaluations

It remains, then, to evaluate $\Sigma$ for the two cases. For Fermions, we know that $v = 1$, so that the sum in Eq. (E.1.7) is $1 + x$. For Bosons, $v$ is infinity; the sum is an infinite (convergent) geometric series. The sum of such a geometric progression is $1/(1 - x)$. Thus, we have the following:

**Fermions**

$$\Sigma = 1 + x$$

**Bosons**

$$\Sigma = (1 - x)^{-1}$$

the logarithm is:

$$\ln\Sigma = \ln(1 + x)$$

$$\ln\Sigma = -\ln(1 - x)$$

upon differentiating

$$\frac{d\ln\Sigma}{dx} = \frac{1}{1+x}$$

$$\frac{d\ln\Sigma}{dx} = \frac{1}{1-x}$$

so that

$$x\frac{d\ln\Sigma}{dx} = \frac{x}{1+x}$$

$$x\frac{d\ln\Sigma}{dx} = \frac{x}{1-x}$$

and $\bar{n}$ is then given by

$$\bar{n} = \frac{1}{x^{-1}+1}$$

$$\bar{n} = \frac{1}{x^{-1}-1}$$

Finally, substituting for $x$ from Eq. (E.1.2)

$$\bar{n} = \frac{1}{e^{(\varepsilon-\mu)/kT}+1}$$

$$\bar{n} = \frac{1}{e^{(\varepsilon-\mu)/kT}-1}$$

These expressions will be recognised as the Fermi–Dirac and the Bose–Einstein distribution functions. However, it is necessary to understand the way in which this idealised model relates to realistic assemblies of Bosons or Fermions. We have focussed attention on a given quantum state, and treated it as if it were apart from the reservoir. In reality, the reservoir is the entire system and the quantum state of interest is *in* that system. The model analysis then follows through so long as the mean energy of a particle, $\mu$, in the system is changed by a negligible amount if a single quantum state is excluded. And this must be so for any macroscopic system.

We now turn to an examination of the meaning of $\mu$ within the spirit of this picture. We said that it was the mean energy lost when a particle is removed from the reservoir, which we now understand to mean the entire system. When a particle is removed, the system remains otherwise unchanged. In particular, the distribution of particles in the other energy states is unchanged — the entropy remains constant. Also, the energy of the various states is unchanged as the volume remains constant. Thus, our $\mu$ is equal to $\partial E/\partial N$ at constant $S$ and $V$, which when compared with the extended statement of the First Law of thermodynamics:

$$dE = TdS - pdV + \mu dN, \tag{E.2.1}$$

indicates that our $\mu$ corresponds to the conventional definition of the chemical potential.

# References

[1] H. B. Callen, *Thermodynamics and an Introduction to Thermostatistics* (John Wiley, 1985).
[2] J. W. Gibbs, On the equilibrium of heterogeneous substances, *Trans. Conn. Acad. Arts Sci.*, **3** (1875–1878) 108. Reprinted in J. W. Gibbs, The Collected Works, Vol. 2, Longman and Green (New York, 1928).
[3] B. Cowan, On the chemical potential of ideal Fermi and Bose gases, *J. Low Temp. Phys.*, **197**, 5–6 (2019) 412–444.
[4] R. P. Feynman, *Statistical Mechanics* (Benjamin, 1972).

# INDEX